Amazon Peasant Societies in a Changing Environment

Amazon Peasant Societies in a Changing Environment

Political Ecology, Invisibility and Modernity in the Rainforest

by

Cristina Adams
University of São Paulo, Brazil

Rui Murrieta
University of São Paulo, Brazil

Walter Neves
University of São Paulo, Brazil

and

Mark Harris
University of St Andrews, Fife, UK

Springer

Editors
Cristina Adams
School of Arts, Sciences and Humanities
University of São Paulo
Brazil

Rui Murrieta
Department of Genetics and
Evolutionary Biology
University of São Paulo
Brazil

Walter Neves
Department of Genetics and
Evolutionary Biology
University of São Paulo
Brazil

Mark Harris
Department of Social Anthropology
University of St Andrews
Fife, United Kingdom

Translated and updated version of:
Adams, C., Murrieta, R., & Neves, W. A. (2006). Sociedades caboclas amazônicas: modernidade e invisibilidade (1a ed.). São Paulo: Annablume.

Cover image taken by Mark Harris in November 2003.

ISBN: 978-1-4020-9282-4 e-ISBN: 978-1-4020-9283-1

DOI: 10.1007/978-1-4020-9283-1

Library of Congress Control Number: 2008936510

Printed on acid-free paper

9 8 7 6 5 4 3 2 1

springer.com

To Guilherme M.M de La Penha,
in memoriam

Preface

I am delighted to be able to provide some comments on the English edition of this book, *Sociedades Caboclas Amazonicas: Modernidade e Invisibilidade*, originally published in Portuguese by Annablume Editora, with support from FAPESP (The State of São Paulo Research Foundation). The book presents a synthesis of the advances in understanding of *caboclos* in the past 20 years, and much of this material has not been available heretofore in English.

This book is appearing in English at a critical moment in the evolution of our understanding of caboclos. For one thing, this body of work has made it possible to get away from the earlier connotations of the term caboclo which was a demeaning term used always for someone below you, as noted by Wagley and Galvão in the 1950s. As research has accumulated, more and more scholars are opting for the use of the term *ribeirinho* instead. This is as it should be, as the term caboclo is in any number of ways intentionally ironic. Once in a conversation with a *colono* from the pre-Amazonia maranhense, he called me over and showed me his pig pen saying 'this is the cattle of the caboclo'. Not all caboclos have such a refined sense of irony, or perhaps they do and we have failed to take note of it.

I will not try to summarize the papers in the volume, as the excellent introduction to the volume does a superb job of doing so. Rather, I want to note what a remarkable shift this volume represents in the discourse on caboclos that preceded it. The understanding of caboclos from the 1950s to the 1980s was strongly driven by notions of cultural ecology (culture shaped by the way people organize to exploit resources), and by notions of syncretism (mixing up of cultures coming into contact). Then a more historical bent entered into the analysis, represented by various authors in this volume, who began to recover the complex interactions of native Amazonians in Directorate (state run) villages, and the subsequent economic conditions faced by these populations following the breakdown of the villages and a more isolated economic life. In the late 1980s and early 1990s came a more human ecological approach, that included bioanthropological, nutritional, and political economic approach that complexified the underlying bases for the way of life of caboclos. It is this direction which is most strongly represented in this book. Whereas before caboclo cultural explanations focused on their isolation, now the analysis showed how connected they have been for a good part of the nineteenth and twentieth century to a global economy, and to regional patterns of economic inequality.

When I first went to the Amazon many decades ago, it was quickly clear to me that the Amazonian caboclos that I came across in new settlement projects along the Transamazon Highway were gifted resource managers. They had higher yields per unit of land and per unit of labor, they had better health, and they even took advantage of available capital and technology better than the allegedly more modern southerners who came to the area. They knew how to hunt, how to recognize the best soils for farming, and how to diversify their portfolio. They even led the way in investing their profits in nearby urban areas in the form of a house, a store, and education for their children—without abandoning their land or their knowledge. What was hard for me to understand at that time, was why so many other colonists, and government civil servants working in the area, could not see that. Even when I presented them with quantitative evidence for the higher performance of caboclos, technical personnel insisted that that could not be—and colonists dismissed caboclos as poor farmers who were more interested in hunting than cultivating their land.

Why this resistance? I have reflected on that over the years, and found that there are many reasons. It takes time for people to learn how to use the resources in a new environment (sometimes as long as one generation), and while a few individuals will reduce that period by borrowing from neighbors, most people resist learning from those whom they have placed in a socially inferior position (i.e. the caboclo). Even if people are willing to learn, it still takes time to learn something as complex as a full array of adaptive strategies to ecosystems as complex as those in Amazonia. People prefer to try what they know, rather than to change, and this attitude delays learning precise knowledge. If we add to environmental knowledge, the knowledge required to understand one's position vis-à-vis the regional political economy – such a task becomes fraught with huge challenges. This was further complicated by the shifts in social and economic relations that accompanied life in Amazonia following the developmentalist efforts that began in the late 1960s and which brought many different actors to the scene, who challenged the traditional political economic order, and attempted to restructure the relations of production.

Is it any wonder that the rural populations of Amazonia, whether caboclo, or immigrant, found themselves in ever more precarious situations—but also provided with some opportunities. As the papers in this volume indicate, one sees places in Amazonia where the health, nutrition, and economic conditions of caboclos have improved—but also too many where it has not. Whereas the market for açaí has exploded globally, local caboclos have not benefited as much as they should have every right to expect. Whereas Brazil now leads the world in beef exports, caboclos have not benefited. Whereas Brazil's soybeans keep growing in world market share, this mechanized crop has not benefited caboclos. In fact, the trickle down benefits of the development efforts of the past 38 years in Amazonia have been miniscule, and people in Amazonia have not shared in whatever profits have been made—and the caboclo least of all among them.

This book points to the need to pay attention to the structural conditions, and the livelihoods, of people in Amazonia— particularly its forgotten rural populations and the marginalized populations in the growing cities. They are a rich resource of knowledge, energy, and appropriate development. They do not lack capacity or desire, they often lack opportunity. The development formulas imposed by outsiders benefit outsiders more often than not, and leave populations further impoverished. A start has to be education and health available to all—but an education that includes knowledge of Amazonia and its people and resources. This has to be an education that values what they know, and the conservation and sustainable use of the region. Without this foundation, development efforts will destroy rather than use and conserve Amazonia. This book offers an important scholarly basis for valuing the caboclo, and for addressing regional development, as if people mattered. I applaud the authors in this volume, for putting people first, and giving us deep understanding of the conditions faced by caboclos.

Terre Haute, IN Emilio Moran

Acknowledgments

We would like to begin by expressing our gratitude to all the families and communities of Amazonia who have helped to make this book a reality. Without their collaboration, involvement and inspiring friendship this initiative would not have been possible. Our special thanks must go to our students and friends who have supported and helped in the long, and many times, tedious process of editing in Brazil: Aglair Ruivo, Nelson Novaes, Henrique Ataíde, Eloise Tonial, Maissa Bakri, Natasha Navazinas and Carolina Taqueda. The immeasurable support of the Department of Genetics and Evolutionary Biology, of the Institute of Biosciences, and the 'Pró-Reitoria de Pesquisa', both of the University of São Paulo, was vital to the realization of the workshop from which this book is derived. To the Foundation for the Support of Research of the State of São Paulo (FAPESP) we owe a large part of the funding for the workshop in June 2002, as well as for the publication of the Brazilian edition of the book. To Emilio Moran, of Indiana University, for having played a fundamental role in the formation of the first generation of Brazilian anthropologists and ecologists dedicated to the study of the riverine populations of the Amazon. In Scotland at the University of St Andrews, Lisa Smith prepared the English language manuscript for publication and Graeme Sandeman designed the map. The translation of the Portuguese originals was made possible by a 'Primer Premio Annual' (2007) de Artículos de Investigación en temas de Seguridad Alimentaria y Nutricional (SAN) from the Iniciativa América Latina y Caribe sin Hambre and the Red de Investigación y Capacitación en Seguridad Alimentaria y Nutricional of the Food and Agriculture Organization (FAO).

Contents

Contributors

Cristina Adams is Assistant Professor at the School of Arts, Sciences and Humanities at the University of São Paulo, Brazil, where she coordinates the Laboratory of Human Ecology. She has a Masters degree in Environmental Science and a PhD in Ecology, both from the University of São Paulo. She was a Visiting Training Fellow at the Department of Anthropology, University of Kent at Canterbury (UK), from 1999–2000. Her main research interests are the adaptability of peasant populations to the neotropical rainforests (Amazon and Atlantic Rainforest), tropical agroforestry systems and slash-and-burn agriculture.

William Balée is Professor of Anthropology at Tulane University. His main lines of research are historical ecology and ethnobotany, with a focus on South America. Among his numerous publications are: *Resource Management in Amazonia* (coeditor with D.A. Posey, 1989), *Footprints of the Forest* (1994) and *Time and Complexity in Historical Ecology: Studies in the Neotropical Lowlands* (coeditor with C. Erickson, 2006).

Eduardo Brondizio is an Associate Professor of Anthropology at the Indiana University, Bloomington. His research focuses on the ways rural populations and their environment co-evolve with shifts in the larger society. Since the late 1980s he has carried out longitudinal ethnography and cross-sectional survey research among Amazonian caboclo and colonist populations with particular attention to historical analysis of land use change, people-forest interaction, and ethnobotany.

Fábio de Castro is a human ecologist with MSc in Ecology (University of Campinas, Brazil) and PhD in Environmental Science (Indiana University, USA). He has collaborated in research projects with NGO's, GO's and academic institutions and published several articles on use and management of natural resources in the Amazon and Atlantic Forest. Fabio is assistant professor at the Centre for Latin American Research and Documentation (CEDLA) in Amsterdam, The Netherlands, and associate researcher at the Anthropological Center for Training and Research on Global Environmental Change (ACT) at IndianaUniversity.

Henyo Barretto Filho received his PhD in social anthropology from the University of São Paulo. He is now the director of the International Institute of Education in Brazil (IEB), a non-governmental organization based in Brasilia.

Célia Futemma has a PhD from the Indiana University. She carried out post-doctoral work at the University of São Paulo and Indiana University. She currently teaches at Federal University of São Carlos, in the State of São Paulo.

Décio de Alencar Guzmán was born and educated in Brazilian Amazonia. He has degrees from the Federal University of Pará, Brazil, and Campinas State University (UNICAMP) São Paulo, Brazil. He has a PhD in History from the *École des Hautes Études en Sciences Sociales*, Paris, France. Since 1994 he has taught Amazonian history and Latin American history at the Federal University of Pará. His interests include the cultural and economic history of Amazonia, and the study of the political culture and cultural history of the South America native peoples and 'mestizo' peoples. His publications include essays in professional journals and books in Brazil, Belgium, Mexico and Peru.

Mark Harris teaches social anthropology at the University of St Andrews. He has conducted fieldwork in the Brazilian Amazon and archival research on a massive rebellion there, known as the Cabanagem, in the 1830s. His publications include *Life on the Amazon* (2000), *Some other Amazonians* (coeditor with Stephen Nugent, 2004), *The Child in the City* (coeditor with Anna Grimshaw, 2000), *Rebellion on the Amazon* (forthcoming, 2009).

Deborah de Magalhães Lima has a PhD in Social Anthropology from the University of Cambridge and is Associate Professor of Anthropology at the Federal University of Minas Gerais, where she is the Coordinator of the Centre for Studies in Quilombolas and Traditional Peoples. She is also Coordinator of Anthropological Research at the Mamirauá Sustainable Development Institute, in the Brazilian Amazon. She has published *Diversidade Socioambiental nas Várzeas dos Rios Amazonas e Solimões* (2005), among other works, and is specialized in the anthropology of traditional peoples in the Amazon.

Rui Murrieta is Assistant Professor of Anthropology in the Department of Genetics and Evolutionary Biology at University of São Paulo, Brazil, where he also coordinates the Human Ecology Division of the Laboratory for Human Evolutionary Studies. He obtained his Ph.D. in Anthropology at the University of Colorado – Boulder, in 2000. He has done extensive research on the nutritional and environmental anthropology of Amazonian and Atlantic Forest indigenous peasants and is co-editor of the books *Amazônia, além dos 500 anos* and *Sociedades Caboclas Amazônicas: Modernidade e Invisibilidade*.

Walter Neves is currently Full Professor at University of São Paulo, where he is ahead of the Laboratory for Human Evolutionary Studies. He received his PhD in Science in 1984 at the same university, after a pre-doctoral training at Stanford and Berkeley, in 1982. He has also developed post-doctoral studies in USA and Brazil, in the areas of Bioarcheology and Ecological Anthropology (in 1985 and 1991/92, respectively). His main fields of investigation are Bioarchaeology and Human Evolution. However, he has dedicated part of his career and attention to several correlated areas of knowledge, like hunter-gatherers archaeology, human adaptation to the rain forest, and megafauna paleontology. Between 1986 and 1992 he created and coordinated the "Human Biology and Ecology Program" of Museu Paraense Emilio Goeldi, in Belém. During that period, he has conducted investigations concerning social and biological adaptations of Amazonian cabloco populations of Marajo Island. In the last few years he has dedicated most of his time on problems concerning the settlement of the Americas.

Stephen Nugent is Professor of Anthropology and director of the visual anthropology centre at Goldsmiths College, University of London. His research has predominantly been with the people of the Lower Amazon. His publications include *Big Mouth* (1989), *Amazon Caboclo Society* (1993), and *Scoping the Amazon* (2007).

Rosely Sanches holds a Masters' degree in Ecology from the University of São Paulo. She currently works at the Instituto Socio-Ambiental (ISA) coordinating projects related to regional planning and indigenous land rights. She is the author of *Caiçaras e a Estação Ecológica de Juréia-Itatins, Litoral Sul, São Paulo* (2005, Annablume, Fapesp).

Hilton Silva is Professor of Biological Anthropology at the Universidade Federal do Rio de Janeiro-UFRJ, Brazil, and Visiting Professor of the Graduate Program in Genetics and Molecular Biology and of the Institute of Philosophy and Social Sciences of the Universidade Federal do Pará-UFPA, Brazil (2007–08). He is currently coordinator of the Biological Anthropology Division, Director of the Laboratory of Bioanthropological Studies on Health and the Environment, and Curator of the Human Evolution exhibit of the Museu Nacional/UFRJ. He is the principal investigator in long term projects about growth, development, health and nutrition of rural populations of the Brazilian Amazon, and about the inter-relations among flora, fauna and human communities in protected areas of mega-biodiversity of the Atlantic Forest in Southeastern Brazil, among others.

Andrea Siqueira is a socio-cultural anthropologist, and has been researching the Brazilian Amazon for the past two decades. Her research interest includes household economy, gender, food, migration, parks and people. Currently she is the Associate Director of the Center for Latin American and Caribbean Studies at Indiana University, Bloomington, Indiana, USA.

Antoinette WinklerPrins is an Associate Professor, and Graduate Program Coordinator, at Michigan State University, Department of Geography. Her research interests sit at the intersection of human and physical geography.

COLOMBIA
VENEZUELA
GUYANA
SURINAM
FRENCH GUIANA
ATLANTIC OCEAN
AMAPÁ
Macapá
Boa Vista
RORAIMA
Branco
Negro
Marajó Island
Belém
São Luis
ECUADOR
Manaus
Santarém
Amazon
Solimões
Tefé
PARÁ
Marabá
MARANHÃO
PACIFIC
AMAZONAS
Juruá
Purus
Madeira
Xingu
BRAZIL
ACRE
Porto Velho
Rio Branco
RONDÔNIA
Tapajós
Araguaia
Tocantin
TOCANTINS
PERU
Cuiabá
GOIÁS
Brasília
MATO GROSSO
BOLIVIA
Main settlements
International boundaries
0
500 km

List of Figures

List of Tables

List of Boxes

Introduction

Cristina Adams, Rui Murrieta and Walter Neves

Abstract Amazonia's contemporary "non-urban" societies can be divided, by and large, into three groups: on the one hand, Amerindian societies and "traditional" or historical peasants (*caboclos*), originated from the Amazonian region's colonial incorporation; and, on the other, the neo-peasants who, from the mid-1970s on, have migrated into the region as part of governmental territorial occupation policies. Despite the great differences between them, these societies hold in common a relative socio-political "invisibility". The Amerindian' invisibility results from the fact that they are absorbed into the representation of the Amazon as a tropical ecosystem, a kind of super-nature; the historical peasants or *Caboclos* are "invisible", in their turn, because they represent the failure of past national integration efforts; and, finally, the neo-peasants, for they are excluded from the developmentist agendas both of the extractive and agribusiness sectors.

Keywords Caboclo identity · Amazonian anthropology · History · Environment

The Amazon has traditionally been portrayed as a tropical forest of continental dimensions or as the territory of the archetypal 'primitive other' (Slater, 1996). Fundamentally, it is pictured as a natural domain in which the social is an invader (Nugent, 1993). There is a clear asymmetry between the social system and the natural system, with the former subordinate to the latter. This view of the Amazon as an essentially natural domain is based on two basic presuppositions about Amazonian societies. The first is that during the dissolution of the indigenous societies and the emergence of the caboclo, nature was the only factor that remained a constant. The second refers to the connections between pre and post-colonial indigenous social structures and supposedly limiting environmental factors. In other words, despite strong evidence of the existence of complex societies in the pre-colonial period (Roosevelt, 1989, 1994; Porro, 1996; Neves, 2000; Heckenberger et al., 2003; Hornborg, 2005), the current state of these groups is always seen as being limited by natural rather than historical factors (Harris, 1998a; Nugent, 1993; Pace, 1998).

C. Adams et al. (eds.), *Amazon Peasant Societies in a Changing Environment*,
DOI 10.1007/978-1-4020-9283-1_1, © Springer Science+Business Media B.V. 2009

Contemporary 'non-urban' Amazonian societies can be broadly divided into indigenous societies; 'traditional' or historical peasant societies (caboclos), the fruit of colonial incorporation of the Amazonian region; and the neo-peasantry that has been migrating into the region since the 1970s in the wake of government policies to colonize Amazonia (Browder & Godfrey, 1997; Bunker, 1984, 1985; Moran, 1981; Nugent, 1993). Despite the considerable differences between these societies, they share the same relative socio-political invisibility. The Indians, because they are steeped in the representation of the Amazon as a tropical ecosystem; the historical peasantry, because they represent the failed efforts of the past at national integration; and the neo-peasantry because they are excluded from the developmentalist project of highly capitalized extractivist industries (Nugent, 1997).

The historical peasant societies, the focus of this book, occupy a problematic place within both the conceptual framework of anthropology and studies of Amazonian development. For anthropology, caboclo societies are the direct antagonists of indigenous societies, veritable spearheads of colonialism and, later, of national society (Brondízio & Siqueira, 1997; Bunker, 1984; Galvão, 1955; Lima, 1992; Murrieta, 2000; Parker, 1985a; Ross, 1978; Wagley, 1955). In a region characterized by so many 'genuine others', but particularly the archetype of the 'noble savage' (Nugent, 1993; Slater, 1996; Viveiros de Castro, 1996) caboclo societies fall outside the theoretical framework of anthropology (Nugent, 1993). On the rare occasions in which the caboclo is actually portrayed, it is usually as an 'inauthentic or pathological' other who cannot serve as an object of anthropological analysis as the very existence of caboclo society subverts the formal distinction between other and observer. The caboclos are, according to the representation, 'falsified others', not only because they derive from European conquest as opposed to the 'local societies', but also because their very existence attests to the pernicious influence of 'civilization' (Nugent, 1993).

However, the 'original others' against which the caboclos are so unfavorably compared are themselves the result of historical processes of colonization (Balée, 1993; Moreira Neto, 1988; Porro, 1996; Viveiros de Castro, 1996). Indeed, the continuity that exists between the indigenous and caboclo identities is much more complex than normally considered. It involves the indigenous societies, the objects of a dramatic conquest, and the emergence of an Amazonian peasantry, fruit of the former's deterioration (Leonardi, 1999; Nugent, 1993; Parker, 1985a, 1985b). Even the more recent process of the incorporation of north-eastern Brazilian immigrants into Amazonian societies has not always been peaceful and has involved exchanges, abandonments, migrations, adaptations, innovations, inventions and various forms of reciprocal acculturation (Cunha & Almeida, 2002; Leonardi, 1999; Moreira Neto, 1988; Nugent, 1993; Oliveira Filho, 1979; Parker, 1995b; Santos, 1980; Weinstein, 1993; Wagley, 1955).

Another aspect rarely considered in the related literature is that of the Brazilian racial ideology. Few writers (Figueiredo, 1999; Lima, 1992; Motta-Maués, 1989) have considered the question of caboclo ethnic and cultural identity within the broader context, showing how their invisibility can, in part, be explained by the debate on miscegenation which preoccupied Brazilian intellectuals at the end of the nineteenth and beginning of the twentieth centuries. The fact that they did not inhabit the southeast of Brazil, where the immigration policies might have enjoyed some success in 'whitening' the mestizo population, basically disqualified the caboclos from the process of racial purification (Nugent, 1993). Furthermore, little effort has been made to understand and bring to light the role played by the Amazonian intellectual elite and the modernist movement in constructing the imagery of contemporary Amazonian cultural 'types' (Figueiredo, 1998, 1999). As such, the social invisibility that characterizes caboclo societies also manifests itself in the negligence of academics to the contribution Amazonians themselves have made to this process (Figueiredo, 1998, 1999).

The virtual invisibility of Amazonian caboclo societies can also be partly explained by their diverse and non-specialized lifestyles. Their resource base consists of the forest and a river system characterized as homogeneous in certain scales of scientific representation (humid tropics, *várzea* [floodplain] and terra firme [uplands]), but which actually harbors a vast array of largely misunderstood heterogeneities (Fraxe, 2004; Murrieta, 2000; Murrieta, Dufour, & Siqueira, 1999; Nugent, 1993, 1994; Raffles & WinklerPrins, 2003; Raffles, 2001; WinklerPrins, 2001, 2002). In addition, many of the resources necessary to peasant reproduction are not traditionally valued by economists (forest, alluvial soil, rivers), thus pushing the caboclo economy to the margins of technocratic, capital-driven 'economic development'. For the developmentalists, caboclo producers belong to an essentially informal economy, where the sectorial activity and class boundaries are not clearly demarcated (Bunker, 1984; Fraxe, 2000; Nugent, 2003; Pace, 1998; Parker, 1985a). Even the rural/urban distinction within caboclo societies themselves is somewhat blurred, as has been discussed recently concerning other rural Brazilian societies (Abramovay, 2003; Browder & Godfrey, 1997; Cleary, 1993; Fraxe, 2000; Nugent, 2003).

Caboclo Identity

The formation of the caboclo identity resides within processes defined more by externalities (global economic transformations) than by local cultural continuities (Leonardi, 1999; Nugent, 1993). The context of violence and domination in which their identity was forged has made the caboclos define themselves in opposition to powerful outsiders (Harris, 1999; Slater, 1997). According to Harris (1999), as they live by a short-term logic and far from the urban centers of power, caboclos combine opposition and indifference in their relationship with the local elite, hoping to avoid or at least diminish domination through an apparently anarchic social life. The fact that the caboclo societies, unlike the peasant societies with which anthropology traditionally deals, have no clear pre-capitalist antecedents (if we chose to disregard their indigenous origins as a continuity) makes a historical approach even more difficult (Nugent, 1993, 1997).

The complexity of the situation has been further aggravated by the drastic changes of the last thirty years in Amazonia. A proliferation of class distinctions, the strangulation of the political structure of patronage by the expansion of the capital, the growth of the urban contingent of the caboclo population, the impact of the transport network and increased deforestation are just some of the phenomena observed in recent times. Recognizing the significance of caboclo societies requires that they be considered as *societies* within this context of historical change, and therefore subject to the same dynamic which has incorporated other 'peripheries' of the domain of capitalist politico-economic systems (Brondízio & Siqueira, 1997; Murrieta et al., 1992; Pace, 1998; Schmink, 2003).

The inclusion of caboclos within anthropological discourse has proved somewhat ambiguous: on one hand, the growing legitimacy of the environmentalist/ecological argument (Murrieta, 2000; Nugent, 1993) has seen their insertion in such issues as the use and management of resources within the context of a highly fetishized Amazon ('lungs of the earth', 'genetic bank', 'wellspring of potential miracle drugs', 'air-conditioner of the planet'), while, on the other, caboclo societies continue to be portrayed in a negative light. From the anthropological perspective on indigenous populations, caboclos represent at once the 'left-overs' of degraded indigenous societies and an immediate threat (land invasions) to those Indian societies that managed to make it through the catastrophe of colonization. Nationally, the caboclo represents an unfinished project of the creation of a Brazilian culture that broke with its European, African and Indigenous antecedents (Motta-Maués, 1989; Nugent, 1993, 1997).

Theoretical Lines in the Study of Caboclo Societies

For the majority of contemporary academics, the cornerstone in the anthropological study of caboclo societies continues to be the work of Charles Wagley and Eduardo Galvão. These studies, given their American origin, clearly descend from 'culturalist' studies, Boasian historical particularism and the cultural ecology of Julian Stewart. However, little attention is paid to the local output by such scholars as José Veríssimo and Dalcídio Jurandir, among others, who greatly influenced this supposedly spontaneous generation of professional academics (Bezerra Neto, 2002; Figueiredo, 1998, 1999). These writers were pioneers in casting the mould of the regional literature and declared proponents of studying modern Amazonian identity; yet they have been made invisible by other forms of imperialism, namely scholarly.

Originating from and influenced by Stewardian cultural ecology and the Boasian culturalism of Charles Wagely and Eduardo Galvão, there is a whole cast of foreign and national researchers who have devoted themselves to studying caboclo societies, such as Emilio Moran, Lourdes Furtado, Eugene Parker,

Eric Ross, Richard Pace, Angélica Motta-Maués and Heraldo Maués[1]. The ecological vein of this lineage produced some of the markedly ecologically deterministic and ahistorical hypotheses that dominated academic output for over two decades (Harris, 1998a; Neves, 1991; Nugent, 1993). In the critical appraisal of European and Brazilian authors, this first generation of American anthropologists and their Brazilian pupils who wrote about caboclos upheld a functionalist notion of culture, portraying caboclo society as a 'model' under which the rural populations of the Amazon basin have been catalogued ever since. This view contradicted the idea of a historical identity, which was materially produced and connected to externalities (Harris, 1998a; Murrieta, 2000; Nugent, 1993, 1997; Pace, 1998).

In the 1970s, the work of Emilio Moran (1974), despite also emphasizing environmental limitations just like his antecedents, diverged from this approach by considering the caboclo social system to be the most important factor in the environmental and socio-political context of the post-colonial Amazon. For Moran (1974), the caboclo was a cultural 'type' that emerged as a result of the 'tupinization' of the Iberian and local non-Tupi cultures. Though 'tupinization' would imply a historical process, Moran does not propose any causal explanation for the occurrence of change and therefore no theory of socio-economic transformation (Harris, 1998a: 89).

Despite their obvious evolutionist and ecological inspiration, researchers like Eugene Parker express a more embracing recognition of the importance of history in the formation of the caboclo societies. Instead of Moran's 'tupinization' (1974), Parker uses the term 'caboclization' to refer to the events and conditions that destroyed a large portion of the Amerindian societies, transformed those that were left, and spawned the emergence and consolidation of caboclo culture in the nineteenth century Amazon. Perhaps a happy medium between the works of these two authors would be that of Ross (1978). This author has best delineated the set of different historical and environmental factors that influenced the formation of the contemporary aspects of the historical Amazonian peasantry (Harris, 1998a).

In general, for the American anthropologists of this first period, caboclo culture was consolidated in the twentieth century, largely in the form of family units living on the river floodplains and on the banks of other waterways and lakes, practicing smallholder family agriculture allied with fishing and hunting. Nevertheless, once this way of life became established, the impression was that the system crystallized within an ahistorical reality cut off from external pressures (Harris, 1998a).

After this first phase of studies on Amazonian caboclos, other lines of enquiry began to emerge in American anthropology, such as political ecology and institutional analysis, which sought to approach the rural smallholder

[1] Angélica Motta-Maués and Heraldo Maués are the authors of the two most important anthropological studies of caboclo societies in Portuguese from the last decade of the twentieth century (Motta-Maués, 1993; Maués, 1995).

(whether Indian, caboclo or migrant settler) as an active agent capable of taking decisions and changing his situation in a dynamic way (Bunker, 1984; Chibnik, 1994; Pace, 1998; Schmink, 1985). In addition, theories of cultural and human ecology were revised to adopt social, economic and political contexts in their analyses, which now included caboclo societies. The initial attempts at the cultural typification of the caboclo were replaced by approaches that looked to understand the way these Amazonian populations use the land and its natural resources, and how this relates with political and environmental questions of management (Castro, 1999; Chibnik, 1994; Futemma, 2000; Furtado, 1993; Lima, 1992; McGrath, Calabria, Amaral, Futemma, & Castro, 1993a; McGrath, Castro, Futema, Amaral, & Calabria, 1993b; McGrath, Castro, Câmara, & Futemma,1999).

Another branch of ecological studies that has been recruiting a growing number of sympathizers is historical ecology (Balée, 1995, 1998; Crumley, 1994). This line, represented by the works of William Balée, among the indigenous groups (1992, 1993, 1995, 1998), and Hugh Raffles among caboclos (Raffles, 2001; Raffles & WinklerPrins, 2003), stresses the adoption of a historical approach with a view to moving beyond the debate on the reification of limiting environmental factors and landscape homogenization.

In Britain, on the other hand, a group of academics heavily influenced by social theory, particularly a British version of structural Marxism, has been leveling deeply incisive criticism against the American academic tradition on the Amazon. Chief amongst these are Mark Harris (1996, 1998a, 1998b, 1999, 2001), David Cleary (1993) and, especially, Stephen Nugent (1993, 1994, 1997)[2]. The Amazon that interests the British anthropologists is inter-connected by trade, credit, migration, exchanges, conflict, the search for commodities, and an enormous group of people engaged in the informal economy and therefore beyond the reach of the state (Harris, 1998a).

Meanwhile, a whole new generation of Brazilian researchers has developed its own scholarly trajectory in relation to these two academic traditions. Among its members we can see, besides a visible interest in environmental and microeconomic problems, a clear manifestation of concern with the interactions between these local levels and the wider political and economic context that has enveloped the region for the last 500 years. Comprising this generation are those that emphasize management practices and land tenure and identify the caboclos as agricultural/rural producers (Adams, Murrieta, & Sanches, 2005; Brondízio & Siqueira, 1997; Fraxe, 2004), those who adopt the analytical lens of political ecology and institutional analysis (Castro, 1999; Futemma, 2000; Lima, 1992; Murrieta et al, 1999; Siqueira, 1997; Winkler-Prins, 2001, 2002), those who concentrate on the socio-economic and political identity of the caboclo (Alencar, 1994; Cunha, 2001, 2002; Furtado, 1993;

[2] Though American, Stephen Bunker (1984) could also be considered within this school.

Furtado & Quaresma, 2002; Lima, 1992, 1999; Lima & Alencar, 2000; Lima & Pozzobon, 2001; McGrath et al., 1993a; 1993b; Simonian, 1995), those who focus on processes related to daily practices and subjacent cultural motivations (Murrieta, 2000, 2001a, 2001b; Harris, 2001), those that accentuate the biological patterns present in the processes of change caboclo populations have been going through (Adams, 2002; Silva, 1995; Silva & Eckhardt, 1994; Silva, Crews, & Neves,1995; Siqueira, 1997) and, lastly, those who verticalize the functionalist ecological models under the orientation of behavioral ecology (Begossi, 2004)[3].

Though largely ignored, the works of such historians as Barbara Weinstein (1993), Warren Dean (1987), Roberto Santos (1980), Vicente Salles (1971) and Arthur César Ferreira Reis (1953) are enormously important in this new context. From different perspectives, these writers have sought to describe and analyze the set of factors that have molded the 'profile' of the historical Amazonian peasantry. More recently, another new generation of researchers has successfully managed to reconcile different hypotheses and theoretical perspectives from Europe and North America in an interesting criticism that underscores the factual and interpretational lacunas that need to be filled (Bezerra Neto, 2002; Bezerra Neto & Guzmán, 2002[4]; Figueiredo, 1993, 1998, 1999; Gondim, 1994; Guzmán, 1997, 1998; Meira, 1993, 2006).

The convergence of these different intellectual approaches and traditions, not to mention personal experiences, upon a single 'object', namely the historical or caboclo peasantry, creates a unique opportunity to produce new questions, exchanges and syntheses. Expanding our interpretation of caboclo societies in this way, beyond a folk culture model or 'pristine' systems of intervention in the environment, we can produce a general framework for explaining and interpreting the coalition of historical forces and their contribution to the Amazonian social reality. In this manner, the definition of caboclo will neither exclude nor simplify the ambiguity and complexity that pervade it, nor naively seek to divorce the historical Amazon from the Amazon of western expansionism and all the resistance, tensions and negotiations that go with it.

Based on these presuppositions, in the year 2000 Cristina Adams (School of the Arts, Sciences and Humanities/University of São Paulo, Brazil) and Mark Harris (University of St Andrews, UK) proposed holding a discussion forum

[3] In this volume organized by Alpina Begossi, she and her collaborators conduct comparative analyses between the caiçara and caboclo populations. Various dimensions of the ecological and socio-economic systems of these groups are approached, though the authors proceed with what is basically varying degrees of a functionalist ecological analysis.

[4] In this work, Bezerra Neto and Guzmán present the ideas of various Amazonian historians of the new generation. Despite the disparities in academic quality among the articles themselves, this is without doubt a praiseworthy initiative in presenting the state-of-the-art of Amazonian historiography.

with the small (but growing) group of researchers dedicated to the study of caboclo societies. The main goal of the meeting was to review critically the anthropological production on these societies and to discuss the general state of the research and the possibility of drawing up a new agenda for work on the theme at the turn of the millennium. Adams and Harris' intention was to assemble a group small enough for the discussions to be productive, but large enough to cover all the existing approaches in the human sciences and related areas.

The idea materialized in 2002 with the event 'Amazonian Caboclo Societies: Modernity and Invisibility', held at the Biosciences Institute of the University of São Paulo (USP). The event was divided into two parts: a closed workshop with specially invited researchers, and two days of open lectures at which some of the material discussed was presented to the general public. Organized by Cristina Adams, Rui Murrieta and Walter Neves, the event was made possible by funding from Fundação de Amparo à Pesquisa do Estado de São Paulo – Fapesp (the São Paulo Foundation for Research Support) and the Research Rectory at USP.

The papers presented at the workshop 'Amazonian Caboclo Societies: Modernity and Invisibility' are collected in this book and represent the diversity of new approaches to the study of caboclo societies.

Summary of the Book

The first chapter in this volume, which opens the section Identity, History and Society, is 'Utopias and Dystopias on the Amazonian Social Landscape', by Stephen Nugent. The author begins his chapter by drawing attention to the invisibility of these societies in the anthropological literature of the 1960s and 70s. For Nugent, the invisibility of caboclo societies (historical peasantry, to use his terminology) in that period is a result of four factors: the idealization of the Amazonian landscape as pure nature; the fact that the historical peasants never adopted plantation agriculture; the nature of the Amazonian frontier; and, last but by no means least, the fact that caboclo agrarian systems are neo-colonial 'experiments' largely based on foreign involvement. In relation to a central point visited to a lesser or greater extent in most of the articles in this volume, Nugent devotes the closing pages of his chapter to discussing the ahistoricity that typifies much of the anthropological production on Amazonian societies. Nugent seems to entertain no doubts that the main 'villain' behind the Amazonist tradition is the ideological naturalization of the view that the human populations that inhabit the region are victims. In other words, the anthropological literature on the historical Amazonian peasantry inherited the same bias extensively applied to the local indigenous populations: the search for a 'natural society', practically untouched by the political economy, whose structure, social organization and subsistence strategy ought to reflect, first and foremost, the environmental factors specific to the surroundings, no matter how limiting they may be.

As part of the section 'Identity, History and Society', William Balée shows the influence of colonization and penetration by the European system of finance and commerce on changes in the native vocabulary, and, subsequently, the knowledge of the Amazonian landscape and associated biota. The author concerns himself in particular with the case of cocoa and the way its denominations and conceptions gradually transformed in the indigenous languages thanks to the importance of the commodity on the export agenda of the eighteenth century. In reality, Balée shows us that the socio-environmental context in which the caboclo and indigenous societies formed is highly complex, thus reminding us of the importance of the historical aspect in our analysis of both.

Décio Guzmán's chapter analyzes the process of miscegenation on the Rio Negro during the eighteenth and nineteenth centuries. After presenting some background to the components in this miscegenation (the Indian, European and African), the author focuses on the issue of the inter-racial marriage policies decreed by the Portuguese Crown as one element in a package of projects geared towards exploiting the human resources of Portuguese America. Like Nugent, Guzmán holds that one of the biggest impediments to advances in the study of Amazonian caboclo societies is the belief that they somehow correspond to the Stewardian ideal of the 'natural population', that is, to independent, self-reproducing and self-regulating systems. This conception has hindered a more accurate understanding of these societies as the product of historical transformations involving internal and external tensions.

Though Mark Harris' chapter, 'Ambivalent Present: an Amazonian way of being in time', does not formally figure as a historical analysis, it presents undeniable affinities with the two previous contributions (particularly for its deconstructivist vocation). Harris' core argument is that the caboclos are modern because of their constant renewal of the past in the present, a strategy that has granted them enormous reproductive success (socially and biologically) and which has been critical to their adjustment to unstable economic and political conditions and the background of general socio-cultural collapse. For the author, resilience and flexibility are the key traits of the riverine peoples. Throughout the text, Harris dialogues, sometimes implicitly, sometimes explicitly, with two earlier ways of referring to the origination of these populations: Parker's process of 'caboclization', formulated in the mid 1980s, and Nugent's Boom-Bust theory from the early 1990s. For the author, by imposing categories and abstract concepts with a view to constructing collective entities like a caboclo 'culture' or 'identity', we end up losing the richness of the subject of study, namely the heterogeneity, ambivalence, ideology of 'mixing' and 'openness' to the new, that emerge from the analysis of specific biographies in their respective socio-economic contexts. Harris argues that these riverine societies can be characterized by their ongoing capacity to negotiate the conditions of the *present*. Scant attention is given to preserving the past, whether materially or ideologically speaking. 'Riverine' identity is the product of what these people are in the present, and is in contrast to what they were in the recent past.

With the article 'Traditional peoples: introduction to the political ecology critique of a Notion', Henyo Barretto Filho opens the section 'Sustainability and Development Policies' by discussing the controversial matter of whether or not traditional populations (including various categories of 'caboclo') should remain in conservation units. Following this idea, Barretto Filho presents and discusses an issue that is fundamental in this context: how does one define and characterize 'traditional peoples'? In the specific case of the Amazon, indigenous and caboclo populations are usually classified in this category because, on one hand, their cultivation practices do not hamper the proper functioning of the regenerative system of the tropical rainforest, and, on the other, because the impact of their economic activities are no worse than small-scale, short-lived and infrequent natural disturbances. Barretto Filho vehemently questions these aphorisms, reminding us that a large portion of the Amazonian forest of today can be seen as an immense 'cultural forest', to use the term introduced to ecological anthropology by William Balée at the end of the 1980s. If human populations that inhabited the Amazon before the arrival of the Europeans managed to alter the hylea (dense tropical forest) so extensively, albeit largely positively (using biodiversity as their assessment criteria), can we really say with any certainty that the environmental impact of these populations is as slight as commonly stated? For the author, there is no precise way to conceptualize the category 'traditional peoples', much less formulate a scientific response to the question of whether human presence should be allowed in environmentally protected areas. It is, he says, a matter of 'an ideological construct whose power resides precisely in the general nature of its meaning and fluctuating use'. Furthermore, by branding these groups traditional and betting on their non-monetary strategies for economic subsistence as the passport to the future survival of humanity (in terms of biodiversity), do we not risk turning them into hostages of an ahistorical definition, galvanizing the potential processes of socio-cultural change to which they are entitled? Emphasizing the spatial over the temporal, Barretto Filho concludes by suggesting that we substitute the term 'traditional population' for 'resident population' in discussions on the permanence (or not) of human populations in protected areas, thus defending the inclusion of all 'traditional' and 'modern' peoples that are, in this context, struggling to survive and reproduce.

Next in this section, Deborah de Magalhães Lima analyzes the economy of the domestic units of the Mamirauá Sustainable Development Reserve in the Upper Amazon, not with the aim of assessing or perhaps even legitimizing traditional strategies for the use of local natural resources as self-sustainable, but with a view to discussing the concept of self-sustainable development against the socio-economic background that emerges from her analysis. Firstly, she discusses the conceptual fuzziness that surrounds the expression 'sustainable development', defending the idea that environmentally protected areas could help clear up this vagueness, as they are being constructed empirically and without pre-defined models or schema. After a detailed presentation of the volumes and composition of the incomes of various communities and settlements within the geographical perimeters of the Mamirauá reserve, based on studies conducted in 1991 and 1994/95, the author demonstrates the enormous variability in incomes both within and among the riverine communities of the Upper Amazon floodplains, whose production is organized around the ideal of the economically autonomous household. The author argues that the bonds of kith and kin are just as crucial to the survival of these populations as the natural resources on which they depend. To support this contention, the author presents valuable data for the Mamirauá reserve, collated under a rigorous research plan, and compares it with data from other caboclo societies, particularly from the Lower Amazon and Amazon Estuary. When taken together, these studies, which cover a considerable area of the Amazon from the Upper Solimões River to its mouth reveal that the annual income of the riverine domestic units varies from five hundred to three thousand dollars, with those of Mamirauá much closer to the lower end of the scale. For Lima, the most palpable effect of the environmental conditions at Mamirauá on the local populations is their constant sense of risk and incessant search for better living conditions. In this context, the emphasis of Amazonian development projects place on environmental variables might stimulate the low levels of production and consumption of the so-called traditional populations, but not question the social inequality represented by the unsustainable levels of consumption in other sectors of society, if not entire societies. As such, the author contends that sustainable development rests, albeit inexplicitly, on social inequality.

Fábio de Castro's chapter provides a thorough analysis of the diversity of economic strategies among the riverine domestic units (or family units, as de Castro prefers) of the Lower Amazon River, units that depend on the floodplain resources for their survival. The empirical data used in the analysis was collated using *Censos Estatísticos Comunitários* (CEC) [Community Statistical Censuses], a participatory methodology based on community meetings, for 8,570 domestic units across 172 communities in the focus region. The economic strategies of the population of the Lower Amazon floodplain basically combine four main activities: fishing, agriculture, cattle raising, and paid labor and pensions from the state. While fishing is the core commercial activity of these communities, agriculture is more important to their subsistence. The best part of de Castro's article is devoted to analyzing the combinations of these four activities in different communities, and from domestic unit to domestic unit within the same community, to ascertain what factors influence most in the distinct combinations of these four economic pillars. The author shows that the importance of each of these economic activities varies widely in terms of local production repertoires, generating a great deal of heterogeneity among and within these communities. One way or another, in both cases, the determinants that carry the most weight in these variations are the economic ends of the activity, the structure of each family unit and the level of access to resources. The socio-economic panorama that emerges from this exercise is one of enormous heterogeneity as opposed to the homogeneity suggested by various authors, such as Lourdes Furtado (1993).

Eduardo Brondízio opens the section 'Resource Management' with the article 'Agricultural Intensification, Economic Identity and Invisibility among Amazonian Rural Smallholders: a comparative perspective on caboclos and settlers'. The focus of his contribution is to demonstrate the erroneous interpretation that prevails concerning the agricultural production systems of the Amazonian rural smallholder, especially when analyzed through the lens of intensification, and of how this equivocal judgment stems from a depreciative view of these social groups and their contributions to the regional economy. Both patterns of land use – caboclo and settler – are often based on the co-existence of intensive and extensive activities that together minimize risk and ensure the consolidation of the rural properties, as well as their expansion of the market-oriented activities. Brondízio also argues that, whether caboclos or settlers, the smallholders are actively engaged in the regional economy, they respond rapidly to market incentives, and that the commercial movement engendered regionally by their production is highly significant. The author suggests the term 'small producers' be used to refer to these populations, as proposed by Robert Netting in Europe, as a replacement for 'peasants', which carries pejorative connotations. According to Brondízio, the adoption of the term 'small producers' would go a long way toward creating a more positive socio-economic identity for these populations. In short, all of the empirical and conceptual material presented in the article is rallied to this central argument: that redefining the identity of these Amazonian populations as small rural producers would be a significant step toward overcoming the prejudices incorporated into regional and national society (and that includes the national and international development agencies and academia itself).

In the article 'Use and Access to Forest Resources: the caboclos of the Lower Amazon and their socio-cultural attributes', Célia Futemma further contributes to the issue of the use of natural resources by riverine populations, presenting what she calls an institutional analysis of the access to, and use of, forest resources by the Patos riverine community in the Lower Amazon. Futemma pays particular attention to the role social networks, particularly those of kith and kin, play in the provision of access to floodplain and upland resources. Another core aspect of the author's analysis is the flexible role of formal land ownership (not shared by all of the domestic units of the community) in enabling access of all – to varying degrees, depending on social bonds – to the resources vital to the material survival of the Patos residents. The author concludes that the formal rules of the state (which designed and promoted the agrarian reform implanted in the area) have been overlain by a system of informal local rules for access to and use of forest resources designed to accommodate the social and landscape diversity in a manner that reduces disparities among the rural smallholders of Patos.

In the section 'Gender and Daily Life', Andréa Siqueira presents an ethnographic account of five case studies on household dynamics in five distinct family units she studied in the region of Ponta de Pedras, Ilha de Marajó, Pará, during the early 1990s, with emphasis on the balance of power between men and women. From these cases Siqueira deduces that the women's influence in domestic decision-making is affected by the following factors: type of rights they have over the land (ownership), their effective economic contribution, and education (not necessarily formal). In these households, the decision-making can either be centered on the man, the 'head of the household', or shared and negotiated between the couple. Siqueira highlights the fact that understanding and valorizing women's role in this context and attaining a thorough comprehension of the dynamic of the domestic units are fundamentally important to the success of any interventionist projects that envisage improving the material quality of life of these riverine households.

Verticalizing some of the concerns raised by Siqueira, Rui Murrieta and Antoinette WinklerPrins offer a detailed ethnographic analysis of the motivations behind the initiative of a group of women kinsfolk of the São Benedito community, on Ituqui Island, in the Lower Amazon, to cultivate gardens and vegetable plots. The authors begin by discussing a certain dichotomy that exists in studies of a socio-cultural bent, on one hand, and of an agroforest or ecological persuasion, on the other, when it comes to caboclo societies in general and their subsistence practices in particular. Through a personalized and constantly multi-vocal narrative, Murrieta and WinklerPrins show how these women's cultivation practices are an intricate product of social, ecological, economic and emotional motivations. Among herbs, trees and flowers, social roles are contested, affinities and alliances cemented, agricultural practices tried and tested, thus creating a forum in which these women can express their subjectivities. The authors endeavor to show how broader and more concomitant knowledge of these social and material, personal and subjective circumstances and all their practical ramifications is central to a genuine understanding of the motivational elements present in the actions of the women of São Benedito. Consequently, further integration between studies of economic/ecological stress with those of socio-cultural inclination is also essential to attaining that level of understanding.

In the section 'Diet and Health', Cristina Adams, Rui Murrieta and collaborators present a detailed analysis on caloric and protein ingestion among Amazonian riverine communities based on the authors' own stock of thirteen years work of quantitative data on dietary features. The five communities featured in the study are located in the Estuary (Marajó Island) and Lower Amazon River (Ituqui Island). Using this unique data bank on Amazonian riverine populations, the authors conduct a minute analysis of the influence of various local factors on the alimentary intake of these populations: ecosystem, micro-environment, seasonality, social history and organization. As a brief overview, we could say that the main calorie sources on the estuarine floodplain are manioc flour and açaí, while further up-river, this function is fulfilled by manioc flour, fish and sugar. As for proteins, fish is the main source, regardless of the ecosystem studied. Each of the five communities ingests more than 100% of the minimum protein intake recommended by the FAO. In terms of calories, however, only one of the communities – that located in the estuary – revealed a caloric energy intake in-line with the internationally recommended minimum. The authors close their article by discussing the anthropological implications of the historical longevity and ubiquity of the association between manioc flour and fish as the staple of riverine sustenance.

Continuing with the issue of the material bases of the survival of caboclo societies and their biosocial consequences for these riverine communities, Hilton Silva, in his paper entitled 'Socio-ecology of Health and Disease: the effects of invisibility on caboclo populations in the Amazon', presents a judicious biomedical study of four riverine communities – two floodplain settlements and two upland settlements –, associating nutritional and parasitological anthropometric analyses for children and adolescents in these communities. The Caxiuanã community is located along a blackwater river system inside the National Forest of the same name, some 400 km southwest of Belém. Subsistence in Caxiuanã rests upon slash-and-burn agriculture, fishing, açaí extractivism and hunting. In Ilha de Ituqui, located 900 km from Belém, a region covered by various contributions to this volume, the author studied two communities: Aracampina and Santana. Silva's biomedical research reveals extremely high rates of malnutrition in all four communities, ranging from acute to chronic. More frightening still are the stupendous rates of infection by intestinal parasites in these communities. Silva's hope is that this data can serve as groundwork for the formulation of integrative and participative public health policies and spur wide ranging sanitary interventions to improve the quality of life of the Amazonian riverine populations studied and many others besides.

In the final section of the book, we present our conclusions, analyzing and enumerating the chapters' contributions to research and interventions in Amazonian caboclo societies.

Acknowledgement The authors would like to thank Célia Futemma for her critical contributions to the first draft.

References

Abramovay, R. (2003). *O futuro das regiões rurais*. Porto Alegre: UFRGS Editora.

Alencar, E. F. (1994). Gênero e trabalho nas sociedades pesqueiras. In L. G. Furtado, W. Leitão, & A. F. De Mello (Eds.), *Povos das águas* (63–81). Belém: Museu Paraense Emilio Goeldi/Coleção Eduardo Galvão.

Adams, C. (2002). *Estratégias adaptativas de duas populações caboclas (Pará) aos ecossistemas de várzea estuarina e estacional: Uma análise comparativa*. Tese (Doutorado) – Departamento de Ecologia, Universidade de São Paulo, São Paulo.

Adams, C., Murrieta, R. S. S., & Sanches, R. A. (2005). Agricultura e alimentação em populações ribeirinhas das várzeas do Amazonas: Novas perspectivas. *Ambiente & Sociedade*, 8, 1–22.

Balée, W. (1992). Indigenous history and Amazonian biodiversity. In H. K. Steen & R. P. Tucker (Eds.), *Changing tropical forests: Historical perspectives on today's challenges in Central & South America* (pp. 185–97). Durham: Forest History Society.

Balée, W. (1993). *Footprints of the forest. Ka'apor ethnobotany – The historical ecology of plant utilization by an Amazonian people*. New York: Columbia University Press.

Balée, W. (1995). Historical ecology of Amazonia. In L. E. Sponsel (Ed.), *Indigenous people and the future of Amazonia: An ecological anthropology of an endangered world* (pp. 97–110). Tucson/London: University of Arizona Press.

Balée, W. (1998). *Advances In Historical Ecology*. New York: Columbia University Press.

Begossi, A. (Ed). (2004). *Ecologia de pescadores da Mata Atlântica e da Amazônia*. São Paulo: Hucitec/Nepam/Nupaub/Fapesp.

Bezerra Neto, J. M. (2002). Os males da nossa origem: O passado colonial através de José Veríssimo. In J. M. Bezzera Neto & D. D. A. Guzmán (Eds), *Terra matura: Historiografia & história social na Amazônia* (pp. 39–65). Belém: Paka-Tatu.

Bezerra Neto, J. M., & Guzmán, D. D. A. (Eds.). (2002). *Terra matura: Historiografia & história social na Amazônia*. Belém: Paka-Tatu.

Brondízio, E. S., & Siqueira, A. D. (1997). From extractivists to farmers: Changing concepts of caboclo agroforestry in the Amazon Estuary. *Research in Economic Anthropology*, *18*, 233–79.

Browder, J. O., & Godfrey, B. J. (1997). *Rainforest cities: Urbanization, development, and globalisation of the Brazilian Amazon*. New York: Columbia University Press.

Bunker, S. (1984). Modes of extraction, unequal exchange and the progressive underdevelopment of an extreme periphery: The Brazilian Amazon, 1600–1980. *American Journal of Sociology*, *89*, 1017–64.

Bunker, S. (1985). *Underdeveloping the Amazon*. Urbana: University Of Illinois Press.

Castro, F. D. (1999). *Fishing accords: The political ecology of fishing intensification in the Amazon*. Ph.D. Dissertation, School of Public and Environmental Affairs, Indiana University, Bloomington.

Chibnik, M. (1994). *Risky rivers: The economics and politics of floodplain farming in Amazonia*. Tucson/London: University Of Arizona Press.

Cleary D. (1993). After the frontier: Problems with political economy in the modern Brazilian Amazon. *Journal of Latin American Studies*, *25*, 331–349.

Crumley, C. L. (1994). *Historical ecology: Cultural knowledge and changing landscapes*. Santa Fé: School of American Research Press.

Cunha, M. C. (2001). *Biodiversidade na Amazônia brasileira*. São Paulo: Editora Estação Liberdade/Instituto Sociambiental.

Cunha, M. C., & Almeida, M. B. (2002). *Enciclopédia da floresta, o Alto Juruá: Práticas e conhecimentos das populações*. São Paulo: Companhia das Letras.

Dean, W. (1987). *Brazil and the strugle for rubber: A study in environmental history*. Cambridge: Cambridge University Press.

Figueiredo, A. M. (1993). Pajés, médicos & alquimistas: Uma discussão em torno de ciência e magia no Pará oitocentista. *Cadernos do CFCH, Belém*, *12*, 41–54.

Figueiredo, A. M. (1998). Letras insulares: Leituras e formas de história no modernismo brasileiro. In S. Halhoub & L. A. M. Pereira (Eds.), *A história contada: Capítulos de história social da literatura no Brasil*. Rio de Janeiro: Nova Fronteira.

Figueiredo, A. M. (1999). *A cidade dos encantados: Pajelanças, feitiçarias e religiões afro-brasileiras na Amazônia, 1870–1950*. Dissertação (Mestrado) – Departamento de História, Universidade de Campinas.

Fraxe, T. J. P. (2004). *Cultura cabocla-ribeirinha: Mitos, lendas e transculturalidade*. São Paulo: Annablume.

Furtado, L. G. (1993). *Pescadores do Rio Amazonas: Um estudo antropológico da pesca ribeirinha numa área amazônica*. Belém: MPEG.

Furtado, L. G., & Quaresma, H. D. A. B. (Eds.). (2002). *Gente e ambiente no mundo da pesca artesanal*. Belém: MPEG.

Futemma, C. (2000). *Collective action and assurance of property rights to natural resources: A case study from the Lower Amazon region, Santarém, Brazil*. Ph.D. Dissertation, School of Public and Environmental Affairs, Indiana University Bloomington.

Galvão E. (1955). *Santos e visagens*. São Paulo: Companhia Editora Nacional.

Gondim, N. (1994). *A invenção da Amazônia*. São Paulo: Editora Marco Zero.

Guzmán, D. (1997). *Histórias de brancos: Memória e historiografia dos índios Manao do Rio Negro (séculos XVIII–XX)*. Dissertação (Mestrado) – IFCH-Unicamp, Campinas.

Guzmán, D. (1998). *Les chefferies indigènes du rio Negro à l'époque de la conquête de l'Amazonie, 1650–1750: Le cas des indiens Manao. Mémorie de dea en histoire et civilisations*. Paris: Ehess.

Heckenberger, M. J., Kuikuro, A., Kuikuro, U. T., Russell, J. C., Schmidt, M., Fausto, C., et al. (2003). Amazonia 1492: Pristine forest or cultural parkland? *Science*, *301*, 1710–14.

Harris, M. (1996). *People of the Amazon floodplain: Kinship, work and exchange in a caboclo community near Óbidos, Pará, Brazil*. PhD Dissertation, Department of Anthropology, University of London, London.

Harris, M. (1998a). What it means to be caboclo: Some critical notes on the construction of Amazonian caboclo society as an anthropological object. *Critique of Anthropology*, *18*, 83–95.

Harris, M. (1998b). The rythm of life in the Amazon floodplain: Seasonality and sociality in a riverine village. *Journal of the Royal Anthropological Institute*, *4*, 65–82.

Harris, M. (1999). The Brazilian floodplains: Where cholera does not kill caboclos. In S. Day, E. Parataxiarches, & M. Stewart (Eds.), *Lilies of the field. Marginal people who live for the moment* (pp. 196–211). Oxford: Westview Press.

Harris, M. (2001). *Life on the Amazon: The anthropology of a Brazilian peasant village*. London: British Academy.

Hornborg, A. (2005). Ethnogenesis, regional integration, and ecology in prehistoric Amazonia: Toward a system perspective. *Current Anthropology*, *46*, 589–620.

Leonardi, V. (1999). *Os historiadores e os rios: Natureza e ruína na Amazônia brasileira*. Brasília: Paralelo 15/UnB.

Lima, D. M. (1992). *The social category caboclo: History, social organization, identity and outsider's social classification of the rural population of an Amazonian region*. PhD Dissertation, King's College, Cambridge University, Cambridge.

Lima, D. M. (1999). A construção histórica do termo caboclo: Sobre estruturas e representações sociais no meio rural amazônico. *Novos Cadernos do NAEA*, *2*, 5–32.

Lima, D. M., & Alencar, E. F. (2000). Histórico da ocupação humana e mobilidade geográfica de assentamentos na várzea do Médio Solimões. In H. Torres & H. Monteiro (Eds.), *Populações e meio ambiente* (pp. 133–61). Brasília: Senac/Associação Brasileira de Estudos Populacionais (ABEP).

Lima, D., & Pozzobon, J. (2001). Amazônia socioambiental: Sustentabilidade ecológica e diversidade social. In I. C. G. Vieira, J. M. C. Silva, D.Oren, & M. A. D'incao (Eds.), *Diversidade biológica e cultural da Amazônia* (pp. 195–251). Belém: Museu Paraense Emílio Goeldi.

Maués, R. H. (1995). *Padres, Pajés, Festas e Santos: Catolicismo Popular e Controle Eclesiástico*. Belém: Sejup.

McGrath, D. G., Castro, F., Câmara, E., & Futemma, C. R. (1999). Community management of floodplain lakes and the sustainable development of Amazonian fisheries. In C. Padoch, J. M. Ayres, M. Pinedo-Vasquez, A. Henderson (Eds.), *Várzea: Diversity, development, and conservation of Amazonia's whitewater floodplains* (pp. 59–81). New York: The New York Botanical Garden.

McGrath, D. G., Calabria, J. A., Amaral, B. D., Futemma, C. R., & Castro, F. (1993a). Varzeiros, geleiros e o manejo dos recursos naturais na várzea do Baixo Amazonas. *Cadernos do NAEA, 11*, 91–125.

McGrath, D. G., Castro, F., Futema, C. R., Amaral, B. R., & Calabria, J. A. (1993b). Fisheries and the evolution of resources management on the Lower Amazon varzea. *Human Ecology, 21*, 167–95.

Meira, M. (1993). *O tempo dos patrões: O extrativismo da piaçava entre os índios do rio Xie, Alto rio Negro.* Dissertação (Mestrado) – IFCH-Unicamp, Campinas.

Meira, M. (2006). Índios e brancos nas águas pretas: Histórias do Rio Negro. In L. C. Forline, R. S. S. Murrieta, & I. C. G. Vieira (Eds.), *Amazônia além dos 500 anos* (pp. 183–226). Belém: Museu Paraense Emilio Goeldi.

Moran, E. F. (1974). The adaptive system of the Amazonian caboclo. In C. Wagley (Ed.), *Man in the Amazon* (pp. 136–59). Gainsville: University of Florida Press.

Moran, E. F. (1981). *Developing the Amazon.* Bloomington: Indiana University Press.

Moreira Neto, C. A. (1988). *Índios da Amazônia: De maioria a minoria* (1750–1850). Petrópolis: Vozes.

Motta-Maués, M. (1989). A questão étnica: Índios, brancos, negros e caboclos. *Estudos e Problemas Amazônicos.* Belém: Seduc/Idesp.

Motta-Maués, M. (1993). 'Trabalhadeiras' e 'Camarados': Relações de Gênero, Simbolismo e Ritualização numa Comunidade Amazônica. Belém: UFPa.

Murrieta, R. S. S. (2000). *The dilemma of the 'chibé'-eater: Food choices, ecology and everyday life among peasant communities in the Lower Amazon, Brazil.* PhD Dissertation, Department of Anthropology, University of Colorado, Boulder.

Murrieta, R. S. S. (2001a). Dialética do sabor: Alimentação, ecologia e vida cotidiana em comunidades ribeirinhas da Ilha de Ituqui, Baixo Amazonas, Pará. *Revista de Antropologia, 44*, 39–88.

Murrieta, R. S. S. (2001b). A mística do pirarucu: Pesca, ethos e paisagem em comunidades rurais do Baixo Amazonas. *Horizontes Antropológicos, 7*, 113–30.

Murrieta, R. S. S., Dufour, D., & Siqueira, A. D. (1999). Food consumption and subsistence in three caboclo populations on Marajo island, Amazônia, Brazil. *Human Ecology, 27*, 455–75.

Murrieta, R. S. S., Brondizio, E. S., Siqueira, A. D., & Moran, E. (1992). Estratégias de Subsistência da Comunidade de Praia Grande, Ilha de Marajó, Pará, Brasil. *Bol. Mus. Para. Emílio Goeldi, série Antropologia, 8*, 185–201.

Neves, E. (2000). O velho e o novo na arqueologia amazônica. *Revista USP, 44*, 86–111.

Neves, W. A. (1991). *Origens, adaptações e diversidade biológica do homem nativo da Amazônia.* Belém: Museu Paraense Emílio Goeldi.

Nugent, S. (1993). *Amazonian caboclo society: An essay on invisibility and peasant economy.* Oxford: Berg.

Nugent, S. (1997). *Big mouth: The Amazon speaks.* London: Fourth Estate.

Nugent, S. (1997). The coordinates of identity in Amazonia: At play in the fields of culture. *Critique of Anthropology, 17*, 33–51.

Oliveira Filho, J. P. (1979). O caboclo e o brabo: Notas sobre duas modalidades de força de trabalho na expansão da fronteira amazônica no século XIX. *Encontros com a Civilização Brasileira, 11*, 101–140.

Pace, R. (1998). *Amazon town: Gurupá revisited.* London/Boulder: Lynne Rienner, 1998.

Parker, E. (1985a). The Amazon caboclo: An introduction and overview. In E. Parker (Ed.), *The Amazon caboclo: Historical and contemporary perspectives* (pp. xvii–li). Studies in Third World Societies, V. 32, Williamsburg: College of William and Mary.

Parker, E. (1985b). Caboclization: Transformation of the amerindian in Amazonia, 1615–1800. In E. Parker (Ed.), *The Amazon caboclo: Historical and contemporary perspectives* (pp. 1–50). Studies in Third World Societies, V. 32, Williamsburg: College of William and Mary.

Porro, A. (1996). *O povo das águas*. São Paulo: Edusp.

Raffles, H. (2001). *In Amazonia: A natural history*. Princeton: Princeton University Press.

Raffles, H., & WinklerPrins, A. M. G. A. (2003). Further reflections on Amazonian environmental history: Transformations of rivers and streams. *Latin American Research Review*, *38*, 165–87.

Ross, E. (1978). The evolution of the Amazon peasantry. *Journal of Latin American Studies*, *10*, 30–62.

Roosevelt A. C. (1989). Natural Resources management in Amazonia before the conquest: Beyond ethnographic projection. *Advances in Economic Botany*, *7*, 30–62.

Roosevelt A. C. (1994). *Amazonian indians from prehistory to present: Anthropological perspectives*. Tucson/London: University of Arizona Press.

Salles, V. (1971). *O negro no Pará*. Belém: Fundação Getúlio Vargas.

Santos, R. (1980). *História econômica da Amazônia*. Belém: Sejup.

Schmink, M. (1985). São Félix do Xingu: A caboclo community in transition. In E. Parker (Ed.), *The Amazon caboclo: Historical and contemporary perspectives* (pp. 143–66). Studies in Third World Societies, V. 32, Williamsburg: College Of William And Mary.

Schmink, M. (2003). No longer invisible, but still enigmatic: Amazonian peasant identities and cosmographies. *Reviews in Anthropology*, *32*, 223–37.

Silva, H. P., & Eckhardt, R. B. (1994). Westernization and blood pressure variation in four Amazonian populations. *Collegium Anthropologicum*, *18*, 81–7.

Silva, H. P., Crews, D. E., & Neves, W. A. (1995). Subsistence patterns and blood pressure variation in two rural caboclo communities of Marajó island, Pará, Brazil. *American Journal of Human Biology*, *7*, 535–42.

Siqueira, A. D. (1997). *The ecology of food and nutrition: Patterns of land use and nutritional status among caboclo populations on Marajó island, Pará, Brazil*. PhD Dissertation, Department of Anthropology, Indiana University, Bloomington.

Simonian, L. T. L. (1995). Mulheres seringueiras na Amazônia brasileira: Uma vida de trabalho silenciado. In L. G. Furtado & H. D. A. B. Quaresma (Eds.), *Gente e ambiente no mundo da pesca artesanal* (pp. 97–116). Belém: MPEG.

Slater, C. (1996). Amazonia as edenic narrative. In W. Cronon (Ed.), *Uncommon ground: Rethinking the human place in nature* (pp. 114–31). New York/London: W. W. Norton & Company.

Slater, C. (1994). *Dance of the dolphin: Transformation and disenchantment in the Amazon imagination*. Chicago/London: University of Chicago Press.

Viveiros de Castro, E. Images of nature and society. *Annual Review of Anthropology*, *25*, 179–200.

Wagley, C. (1955). *Amazon town: A study of man in the tropics*. New York: Macmillan.

Weinstein, B. (1993). *A borracha na Amazônia: Expansão e decadência*. São Paulo: Hucitec/Edusp.

WinklerPrins, A. M. G. A. (2001). Why context matters: Local soil knowledge and management among an indigenous peasantry on the Lower Amazon floodplain, Brazil. *Etnoecológica*, *5*, 6–20.

WinklerPrins, A. M. G. A. (2002). Recent seasonal flooplain-upland migration along the Lower Amazon river. *The Geographical Review*, *92*, 415–31.

Part I
Identity, History and Society

Chapter 1
Utopias and Dystopias in the Amazonian Social Landscape

Stephen Nugent

Abstract This chapter starts by relating the invisibility of *caboclo* societies in the anthropological literature of the 1960s and 1970s. The invisibility of caboclo (or, to use Nugent's terminology, historical peasant) societies was grounded on four main reasons: the idealisation of the Amazonian landscape as strictly natural; the fact that the historical Amazonian peasant has never been incorporated by the plantation; Amazonia's frontier character; and last but not least, the fact that caboclo agrarian systems are neo-colonial 'experiments', significantly based on immigrant practices. Regarding a central point appearing in various degrees in a number of the articles brought together in this volume, Nugent discusses, the a-historicity that typifies a large portion of anthropological production on Amazonian societies, including caboclo. Nugent argues that the central element behind this Amazonist anthropological tradition is the ideological naturalisation of which the local human populations are victims.

Keywords Theory · History · Development · Peasantry · Santarém · Jews · Agrarian Systems

It is a great pleasure to contribute to a volume whose primary aim is to ratify the coherence of a research program in which caboclos and caboclo societies are not merely bystanders of the project of Amazonian anthropology, but occupy a central position. This is not in any sense to disavow the work of many decades in which the importance – historical, ethnographic, economic – of caboclo societies has been acknowledged, merely to underscore the fact that such study now offers a more complete rather than artefactual appreciation of the significance of caboclo societies in the emergence of a modern Amazonia.

S. Nugent (✉)
Department of Anthropology, Goldsmiths College, University of London, London, UK
e-mail: s.nugent@gold.ac.uk

C. Adams et al. (eds.), *Amazon Peasant Societies in a Changing Environment*,
DOI 10.1007/978-1-4020-9283-1_2, © Springer Science+Business Media B.V. 2009

This chapter is offered in the spirit of a workshop presentation, intended to raise issues prompted by earlier research rather than provide recent results of fieldwork. My last visit to Amazonia took place in the 2000, and the work of that period – although it has some bearing on what I have to say below – proceeds in a slightly different direction. It might be worth saying a few words about it at the outset, however, for the reception of that work has some bearing on the way in which research on Amazonia tends to be articulated to a wider audience.

The project of 2000 was the making of an ethnodocumentary about Jewish communities of the Lower Amazon that I undertook with Renato Athias. Part of the intention of the film was to document categories of standard/average/normal Amazonians who, we felt, were subsumed under a crude taxonomy consisting of few categories beyond 'indigenous Amazonians' and 'non-indigenous Amazonians'. Through a series of interviews we asked people of Alenquer, Santarém and Belém to talk about their hyphenated identities. The results, I thought, were massively pleasing: ambiguity, contradiction, uncertainty. Lacking many fora beyond circles of kin and friends to engage in such conversations, the interviewees were very candid in speaking of the mutuality of Amazonian and Jewish identity, past and present.

Screening the film before Europeans in an anthropological/academic milieu, however, it was clear that this was perceived as a film about Jews, not about Amazonia. Well, it was a film about Jews, but Jews in and of Amazonia (with the exception of two people out of two dozen, all were reared in the region, most third or fourth generation). For the audience, this was contrary to expectations, for everyone knows that Amazonia is the land of Indians, predacious colonists, clear-cutters, cattle ranches and environmental insult. That it includes Jews, Japanese, Lebanese, Dutch, French, English, Ukrainians, etc. etc. sits uneasily. It disrupts certainties about the clichés many hold dear. I return to this issue at the end of the presentation.

Although aware of the dangers of descent into whinging about the state of anthropological research into non-indigenous Amazonian societies – what I have referred to elsewhere under the generic caboclo or better, historical peasantries – I think it is worth laboring the point a bit for I feel that even an anthropology of the early twenty-first century – an anthropology that I feel has massively lost its way – has much to offer to the analysis of contemporary Amazonian societies, and paradoxically, it is precisely the neglect of modern Amazonia that holds promise for progressive tendencies within contemporary anthropology: the agrarian transformation – lusotropical style, ethnoecology, historical ecology, New Social Movements, and the maelstrom of race and class.

The utopian aspects of the title refer to these; the dystopian ones refer not so much to Amazonia per se as they refer to the predatory development of modernization studies in Amazonia, the results of which are a literature that, with many notable exceptions, continue to cast Amazonia as an exotic, ahistorical terrain in which the experiments of neo-liberal license are yet to be revealed.

If one were looking for literature on non-indigenous Amazonians in, say, the late 1960s/early 1970s, seeking for some kind of foothold as a naïve and inexperienced researcher in the North, there was precious little. Wagley (1953) (and his collaborations with Galvão) had been available for two decades, alone as a standard monograph on modern Amazonians. Staniford (1964) was narrowly focused on the internal political machinery of an ethnic enclave. Leacock and Leacock (1972) was an unusual study, one approaching the kind of fine-grained ethnography still extolled as the benchmark of anthropological study, but it was consigned to the box: anthropology of religion. Certainly the monographs of the post-WWII period made passing reference to caboclo society (cf. Murphy and Murphy (1974), usually disparaging, and there was work such as that of José Verissimo (1970) that provided a folk-culture type account, but by and large, it would be hard to pull together a literature that might encourage further research.

The reasons for this are various, but those that come first to mind are the following:

- an intransigence from a European and North American point of view vis-à-vis the dislodging of a natural neo-tropical landscape in favor of a more considered social landscape view.
- the fact that the most interesting and compelling accounts of New World peasantries shaped by the capitalist world economy (those of Eric Wolf and Sidney Mintz in Puerto Rico, Mexico and Jamaica 1957) could demonstrate a relationship between agrarian and industrial capitalism based on plantation production quite suppressed in Amazonia. Furthermore, the rise of peasant studies in Southeast Asia, Mesoamerica, Africa and Europe (and it should be noted that even as late as the 1970s, the acceptability of European peasantries as proper objects of anthropological scrutiny was officially contested) was quite provincial. The oddities of the Brazilian trajectory were difficult to include. Even Forman (1975) has almost nothing to say about Amazonia.
- Amazonian peasantries were not post-indigenous peasantries (unlike the Andes region, for example, or Mesoamerica). They were creations of colonial experiments in which labor was largely imported. Additionally, the high point of Amazonian integration in the modern world economy – the rubber era – was extractivist, not plantation based.
- finally – although not conclusively – Amazonia was a frontier in relation to the state as well as the world economy. It was not integral to state formation, and even during the rubber era, the crucial relations were between Amazonia and foreign markets relatively unmediated by the intervention of the state.

These, and other factors, place contemporary Amazonian peasantries in a strange position in relation to anthropological investigation:

- Amazonian peasant societies have non-indigenous as well as indigenous roots (i.e. they are in a crucial sense, widely portrayed as inauthentic anthropological subjects);

- the crucial event in Amazonian integration into the world economy (the rubber industry) was largely driven by remote demand (industrialization outside Brazil), and it should be remembered that the rubber epoch lasted almost 100 years, not just the 25 years of the 'boom';
- Amazonia as a tropical domain has long been extensively encoded by natural historians (Humboldt, Agassiz, Bates, Spruce, Wallace). Nancy Leys Stepan's recent *Picturing Tropical Nature* (2001), an investigation of the persistence of Victorian and pre-Victorian imagery, has no entry for caboclo.

I would like to look at three aspects of non-indigenous Amazonians (*mestiçagem*, extraction, plantations) in order to try to clarify why the marginality of caboclos is so persistent.

In a recent discussion of mestiçagem, Stuart Schwartz and Frank Saloman (1999) observe that:

> Colonial society used the language of 'birth' (a semantic field including genealogy, supposedly inherited characteristics like color and moral disposition, and hereditary status) to discuss what would later be called 'class' and 'race. The unequal parts of society were often spoken of in terms of 'blood'; people of similar 'birth' or 'blood' were suitable mates to each other. Bad or good sources and matches of 'blood' in one's genealogy determined 'purity', which had become an ideological obsession of Spanish Christians well before 1492. 'Blood' and 'purity' flourished as elements of ideology in Spain's wars against the Moors and Sephardim, and also in dealing with other kinds of Iberian diversity including the presence of enslaved Africans. (1999:444)

Thus, people of mixed birth – *mestizos*, caboclos, *mamelucos* 'formed not so much a new category as a challenge to categorization itself' (1999:482), and they conclude that mixed peoples of the Spanish and Portuguese empires represented not so much a new people as a new kind of people. They had neither corporate rights as a people, nor did they enjoy dual cultural citizenship, and 'their basic social locus depended on luck, strategy, and achievement as much as it did on ascription (1999:487).

This exclusion from categorization is, I would argue – and to use an old expression – overdetermined in Brazilian Amazonia, for the simple reason that Amazonia was not the locus of the national debates about race and ethnicity so cogently explored in Schwarcz (1999) (although Amazonia is not wholly excluded: the Museu Goeldi was, after all, the first national museum). Rather, Amazonia was in national terms the homeground of what Ramos (1998) has referred to as the hyper-real Indian whose symbolic potency is both reviled and venerated in national racialist discourse. This super-marginality is still evident in the way peoples of Amazonia (Indians and caboclos/*mestiços*) are referred to in much of the scientific literature: they are inhabitants, occupants, populations. Their categorization is uncertain.

Extractivism: it is persuasively argued – if still perhaps controversial – that the kinds of hunter/gatherer/extractivist/horticultural societies that are widely regarded as typical of Amazonian social formations are, in fact, atypical. Following the work of Lathrap (1968) and Roosevelt (1991) (and others), the argument goes that the archetypal foraging societies of interfluve Amazonia represent either culture groups driven from land adjacent to the main course of the Amazon by sedentary, proto-state social formations in pre-history or refugee societies driven into the interior following conquest. Whatever the merits of these arguments – and I think they are considerable – they are part of a characterization of Amazonia as a foraging landscape that has shaped – critically and acritically – much discussion of Amazonian sociality. The well-known arguments of Meggers (1971) (following Steward 1949), Gross (1975) and later, in the context of modernization, Bunker (1985) all allude to environmental ultimate factors that inform the few proximate factors available for modern scrutiny. The particulars of those debates over the extent of environmental constraint on the development of Amazonian societies will likely remain harshly contested until a better understanding of the pre-history of Amazonia is available, but there is an historic example of extractivism that falls well within the purview of contemporary anthropological research and that is, of course, the rubber industry. As is well known, the rubber industry fits easily into the boom-and-bust historiography of a Brazil simultaneously articulated with a global economy demanding tropical preciosities and cheap raw materials and the development of a major New World nation, yet there are significant differences between the rubber boom and the other well-known booms. Two of these are: regional development somewhat at a remove from central state control and the kinds of backward linkages (or better, absence of backward linkages) associated with this intense export-oriented production. Schwartz himself, referred to above in the discussion of mestiçagem, has provided in his study of the sugar industry of the Northeast a revealing contrast to the developments that transpired in Amazonia under the auspices of the rubber boom, a contrast with direct relevance to the analysis of peasant society in Amazonia. One of the key arguments of Schwartz (1985) is that the relative closure of the archetypal sugar plantations (African slave labor engaged in the production of sugar for external markets) is overstated. Given the costs of acquiring slave labor for sugar plantations, demand for the provision of food, firewood, and other elements crucial to sugar production laid the groundwork for the emergence of a free peasantry effectively functioning as sub-contractors in the plantation enterprise. Thus, far from being self-contained, sugar plantations created the conditions for a different set of relations of production (an extended form of informal subsumption) with the consequent development of an elaborated division of labor as well as the growth of local and regional markets. These kinds of backward linkages are relatively absent when considering the rubber industry. Whereas the sugar-plantation-created peasantry persisted after the collapse of sugar and emergence of new foci of agrarian development elsewhere in Brazil, with the precipitous disappearance of rubber extraction in Amazonia in the second decade of the last century, what remained was a skeletal version of a previously extensive mercantile-production network. This did not perhaps represent the complete economic stagnation so widely referred to in the literature, but there seems little question that the imprints of a hundred years of rubber production were shadowy (despite the persistent monuments in Manaus and Belém). The significance of this in terms of what I have referred to above as the supermarginality of Amazonian peasantries is that what transpired was a kind of extractivist involution. Non-indigenous Amazonians were depicted as occupants of a bereft social landscape. Neither authentic Amazonians (qua Indians), and largely not even of regional origins (a significant majority were *nordestinos*), these former members of the rubber brigades were portrayed as incohesive, predatory hangers-on (and, in a classic example of blaming the poor for their poverty, implicitly and often explicitly charged with responsibility for failing to rise to the challenge of Southeast Asian rubber plantations).

However formative of the modern Amazonian peasantry the rubber industry was, two factors contribute to the peripheralization of its legatees. First, they are rarely invoked as major players in the dynamic of the century long industry (although I think Warren Dean (1987) has done more than most in addressing the true role of labor in the rise and fall of the industry), overshadowed as they are by competing colonial machination versus ecosystemic constraint arguments. Second, the reversion to a pre-mercantile economy of hunter/gatherer/extraction/horticulture (a very rough and in many respects inaccurate gloss) is taken as evidence of intransigent rural defeatism. Although I have argued strenuously against this reading in *Amazonian Caboclo Society* (Nugent 1993) it would be unfair to say that the weight of opinion sides with me in seeing the post-rubber development of Amazonian caboclo society in a positive light. As a footnote, it should be pointed out that the national policies encouraging the unregulated conquest of the Amazonian frontier post-1970 have hardly enhanced the image of the Amazonian peasant, neo- or historical. The *caboclo com mota serra* is a potent symbol.

Finally in this section, plantations:

Wolf (1959) and Mintz (1974) are justly celebrated in their own rights and in collaboration for introducing into anthropology a rigorous historical perspective, especially in relation to the study of peasant societies. I suspect that the impact of their contribution is somewhat played down because of the simplicity with which they framed a mature anthropological problematic: the idea that peasant communities formed an integral part of larger, complex societies (Blok 1992), a seemingly anodyne notion, but one that directly challenges a number of key socio-cultural anthropological conceits. It is sometimes difficult to appreciate, for example, that only a few decades ago most peasant studies in Europe adopted a largely tribalist/corporatist approach: studies of peasantries were studies of peasant communities, not the larger social fields in which peasantries were implicated. (In passing, it should be noted that even the idea of studying 'peasants' was for many anathema as they were not regarded as legitimate objects of anthropological scrutiny).

Mintz's work in Jamaica is particularly relevant to questions raised by Amazonian peasantries (and their relative invisibility) because of the counter-example thrown up by sugar production and trade. He argues – in Mintz (1985, 1974) – that an understanding of the Jamaican peasantry requires an historical contextualization that reaches far beyond the island itself. A similar claim should be made for Amazonian peasantries, albeit with a different set of historical referents.

Like Jamaica, Amazonia's peasantries have their roots less in the colonial displacement of pre-capitalist structures than in the eradication of such structures and their replacement by an apparatus suited to the needs of external powers. Despite the fact that all indigenous peoples of the New World suffered enormous demographic loss in the opening stages of colonization (80 per cent widely accepted; 95 per cent according to Diamond 1997), in both Jamaica (and the Caribbean as a whole) and Amazonia – and unlike Andean and Mesoamerican societies – there is little cultural continuity between pre-historical and historical societies. The vestiges of Amazonian societies with pre-historical continuities are those which managed to evade – largely through retreat to the interior – the full weight of conquest. One consequence is that Amazonian peasantries (like Caribbean ones) are not so much artifacts of the New World, but artifacts of colonial enterprise, and one implication of this is that they are by definition 'modern'. While there are strong reasons to believe that ecological constraints militate heavily against the development of agrarian societies (bearing with them the transformations of socio-political structure so well documented globally), the kind of small-scale peasant livelihood that characterizes so much of the Amazonian social landscape is no less modern for that, yet there persists – even among those who should know better – an idea of normative capitalist development which, if absent, consigns the afflicted peasant society to the dust bin of history. W. Arthur Lewis's acclaimed two-sector model (1955) prescribes a development trajectory that hasn't occurred in Amazonia and likely never will. According to this model expansion of modern sector (industrial) development leads to an increase in labor absorption which, over time, draws workers from the countryside (attracted by higher wages, improved material conditions) and shifts the focal economic activities from rural to urban settings. This matches, broadly, what happened in Europe and North America, but is hardly comprehensive in terms of accounting for development in, say, the Caribbean or Amazonia. This is not on occasion to explore the alternative explanations provided by critics of neo-classical (or neo-liberal, for that matter) theory. It is enough to point out that the diversity of development paths requires something more than the self-aggrandizing theoretical platitudes established in the core, and – I would argue – it is precisely the failure to attend to the particularities of Amazonian development that account for the very low profile of Amazonian social actors, not least within the literature of anthropology, a field supposedly specializing in the examination of the heterogeneity of human and social possibility.

The plantation structure examined by Mintz in Jamaica is highly accessible in terms of understanding, not only because of the deftness with which he recounts the development of the sugar industry and the reciprocal transformation of societies – especially in the UK – importing sugar, but also because the Caribbean model bears some strong similarities with the agrarian transformation in Europe, and not least because it makes sense in terms of vernacular understanding of economic rationality. When we turn to Amazonia, however, we are confronted with an apparently intractable problem – or set of problems. At a trivial – but no less heavily commented on for that – level, can a definition of peasant be extended to include fisherfolk and extractivists – or taxi drivers and boat builders for that matter? There is an ample literature attesting to an environmentally-driven seasonality in terms of Amazonian peasant livelihoods, so those seeking resolution of that minor mystery can be quickly put at ease. A more substantive issue, however, concerns the well-recognized relationship between sedentism and the development of complex societies, an echo of the pervasive effects of cultural ecological arguments codified by Steward and employed subsequently by many others.

The tendency in the anthropological literature has been to treat non-agriculture-based sedentism as an anomaly, the mostly cited example probably being the extractivist societies of pre-contact Northwest America – Tlinglit, Kwakiutl - where extractive resources were sufficient to underwrite the development of a significant degree of political hierarchy and material well-being. Recent archaeological work on the lower Amazon is likely to result – among other things - in similar kinds of claims of anomaly/exceptionalism. But there are other forms of long-term viable sedentism in the Amazonian ethnographic record that are less easily set aside as exceptionalist. These would include, for example, the permanent cultivations at *Tomé-Açu* (Subler & Uhl 1990) and Uhl's case studies of small-scale dairy farms exploiting one of Pará's most unlikely resource bases, degraded pasture/*capoeira*. I mention these only in passing and to draw attention to the micro-level studies that only achieve full resonance in the context of considering the broader picture: the ways in which Amazonian peasantries conduct their livelihoods are not necessarily stipulated by the conventions of received wisdom that constitute the 'textbook' accounts of modern Amazonia.

The ahistoricity of Amazonian studies was mentioned at the outset and a number of examples have been presented in the spirit of recasting the way historical Amazonian societies are dealt with in anthropology.

The underlying causes of the reluctance to reshape the anthropological Amazonian context have been widely discussed in the (general) literature: anthropological investigation is non-reciprocal (anthropologists are interested in anthropological explanations; the subjects of such explanation, much less so); the conditions of existence of the field are inextricably linked with the systematic destruction of the societies of anthropological subjects, however inadvertently: anthropology is a legatee of an expansionist capitalist world economy commencing in the 16th century; and finally, to truncate this list, there are at least three anthropological practices operating under a generic term that obscures differences rather than providing synthesis (assuming that is possible). These three tendencies are, in rough form, the scientific, the interpretivist and the political, and while these divisions do not seem to bear particularly on contemporary Amazonianist research, they inform it from afar; and paradoxically, as the subject matter of classical anthropological investigation disappears (to take one index, of the 6000 extant languages in the world, probably half will disappear in another generation or two) specialist divisions within the field proliferate.

While there are adherents of these three tendencies who willingly tolerate an epistemological pluralism, it is more often the case the agendas are incompatible, and that/those incompatibility/ies do not merely reflect the academic concerns of anthropologists, but the wider milieu in which they operate.

A charitable reading of that incompatibility is that anthropology thrives on eclecticism (and a strong case can be made, viz Wolf). A less charitable reading is that eclecticism has different effects depending on where you are positioned within an academic field such as anthropology (which is open-ended in terms of the subject matter is presumes to drag under the spotlight) and within that world system. As an informant from the core, for example, I am amazed (probably naively) at the simplifications perpetrated under the heading 'globalization', an expression that has implications for the utopia and dystopia of the title, and an expression munificently exploited as a buzz word for new syntheses in the human and social sciences. The utopian possibility hailed in the core (if only in token manner) is one of multi-culturalism, an acquiescent multiculturalism to be sure, but one still alluding to the possibility of a legitimate diversity, while portrayals of the periphery are of dystopia (nationalist excess, corruption, bad management practices, financial collapse, eco-destruction, famine, endemic disease) as though this dystopia is self-contained and at arm's remove (although observe the politics of refugees' reception in Europe and North America at the moment).

The burden of authenticity has been a consistent feature of modern anthropology (are the Dinka and Nuer the same; is an acculturated Indian still an Indian; are there Neanderthal genes floating around among homo sapiens sapiens; is a hypenate Italian-American to be offended by *Sopranos*), and is highly marked in Amazonia where the natural landscape has long held sway above the social. The timeless green hell depiction of Amazonia that continues to inform policy-making, the conceits of the educated public as well as an anthropology still configured by the preoccupations of imperial grace and favor is a relic, albeit a persistent one. Writing from the vantage of Imperial Spanish mestiçagem, Schwartz and Salomon note that:

> Racially compromised liberalism was to have a powerful effect on understandings of peoplehood in the emerging independent states. There is, therefore, much realism in the historic tendency of South America's millions of mixed-origin persons to identify themselves not as a new people called 'mestizos' but simply and decisively as 'the people' (1999: 494).

Typically, utopias are prescriptive and anticipated while dystopias are derivative and harshly real. Caboclo society in Amazonia has traditionally been lumbered with both. From the early phases of conquest during which the pursuit of El Dorado still had plausibility to the current era in which a notion of sustainability premised on the compatibility of predatory social systems and oligarchic biosystems, Amazonian utopias have taken various idiomatic forms: the Rousseauean noble savage and Clastres, (1977) romantic anarchism are two well known versions of the social possibility, and the wise forest-manager and hi-tech extractive complex (such as *Carajás*) represent two versions of the virtuous social system/biosystem interface. The dystopias are represented in the near-extermination of the Indian, the collapse of the rubber industry and the social and biological insults perpetrated on behalf of the development enterprise/National Integration. But these terms – utopia and dystopia – while they may illuminate the margins of the large stage refer to well-known systems, plans, concepts whose salience in the lives of Amazonians has not been confirmed. Certainly there are Amazonian peasants well-versed in the utopian discourse of sustainable development, and those whose idiom is dominated by the Beatles and Xuxa; and those fully accepting of the idea that Amazonia is at the margins of the world; but as the chapters in this volume illustrate there is a vast alternative Amazonia for whom ascriptive labels have minimal relevance. A useful analogy may be drawn – perhaps over-optimistically – between the membership criteria of the MST and those of the Amazonian caboclo. João Pedro Stedile (2002: 84), in answer to the questions:

What is the structure of the MST – how many are involved? How are decisions taken, at local and national level? Begins his response:

> We are a mass social movement, whose principal objective is to gather people for the struggle. How do you join the Sem Terra Movement? There is no membership, no cards, and it's not enough just to declare that one wants to be in the MST. The only way to join is to take part in one of the land occupations, to be active on the ground. That's how we get members.

Stedile is describing one of the most important and astonishing social movements to have emerged in late modern South America, a movement membership in which is gained by participating in land occupations, not by wearing a T-shirt, waving a flag or carrying a card. Similarly, Amazonianness is confirmed not by liking *açaí*, responding to *carimbó* or carrying an INCRA receipt. While there are various official versions of Amazonia, the local/regional versions – like Stedile's account of MST membership – do not depend on formal qualifications, but on those conferred as a matter of course in being an active agent in an active social field.

In some respects, this reading of the criteria of inclusion reveals much. For one thing, it accounts for the virtual absence (or minimal importance) of hyphenated Amazonian identities. While Amazonians may have occasion to cite extra-Amazonian cultural antecedents (e.g. my grandfather was Lebanese), generally speaking this is not an issue. To be from Itaituba is to be Amazonian, whether your mother came from Paris or Rabat; the inclusiveness of Amazonian cultural identity is generous. In another respect, however, these criteria of inclusion run afoul of the protocols of social science, and this volume indicates the dimensions of the problem: on the one hand the contributors are attempting to justify the relevance of social dimensions of Amazonia that have previously evaded disciplinary capture (e.g. anthropologists study Indians, not neo-Amazonians). On the other hand, there is the danger that caboclos – say – will be transmogrified into a neo-tribe.

This is not a trivial problem, for as can be inferred from Stedile's observations, the active subjectivity of Amazonians can easily be shifted to a passive objectivity, into a social field in which the emblematic T-shirt displaces the active social force, land occupation in his case, making a living as a peasant in the case of Amazonian caboclos. The very diversity of forms of Amazonianness – of caboclo possibility – though, is some safeguard against that tendency to reify a stereotypical, intermediate caboclo category lying uncertainly between the precariously situated Indian and the fully modernized citizen of the future.

References

Blok, A. (1992). Beyond the bounds of anthropology. In J. Abbink, & H. Vemeulen (Eds.), *History and culture: Essays on the work of eric wolf.* Amsterdam: Het Spinhuis.

Bunker, S. (1985). *Underdeveloping the Amazon: Extraction, Unequal Exchange and the Failure of the Modern State.* Urbana and Chicago: University of Illinois Press.

Clastres, P. (1977). *Society against the state.* New York: Zone Books.

Dean, W. (1987). *Brazil and the struggle for rubber: a Study in Environmental History.* Cambridge: Cambridge University Press.

Diamond, J. (1997). *Guns, steel and germs: The fate of human societies.* London: Jonathan Cape.

Forman, S. (1975). *The Brazilian Peasantry.* New York: Columbia University References

Gross, D. (1975). Protein capture and cultural development in the amazon basin. *American Anthropologist* 77(3):526–549.

Lathrap, D. (1968). The "Hunting" economies of the tropical forest zone of South America: An attempt at historical perspective. In R. Lee & I. DeVore (Eds.), *Man the hunter.* Chicago: Aldine.

Leacock, R. & Leacock, S. (1975). *Spirits of the deep.* New York: Doubleday.

Lewis, W. A. (1955). *The theory of economic growth.* London: Allen and Unwin.

Meggers, B. J. (1971). *Amazonia: Man in Counterfeit Paradise.* Chicago: Aldine.

Mintz, S. (1974). *Caribbean transformations.* Chicago: Aldine.

Mintz, S. (1985). *Sweetness and Power: the Place of Sugar in Modern History.* New York: Viking Penguin.

Murphy, Y. & Murphy, R. (1974). *Women of the forest* . New York: Columbia University Press.

Nugent, S., 1993. *Amazonian caboclo society: an Essay on invisibility and peasant economy.* Oxford and Providence, RI: Berg.

Ramos, A. R. (1998). *Indigenism: ethnic politics in Brazil.* Madison: University of Minnesota Press.

Roosevelt, A. C. (1991). *Moundbuilders of the Amazon.* New York: AcademicPress.

Schwarcz, L. (1999). *The spectacle of races.* New York: Hill & Wang.

Schwartz, S. (1985). *Sugar plantations in the formation of Brazilian Society, Bahia, 1550–1835.* Cambridge: Cambridge University Press.

Schwartz, S. & Saloman, F. 1999. Introduction. S*outh America: The cambridge history of the native peoples pf the Americas.* Cambridge: Cambridge University Press.

Staniford, P. (1964). *Pioneers of the tropics.* London: Athlone Press.

Stedile, J.P. (2002). Landless battalions: the *Sem Terra* movement of Brazil. *New Left Review*, (second series) 15.
Stepan, N. L. (2001). *Picturing tropical nature*. London: Reaktion Books.
Steward, J. (ed.) (1949). *Handbook of South American Indians*, Washington DC: U.S. Government Printing Office.
Subler, S. & Uhl, C. (1990). Japanese agroforestry in Amazonia: a case study in Tomé-Açu. In A. Anderson (ed.) *Alternatives to Deforestation*. New York: Columbia University Press.
Verissimo, J. (1970). *Estudos amazonicos*. Belém: UFPa.
Wagley, C. (1953). *Amazon town: A study of man in the tropics*. New York: Macmillan.
Wolf, E. (1959). *Sons studies.of the shaking earth*. Chicago: University of Chicago Press.
Wolf, E. & Mintz, S. (1957). Haciendas and plantations in middle America and the antilles. *Social and Economic Studies 6*:411–380.

Chapter 2
Landscape Transformation and Language Change: A Case Study in Amazonian Historical Ecology[1]

William Balée

Abstract The author shows the influence of European trade and colonisation on the changes in Amerindian vocabulary. He uses the analysis to reflect on the knowledge of Amazonian landscape and associated biota. Balée is concerned specifically with the case of cacao and the way in which its denominations were transformed in the Tupí-Guaraní language, thanks to the importance of the commodity in the 18th century Amazonia. Balée shows that the socio-environmental picture in which both *caboclo* and Amerindian societies were placed was very complex, a timely reminder of the importance of a historical approach for the understanding of both.

Keywords Cacao · Linguistic change and continuity · South America · Amerindian languages · Tupí · Lingua geral

Historical Ecology and Amazonian Languages

Historical ecology is a perspective on relations between people and the environment that, in principle, envisions how historical phenomena transform landscapes and how such transformations become conditioned and understood through local knowledge, behavior, and culture over time. The current state of landscape knowledge possessed by folk (caboclo) and indigenous peoples of Amazonia is, in part, a product of history. As the landscapes have changed through time, and continue to change, that knowledge, too, shows increments in some domains, losses in others. Such losses and increments of landscape

W. Balée (✉)
Department of Anthropology, Tulane University, New Orleans, LA, USA
e-mail: wbalee@tulane.edu

[1] This chapter has also been published in slightly different form as an article in *Anthropological Linguistics*, 2003, vol. 45, no. 3, pp. 259–280, as "Historical-ecological influences on the word for cacao in Ka'apor."

C. Adams et al. (eds.), *Amazon Peasant Societies in a Changing Environment*,
DOI 10.1007/978-1-4020-9283-1_3,

knowledge are reflected in vocabulary changes, just as vocabulary can be used as an index, however crude, to knowledge of the past state of Amazonian landscapes.

Knowledge of Amazonian landscapes – in the mental control of one or more people – and based on experiential data, is at least as ancient as the Early Holocene, the presumable time of original occupation. No doubt some of that original knowledge of Amazonian landscapes has persisted, but it cannot be accessed with exactitude, since linguistic reconstruction (of words, technology, biota, and concepts) is not reliable beyond about five or six thousand years (see Kaufman, 1990), given what may be considered a background rate of vocabulary loss and change similar to the more well-accepted notion of a background rate of extinction when referring to biota over long sweeps of evolutionary time.

Archaeological data alone are insufficient to probe fully ancient knowledge of Amazonian landscapes, since knowledge is more than material artifacts: it is to be sure those artifacts, but it is also behavior and cognition, which is partly reflected in real language, including written texts. Amazonia lacks written documentation, of course, before 1500, *but one can utilize methods from historical linguistics in order to begin to build a model of landscape knowledge and the changes it underwent during thousands of years before the European conquest.*

One can demonstrate that within a five thousand year time period, however short from an evolutionary viewpoint, many of the landscapes and the languages associated with these in Amazonia underwent transformations, sometimes of a profound character. The landscape is that portion of the environment codified in language and subject to human intervention. A landscape represents an encounter between space and time, nature and history, biotic communities and human societies, and it is central to the conceptual apparatus of historical ecology. Landscape history is linked to environmental knowledge, and in Amazonia it is marked since the Middle Holocene by two deeply transforming phenomena: (1) the development of a system of swidden agriculture and fallow forest management by indigenous (i.e. pre-European) people and (2) the reconstitution of that system by neo-European expansionism, colonialism, and commercialization of existing landscapes in the New World, including Amazonia.

The focus of this chapter is on the second of these two historical phenomena, specifically on how eighteenth century colonialism, Jesuit missionization, dissemination of a contact language (*Língua Geral Amazônica*), and the penetration of the European world system of commerce and finance changed native vocabulary and hence, how these contact and colonialism transformed local interpretation and knowledge of Amazonian landscapes and associated biota. In particular, the emphasis here is on a single product, one of the *drogas do sertão*, cacao, and how the words and concepts for cacao underwent change in native languages given the fact that cacao and cacao beans were for some time in the eighteenth centuries the principal export commodity of Amazonia.

Linguistic and Cultural Background of the Ka'apor in the Eighteenth Century

The Ka'apor language (also known as *Urubu*, Urubú, and Urubu-Kaapor) is one of about forty languages in Tupí-Guaraní, itself one of ten branches of the Tupí family (Jensen, 1999; Rodrigues, 1999; Rodrigues & Cabral, 2002). Ka'apor is spoken in extreme eastern Amazonia in the Brazilian state of Maranhão, specifically in the Gurupi and Turiaçu River basins, though it has recent historical origins to the west, in the present state of Pará. Eight sub-groups of Tupí-Guaraní have been identified, chiefly in terms of phonological criteria (Jensen, 1999; Rodrigues, 1986; Rodrigues & Cabral, 2002), though there is some disagreement over what languages should be included in each of the subgroups (see Mello, 2002). Ka'apor has been classified in subgroup #8 in three slightly different iterations of this model; for the purpose of consistency, I will be using specifically the revised classification of Tupí-Guaraní proposed by Rodrigues and Cabral (2002), with the caveat that minor revisions in that model may become standard in the future. Subgroup #8 also includes Wayãpi, Guajá, and at least seven other living and dead languages (Jensen, 1999; cf. Mello, 2002).

Beatriz Corrêa da Silva (1997, p. 83) argued that Ka'apor is very close to Wayãpi in terms of phonological criteria. Indeed, Ka'apor informants have told me that based on their contacts with Wayãpi speakers (of the Wayampipuku dialect) in the Casa do Índio near Belém, Pará, they can understand Wayãpi better than other Tupí-Guaraní languages they have heard, in spite of unlike stress patterns in the two languages and a number of Carib borrowings in Wayãpi not occurring in Ka'apor. Detailed evidence from ritual also indicates an intimate association between Ka'apor and Wayãpi cultures that would have existed about three hundred years ago (Balée, 2000). But according to Corrêa da Silva (1997, p. 83), Ka'apor is unlike Wayãpi in certain morphological respects. In fact, she claims Ka'apor is more like Língua Geral Amazônica (henceforth, LGA), a Tupí-Guaraní creole known also as Nhe'engatú ('the good talk', Jensen, 1999, p. 127) and which is classified in subgroup #3, in terms of pronominal prefixes, pronominal system in general, and pronominal marking on verbs (see Corrêa da Silva, 2001).

Corrêa da Silva (1997, pp. 88–89) registered numerous, apparently borrowed, lexical items that were presumably present in Ka'apor before the Ka'apor ancestors became peaceful with Brazilian authorities and society in 1928, such as words for caboclo (Amazonian peasant), Catholic priest [*padre*], comrade [*camarada*] (or non-Indian person), Christian, mother [*mamãe*] (vocative), and father [*papai*] (vocative) [also see Balée, 1994, pp. 29–30 for similar evidence]. The supposition is that LGA was the donor language of these and other borrowed terms in Ka'apor. Wayãpi also underwent LGA influence (Jensen, 1990, cited in Corrêa da Silva, 1997, pp. 86), but perhaps not so much as Ka'apor. Where and how did this influence originate, and what, if any, implications does it have for Ka'apor nomenclature regarding natural things in their environment?

LGA made its appearance in Amazonia some time after the Portuguese founded a fort (*Forte do Presépio*) in 1616 that would become the city of Belém. LGA was based on Tupinambá spoken in the Lower Amazonian, Portuguese colony called the Province of Maranhão e Grão Pará, and it underwent significant Portuguese lexical influence. LGA was partly the linguistic product of marriages between Tupinambá women and Portuguese soldiers and colonists (Corrêa da Silva, 1997, pp. 83–84 et passim; Rodrigues, 1986, pp. 102), and partly the influence of learned Jesuits who brought many aspects of the language with them from coastal Brazil (the uniformity of Tupinambá along the coast of Brazil has been widely noted). The Tupí-Guaraní creole of southern Brazil, Língua Geral Paulista or Tupí Austral, developed in quite parallel circumstances (Jensen, 1999, pp. 127). Jesuit missionaries arrived in the region of the estuary and Lower Amazon in 1636 (Cruz, 1973), and they helped institutionalize LGA in mission settings. By 1655, there were 54 Jesuit missions in Amazonia, mostly along the Amazon River itself and south of it (Leonardi, 1999, pp. 56). LGA became the dominant language in the Brazilian Amazon and would be supplanted for the most part by Portuguese only about two hundred years later, during the rubber boom, beginning in the latter half of the nineteenth century, when hundreds of thousands of monolingual immigrants from northeastern Brazil arrived in the region to take up a life of rubber tapping (Leonardi, 1999, pp. 75; Moreira Neto, 1988, pp. 43–45).

By the time of the rubber cycle in Amazonian history, the Ka'apor as a people had long been isolated from and hostile to rubber tappers and Luso-Brazilian society generally (Balée, 1984). In other words, the Ka'apor were never 'caboclos' per se. Rather, the close of the colonial period of Ka'apor history in the mid to late 1700s helps us to comprehend better the beginning of caboclo history, for it seems to be at this time that the caboclos emerge as a people separate from whatever indigenous roots they had, and the Ka'apor and other indigenous groups—continuing to be indigenous – on the one hand and the Lower Amazonian peasantry on the other then diverge and go their separate ways in historical time down to the present day.

At one time, however, the Ka'apor as a people apparently enjoyed relatively peaceful though probably subordinate relations with representatives of the Iberian metropole, especially Jesuit missionaries with whom they would have been in daily, face-to-face contact at least until the expulsion of the Jesuits from Brazil in 1759 (Azevedo, 1930, pp. 375, cited in Balée, 1988, pp. 157). This is the period that in my view should be understood as being immediately at the eve of the formation of the Amazonian peasantry (see Nugent, this volume, who prefers 'historical peasantry' to the use of the term 'caboclo' for the purpose of describing the extant, native-born, Portuguese-speaking people of Amazonia). This time of hypothesized contact between precursors of Ka'apor society and colonial Luso-Brazilian society constitutes the period shortly before the influx of African slaves into the Lower Amazon (see Chapter 3 by Guzmán, this volume), who came to replace waning indigenous labor and populations. And this contact period occurs just before the coining of the neologism 'caboclo,' which would be used in the following years to refer to the Amazonian masses as distinct from individually named indigenous groups of Amazonia. This time frame is not coincidentally contemporary with the expulsion of the Jesuits and the demise of the Jesuit mission system in the Lower Amazon (and elsewhere) in 1759.

On the basis of ethnohistory, oral history, and linguistic data relating to toponyms, Ka'apor society originated at least four hundred kilometers to the west of their present habitat probably before 1800 in the basin of the Tocantins River (Balée, 1994, pp. 30–32) (see Fig. 2.1). Before 1759, the Ka'apor probably

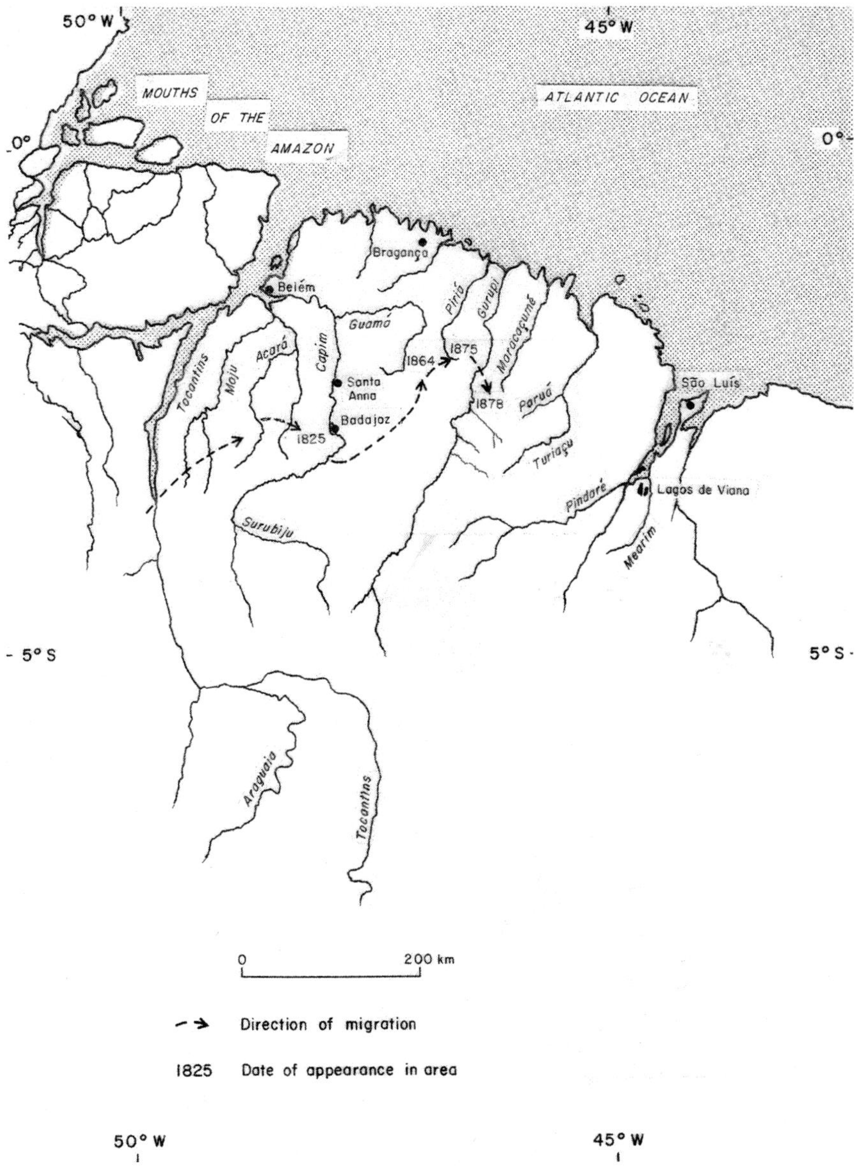

Fig. 2.1 Map of Lower Amazon and the long-term migratory movements of the Ka'apor

lived even farther west, nearer to the Xingu. This is because of a historical connection that seems to have existed between the Wayãpi and the Ka'apor. The Ka'apor and Wayãpi share some esoteric details of a girl's puberty rite that are most likely not due to chance (Balée, 2000, p. 412). The details center on the ant ordeal, which in both societies' cultural practices involves the application of venomous stings on the initiate's skin from the same species of ant (*Pachycondyla commutata*). That ant is called by apparently cognate terms in the two languages (Ka'apor *tapiña' ĩ* and Wayãpi *tapia'i*). This ant ordeal at girl's initiation has not been described for any other pair of Tupí-Guaraní societies, which indicates evidence of shared innovation between Ka'apor and Wayãpi ancestral sociocultural and ritual systems. The ant ordeal, therefore, suggests a historical connection between the two peoples in the comparatively recent past, a connection that further supports their linguistically close pairing in subgroup #8.

Today the Wayãpi and Kaapor live about nine hundred kilometers apart with the estuary of the Amazon River in between them (see Fig. 2.2). But in the early 1700s, the precursors of the Wayãpi lived in the Lower Xingu River basin

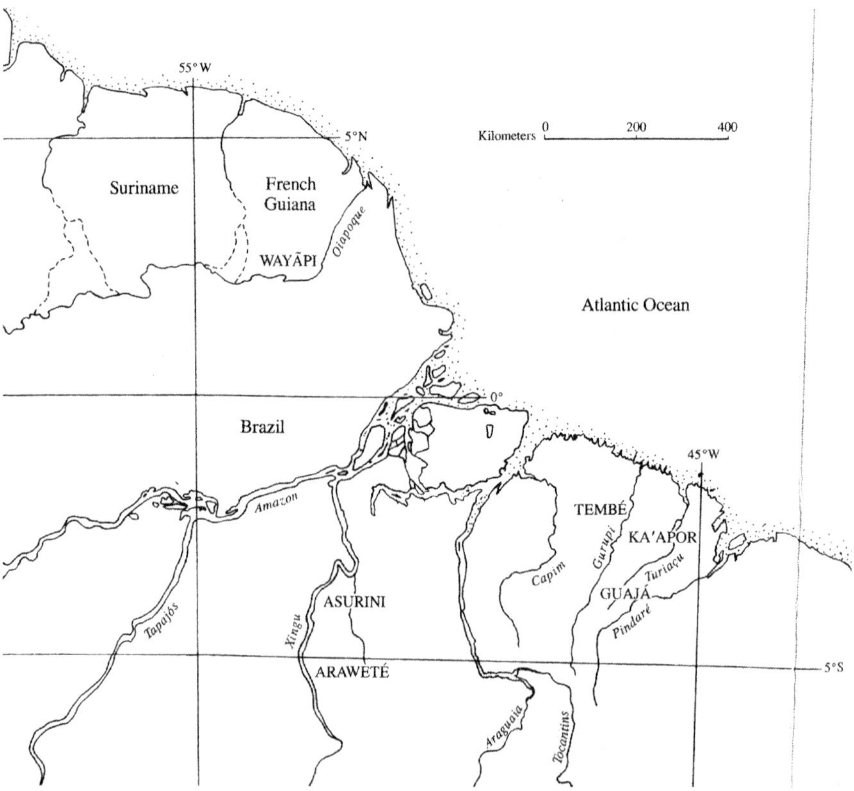

Fig. 2.2 Map of the Lower Amazon showing the location of some Tupi-Guarani people mentioned in the text

and some of them became settled one of three Jesuit missions at that time (Grenand, 1982, p. 20, cited in Corrêa da Silva, 1997, pp. 84–85; Fisher, 2000, p. 46). Evidence from Ka'apor mythology and concepts of species, such as the Brazil nut tree, which are not present in their habitat today, indicates a westerly origin at least as far west as the Rio Tocantins, called *ɨ-takaš ĩ* (i.e. >Smoke River') in Ka'apor (Balée, 1994, p. 25). But because of their close linkages to Wayãpi, both from the perspectives of ritual and language, antecedents of the Ka'apor can be logically placed even farther west than the Tocantins, indeed closer to the Xingu but probably still east of that river, in the early eighteenth century.

The impact of LGA on Ka'apor language and Luso-Brazilian, Jesuit influences on Ka'apor culture, seem to be related to what at one time was the principal commodity extracted by the Iberian metropole from Amazonia: cacao. The word for cacao and the origins, uses, and management of cacao by native peoples in Amazonia in prehistoric times represent an inimitable array of historical-ecological phenomena that allow us to understand Amazonian history within a capsule of a single species and its ultimate effect on the landscape. In this sense, a change in the material and economic landscape—namely, what would become the paramount importance of cacao as a commodity in the export market of the drogas do sertão and the acquisition of native labor to gather it—may have affected the Ka'apor language in the domain of plant nomenclature.

Origins of the Word for 'Cacao' in Various Languages

Where did the word for cacao come from? Mesoamericanist J. Eric S. Thompson indicated that the origins of the words cacao and chocolate are not easily found. There has been a considerable amount of speculation on the subject, but it is to be doubted that any conclusions satisfactory to everyone will ever be reached (1956:107). Written before an explosion of historical-linguistic and epigraphic research in Mesoamerica, Thompson was perhaps too pessimistic, though it must be granted that all linguistic reconstruction as with much of archaeological interpretation must remain speculative, however informed and enlightening.

At least one way of approaching Thompson's problem would be to seek the word where the plant itself originated. This exercise involves consideration of cultural factors, since the cacao of commerce (*Theobroma cacao* L.) is a domesticate.

Two subspecies of cacao are recognized (Cuatrecasas, 1964, pp. 512–513), and the principal subspecies of modern commerce, *T. cacao* ssp. *cacao* (with four formae), was the only domesticated one found in Mexico and Central America at the time of the Hispanic Conquest. Two commercial types are known: *criollo* (*T. cacao* ssp. *cacao*) and *forastero* (which may include other subspecies, all from South America). *Criollo* has 'elongated, ridged, pointed fruits and white cotyledons' while *forastero* has 'short, roundish, almost smooth fruit and purplish cotyledons' (Cuatrecasas, 1964, p. 506; also see Schultes, 1984; Coe & Coe, 1996, p. 27). The *criollo* variety of Mexico and Central America does not grow spontaneously; in contrast, other *forastero* subspecies can be found growing spontaneously in various parts of the Amazon Basin (Huber, 1904 cited in Cuatrecasas, 1964, p. 401; Cavalcante, 1988, p. 63) and the Guianas (Cuatrecasas, 1964, p. 494, map). Indeed, two morphological variants are noted, an Upper Amazon Forastero and a Lower Amazon Forastero (Motamayor, Risterucci, Laurent, Moreno, & Lanaud, 2000). Today *forastero* subspecies and varieties derived from *T. cacao* ssp. *sphaerocarpum* have become the most important in commerce (Gómez-Pompa, Flores, & Fernández, 1990, p. 249), accounting for about eighty percent of world production (Coe & Coe, 1996, pp. 28, 201–202). The pre-contact distribution of many spontaneous varieties in South America and only one, fully domesticated variety in Mesoamerica bespoke the possibility that cacao originated in headwaters of the Amazon, crossed the Andes into northern Colombia, and ultimately made its way to Central America and lands farther north (Cheesman, 1944 cited in Cuatrecasas, 1964, p. 507).

The age-area (or 'least-moves') hypothesis is clearly strengthened by the fact that all twenty-two known *Theobroma* species were originally found in the Amazon Basin and adjoining Guianas and only three (*T. cacao, T. angustifolium,* and *T. bicolor*) have ever grown outside that region. Cuatrecasas (1964, p. 507) confidently asserted, nevertheless, that the first prehistoric cultivation and selection of cacao occurred in Mexico and Central America and subsequent writers have tended to support that claim (e.g. Stone, 1984, p. 69). Gómez-Pompa et al. (1990) presented recent evidence for a possible ancestral form to domesticated cacao, which was noted to be growing in a sinkhole in northern Yucatán. This variety is the rare *T. cacao* L. ssp. *cacao* forma *lacandonica* Cuatrecasas, which was previously only known from the Lacandon Maya area of Chiapas, Mexico (Coe & Coe, 1996, pp. 26–27). Linguistic evidence to date also seems to support an original domestication of cacao in Mesoamerica though the precise language of origin is a matter of dispute.

One account argued for a source of Mayan **kakaw* in Mixe-Zoquean (Justeson, Norman, Campbell, & Kaufman, 1985, p. 59), a putative source of borrowings in Mayan and other Mesoamerican language groups (Campbell and Kaufman, 1976, p. 84). According to this view, many Mixe-Zoquean agricultural terms were borrowed by Mayan and other Mesoamerican language groups, reflecting perhaps the prestige of the proposed first agricultural civilization of the region, the Olmecs. The Olmec civilization might have been associated with speakers of Mixe-Zoquean (Campbell and Kaufman, 1976, p. 84), though this inference too is debatable (Wichmann, 1999). More recently, an argument has been made that cacao is actually a term coined by speakers of Nahuatl, perhaps the people whose capital city was Teotihuacan (Dakin & Wichmann, 2000).

Regardless of which Mesoamerican linguistic group is eventually determined to be the source of the term 'cacao' (i.e., Uto-Aztecan, Mayan, or Mixe-Zoquean), the use of cacao in Classic Maya culture (ca. 200 B.C.–A.D. 600) is now well established. Biochemical evidence for theobromine, one of cacao's characteristic alkaloids, has now been determined to exist on remains of spouted vessels (called 'chocolate pots') in northern Belize that date from 600 B.C. to A.D. 250, i.e., from the time of the Preclassic Maya culture to the beginnings of Classic Maya culture (Hurst, 2001; Powis, Valdez, Hester, Hurst, & Tarka, 2002). By Classic times, cacao is evidently a local crop grown widely in Mesoamerica, including in peripheral areas such as the medium-sized village site of Cerén in El Salvador (Lentz & Ramírez-Sosa, 2002). In other words, it was not evidently a crop only of the elite, but of the common people living on the periphery of urban civilization as well.

After the Spanish conquest of Mesoamerica, with the debut of chocolate in the European marketplace and the rapid conditioning of the Western palate by it, the term cacao became widely diffused to numerous languages worldwide. In Hanunóo of Mindoro Island, Philippines, two of the three words for folk species of cacao exhibit the morpheme *kakaw* (Conklin, 1954, p. 418), no doubt borrowings from Spanish. In the Quichua language of Amazonian Ecuador, all compound names for two species of *Theobroma* (*T. cacao* and *T. subincanum*) incorporate the term *cacao* (Kohn, 2002, p. 432). Many other Amazonian and Lowland South American groups borrowed a term for 'cacao' that entered the continent through Spanish or Portuguese. What is of most interest is why would they, and in particular the Ka'apor, borrow a term for a plant that they already had?

According to the historical-linguistic principle of prestige, whereby in a contact situation goods and services associated with the dominant society that were not previously present in the subordinate society tend to be borrowed by the subordinate society (see Campbell, 1999, pp. 59–60), the word for cacao would not have been borrowed by Ka'apor since it already occurred in their environment, unless cacao had acquired some prestige and economic valorization far above and beyond what it held in native Amazonia. Ethnobotanist Richard Evans Schultes (1984, p. 33) observed that it was difficult to explain why Amazonian Indians would have been motivated to disperse a tree the use of which lay solely in a sweet pulp on which one might suck (also see Coe & Coe, 1996, p. 26 for a similar view). Cacao cultivation in Mesoamerica is probably as old if not older than the Tupí-Guaraní branch of the Tupí family, dating back at least to the beginning of the Common Era and probably much earlier (cf. Alden, 1976, p. 104; see Young, 1994, p. 17), even if the word for cacao may be more recent than its original cultivation (Dakin & Wichmann, 2000). Plant geneticist Charles Clement (1999, p. 201) pointed out that *T. cacao* and its close relative *T. bicolor* (which may grow spontaneously in the Maya lowlands unlike *T. cacao*, though it is of lesser quality and desirability (Thompson, 1956, p. 107) were probably semi-domesticated crops grown as stimulants in the Upper Amazon during late prehistoric times. But the use of cacao beans as stimulants is seldom found outside the Upper Amazon. The Kofán of the Ecuadorian Amazon toast and eat the beans of *T. bicolor* (which they term *mak'av* ɨ-) [Pinkley, 1973, p. 69] as do the Lowland Quichua of Ecuador (Eduardo Kohn, 2001, personal communication). The practice of toasting these beans before consuming them seems fairly widespread in the Upper Amazon, despite the avowedly low quality of the beans and fruit when compared to other species of *Theobroma* (Cavalcante, 1988, p. 66). In any case, no prehistoric Amazonian groups are known to have made chocolate (Schultes, 1984, p. 33; Stone, 1984, p. 69; Gómez-Pompa et al., 1990, p. 249).

Rather, almost everywhere outside the Upper Amazon, native Amazonians have eaten only the sweet, white pulp around the beans and then discarded the beans; in some cases, the pulp around the beans has been made into a non-fermented wine (Coe & Coe, 1996, p. 26). Given the low aboriginal prestige of cacao in the Amazon region, the directionality of borrowing of the term is probably not, basically, Amazonia→Mesoamerica, but rather, the reverse seems much more probable now. It is unlikely that Mixe-Zoquean speakers, who may have been already associated with complex, intensive agricultural society, would have borrowed an Amazonian term for a semi-domesticated (or perhaps even wild) crop that had not yet developed uses as chocolate. And the possibility remains that the development of chocolate production in Mesoamerica began with *criollo* trees that had arisen from spontaneous mutations and subsequent genetic drift along the isthmus of Panama, not far from the northernmost edge of the presumed, original distribution of cacao (Purseglove, 1969, cited in Young, 1994, p. 14–15). It is possible therefore that cacao was not dispersed into Mesoamerica by humans and was part of the original distribution of wild forms of cacao, such as the forma *lacandonica* (Gómez-Pompa et al., 1990, p. 249), but this remains controversial (Stone, 1984; Young, 1994, p. 14).

The first European observation of cacao occurred in 1502 along the northern coast of Honduras, on Columbus' fourth voyage (Alden, 1976, p. 104). Rapidly, the chocolate drink made from it became highly esteemed in Europe (Alden, 1976, p. 109), and it became well known to explorers as a valuable export crop. Cacao plantations begin in Ecuador and Venezuela by the late 1500s and early 1600s. The crop therefore may have been recognizable to Spaniard Cristoval de Acuña, who noted in 1641 that in some places groves of cacao trees along the Amazon River were so thick that the wood could serve to lodge an entire army (1963:76). Cacao exports from the Amazon were reported by 1678–1681, and these beans were being collected from spontaneously occurring trees, not plantation trees (Alden, 1976, pp. 114–115). By about 1725, a cacao boom started in the Amazon, and cacao becomes the dominant export staple of the region (Alden, 1976, p. 118; cf. Hemming, 1987, p. 43). By the mid-1700s, different regions of Brazil exported distinctive commodities to Lisbon. 'The Rio fleet shipped gold, hides and silver; Pernambuco sent wood and sugar; and the fleets of the north [i.e., lower Amazon], of Grão Pará and Maranhão carried cacao ...' (Maxwell, 1973, p. 5). The cacao export sector of the eighteenth century Luso-Brazilian economy was perhaps minor compared to gold in Minas Gerais and Rio de Janeiro and later coffee in São Paulo (Baer, 1995, pp. 15–19), but it seems in many ways to be the precursor of the rubber export economy of the nineteenth century as concerns the Amazon region.

The cacao export sector of the colonial Amazonian economy fell under the control of Jesuit missionaries, who induced Indians under their tutelage to collect cacao in the interior from spontaneous trees, whereas significantly less cacao came from plantations (Alden, 1976, pp. 121–122; Hemming, 1987, p. 43; Coe & Coe, 1996, pp. 194–195). These spontaneous trees were most likely from *Theobroma cacao* and not from nondomesticated species of *Theobroma*. Although *Theobroma speciosum* Willd., a nondomesticated and very widespread cacao species known regionally as *cacauí* (little cacao) produces edible pulp and seeds from which chocolate can and has been made, its fruiting season is only between February and April, hardly enough time to qualify as a major export crop. Amazonian *Theobroma cacao*, in contrast, can be found for sale in all months except September to December at the market in Belém (Cavalcante, 1988, p. 64).

Remarkably, as a percentage of the total exports from the Lower Amazon during 1730–1755, cacao alone ranges between 43.5 and 96.6%, with the highest proportion of total exports from that region occurring in the years 1730–1745 (Alden, 1976, p. 118). The cacao trade begins to decline in the 1740s and 1750s, and this coincides with native population declines due to smallpox and measles epidemics widely reported during the period 1743–1750 (Balée, 1984, pp. 34–35; Hemming, 1987, p. 43; Moreira Neto, 1988, pp. 23–24). African slavery revived the trade after the 1750s, such that what is now the Brazilian state of Pará was exporting 715–850 tons of cacao per year, which constituted about ninety percent of the total from Brazil (Hemming, 1987, p. 43). Even after the expulsion of the Jesuits from the Portuguese Empire in 1759–1760, most of the export from the Amazon still came from collecting expeditions rather than from cultivated trees (Alden, 1976, pp. 123–124), and cacao would not become a dominant export crop from the Brazilian state of Bahia until the late nineteenth century (Baer, 1995, p. 19).

The impact in Amazonia of a cacao export economy combined with Jesuit control seems to have affected native languages. Indeed, the significance of the cacao export sector in the Lower Amazon cannot be overestimated in terms of its effects on local indigenous societies and their languages that were involved in it. In 1743, cacao is clearly the most important of all the drogas do sertão (the various forest and garden products from Amazonia that were shipped to Europe for a variety of purposes: food, spice, medicine, oil, hides, skins, timber, waxes, gums, and so on—see Cleary, 2001, pp. 83–84) for at that time cacao beans were observed to be circulating as money among the Amazonian peasantry (not dissimilar to the way cacao beans had served a monetary purpose in Aztec markets) and cacao beans owned were figured into calculations of an individual's wealth (Bruno, 1966, p. 59). In the colonial era, cacao had assumed an importance it had not been before known in aboriginal Amazonia.

Cacao was a central commodity in the 'Jesuit century' as David Block (1994, p. 98) has so aptly described the eighteenth century in eastern Bolivia, which can apply with slight modifications also to Amazonian Brazil, coastal Brazil, and the mission zones of Paraguay and Argentina. The Jesuits introduced cacao into the Mojos Plains of eastern Bolivia (also known as the Beni), where it had not even existed in the wild before, as an export crop (Block, 1994, p. 98). Cacao was probably not typically planted in pre-Columbian Amazonia, but the Jesuits, using native labor, cultivated it successfully in the vicinity of their Amazon missions (Aubertin, 1996, p. 32; Bruno, 1966, p. 61). Indeed, had it not been for the Jesuits, the ability of Amazonia to meet European demand for chocolate would not have been met (Aubertin, 1996, p. 33). The drogas do sertão—a bona fide term for what today one might call TFPs (tropical forest products—see Cleary, 2001, p. 83–86)—constitute a very long list of wild and cultivated plant and animal materials, but in terms of economic impact both in the Amazon and in the European marketplace, cacao was at the top of this list (Di Paolo, 1985, p. 76).

The Jesuits used LGA, a creole language partially derived from Tupinambá, in their missions. Many LGA vocabulary items are borrowed from Portuguese. In cases of language contact, vocabulary items for native plants, animals, and landscape features are most often borrowed by the dominant or prestige language and vocabulary items related to politics, religion, and finance are most often borrowed by the subordinate or nonprestige language, based on contrasting principles of prestige (i.e., luxury loans) or need (Campbell, 1999, pp. 59–60). Cacao is a native Amazonian plant, so by the principle of need, it is reasonable to argue that the term came originally from some Amazonian language. Cacao is a Portuguese word borrowed from Spanish *cacao* that was in turn borrowed from a Mesoamerican language, where the plant first attained preeminence in terms of world commerce. Controversial evidence suggests that cacao can be reconstructed in Proto-Mixe-Zoquean, which dates from about 3,500 B.P. (Campbell, 1999, p. 349; also see Campbell and Kaufman, 1976; Justeson et al., 1985) and was plausibly associated with the ancient Olmec civilization of the isthmus of Tehuantepec, as **kakawa* (Campbell and Kaufman, 1976, p. 84) The possibility remains that Proto-Mixe-Zoquean borrowed the term from an Amazonian language on the basis of need, if the crop indeed originated there (though perhaps not as a fully domesticated crop) as biogeographic evidence suggests. But the prestige principle and the known time frame militate against that hypothesis.

Cacao Words and Tupí-Guaraní Languages

In several Tupí-Guaraní languages of Amazonia, *Theobroma cacao* L. ssp. *sphaerocarpum* is referred to by words that seem cognate by inspection (Table 2.1), the exceptions being Ka'apor (because of an initial *k*—see below), Parintintin, and Wayãpi. What is puzzling is that the other, seemingly cognate words, resemble the word *cacao* in their phonetic shape. At least some of these languages might be

Table 2.1 Words for cacao (Theobroma cacao L.) in several Tupi-Guaraní languages

Language	Subgroup #	Term[1]	Gloss[2]	Source
Araweté	5	*aka-'i*	L-tree	Balée, field notes, 1985
Assurini do Xingu	5	*aka-'ɨ-wa*	L-tree	Balée, field notes, 1986
Guajá	8	*ako'o-'ɨ*	L-tree	Balée, field notes, 1989
Ka'apor	8	*kaka*	L	Balée, field notes, 1985
Tembé	4	*aka-'ɨ-w-ete*	L-stem-true	Balée, field notes, 1986 (Cf. Boudin, 1978)
LGA	3	*kakáu*[3]	L	Stradelli 1929
Parintintin	6	*ñumi-*	L	Betts 1981
Wayãpi	8	*walapulu*	L	Grenand, 1989

[1]Hyphens indicate morpheme boundaries.
[2]The L refers to a literal, monomorphemic, essentially nonpolysemous plant term (see Balée & Moore 1991). English plant morphemes that heuristically meet this criterion would be "oak," "maple," "pine."
[3]This term refers to the fruit of the cacao tree only.

presumed to have had little if any influence from LGA, especially Guajá (a language of hunter-gatherers who have only been in contact since the 1970s) and Araweté (a language of trekkers only in contact also since the 1970s). But both these languages have a word for comrade (Guajá *kamarar*; Araweté *kamara* – Balée, field notes), borrowed evidently from medieval Portuguese *camarada*; corresponding borrowed terms are also known from Ka'apor (*kamarar*) and LGA (*kamarára*) [Corrêa da Silva, 1997:89], though the range of meaning among them is somewhat divergent, since at least in Guajá *kamarar* refers to the Ka'apor people, whereas in the other languages mentioned the cognate term refers to non Indians, or is even, in the case of Araweté, a personal name for a man.

These five languages are in three different subgroups (nos. 4, 5, and 8) of the eight recognized subgroups of the Tupí-Guaraní branch of the Tupí family (Jensen, 1999; Rodrigues and Cabral, 2002). The phonological structure of the terms apart from the word in Ka'apor in Table 2.1 does not suggest borrowing among the different languages. It is possible that Ka'apor has conserved an initial *k in the word for cacao and that the initial consonant was deleted in Araweté, Assurini do Xingu, Guajá, and Tembé. The proto-term may have been *kaka, and this would be far older than the cacao export economy of lower Amazonia in the 1700s. But this hypothesis seems unlikely. The principle of prestige would tend to preclude a nonprestige language from borrowing a term for a native plant that was not of commercial or agricultural importance. The nondomesticated, widely occurring cacao species, *Theobroma speciosum* Willd., is either designated by the same term (as in Assurini do Xingu and Guajá) or it is linguistically marked as though it is perceived as being a close relative of domesticated cacao (from the point of view of nomenclature, not classification per se) (see Table 2.2).

Table 2.2 Words for non-domesticated cacao (Theobroma Speciosum willd.) in several Tupi-Guaraní languages

Language	Subgroup #	Term[1]	Gloss[2]	Source
Araweté	5	*aka-á-wi'i*	L-fruit-thin-stem	Balée, field notes, 1985
Assurini do Xingu	5	*aka-'ɨ-wa*	L-stem	Balée, field notes, 1986
Aurê and Aurá[3]	8	*aka-ú*	L-large(?)	Balée, field notes, 1987
Guajá	8	*ako'o-'ɨ*	L-stem	Balée, field notes, 1989
Ka'apor	8	*kaka-ran-'ɨ*	L-false-stem	Balée, field notes, 1985
Tembé	4	*aka'u-'ɨ-w*	L-stem	Balée, field notes, 1986, (Cf. Boudin 1978)
Wayãpi	8	*aka-'ɨ-w*	L-stem	Grenand, 1989
Proposed Reconstruction	–	**ako'o-'ɨ-ß*	L-stem	–

[1]See footnote 1, Table 2.1.
[2]See foonote 2, Table 2.1.
[3]Aurê and Aurá, the only known speakers of a newly recorded Tupí-Guaraní language originally spoken between the Xingu and Tocantins Rivers (Jensen, 1999:128; Mello, 1996).

In other words, in keeping with the prestige principle, one would not anticipate borrowing of terms for nondomesticated, seemingly unimportant plants (though nondomesticated cacao, especially *Theobroma speciosum* Willd., like its domesticated congener, does have a sweet, edible pulp, and people gather it for that purpose). But that evidently happened in Ka'apor. The Ka'apor words for *Protium* trees (Burseraceae), *Lacmellea* trees (Apocynaceae), and *Mabea* trees (Euphorbiaceae), all of which are found in high forest and are never cultivated per se, seem to have been borrowed also from LGA (Balée, 1994). It is plausible that products from these trees were part of the drogas do sertão transoceanic trade; *Protium* trees, for example, exude a resin that is highly prized as boat caulking, and caulks were one of the Amazonian drogas do sertão.

Wayãpi, which like Ka'apor is from subgroup #8, denotes domesticated cacao as *walapulu*, clearly a borrowing from one of several Carib languages in the Guianas (Grenand, 1989). Yet the Wayãpi term for nondomesticated cacao, *T. speciosum*, is *aka-' ɨ-*, an apparent cognate with the terms, aside from the Ka'apor term, in Table 2.2. Françoise Grenand (1989:121) gives the etymology as *aka* < *ãkã* >head' and suggests also a comparison with LGA *kakao-'i* >little cacao.' Her etymology of >head' seems problematic, however, since as in Wayãpi, the vowels are also not nasalized in the cognate terms in the five other Tupí-Guaraní languages in Tables 2.1 and 2.2. It seems unlikely that nasalization for this wild cacao word would have been dropped in all of them just as deletion of initial *k* in the cacao word in three different subgroups of Tupí-Guaraní also seems unlikely. Initial consonant loss is, moreover, less common than initial vowel loss (Campbell, 1999, pp. 32–33). On the basis of this mounting evidence, one can logically argue that (a) the original term in Wayãpi for nondomesticated cacao is a sequence of a literal morpheme (*aka*) and a term meaning 'stem' or 'tree' ('ɨ-); (b) the Ka'apor terms for cacao and nondomesticated cacao are most likely to have been borrowed; and (c) the donor language for the cacao terms in Ka'apor was LGA.

The LGA term for cacao fruit is *kakáu* (Stradelli, 1929). In LGA, dipthongs may occur in word final position. The combination of two vowels in principle represents two syllables (Taylor, 1985, pp. 11–12). But in LGA one does not canonically find the following

$$\underline{\mathbf{V}}[+\mathbf{high}, +\mathbf{back}, +\mathbf{vocalic}]\#$$

This combination of phonemes is otherwise common in Portuguese, as in /*páu*/ >wood, tree' and /*kakáu*/ >cacao.' It can be therefore proposed that the directionality of borrowing was LGA→Ka'apor, and not the reverse. Ka'apor retained the initial *k* when it borrowed the term, and phonological substitution (in this case, by deletion of final vowel or apocope – Campbell, 1999, pp. 32, 61) accounts for the absence of the unstressed final, high back vowel in Ka'apor *kaka*. Ka'apor also extended the root lexeme's semantic range to nondomesticated cacao, analogous to the extension noted by Grenand above (1989, p. 121) for LGA. The reason why the term for nondomesticated cacao persisted in Wayãpi is that perhaps Wayãpi was less affected by missionization influences than Ka'apor and because nondomesticated cacao was not an item of prestige, whereas domesticated cacao was a prestigious commodity thanks to Jesuit and Luso-Brazilian valorization and cultivation of it. It is striking nevertheless that the Wayãpi have a Carib loanword for domesticated cacao; it is possible from this evidence, and from the other evidence related to a strong ethnic and linguistic connection between the Ka'apor and Wayãpi, to speculate that Wayãpi once had a term like *kaka* in Ka'apor and exchanged this for walapulu at a later date, after they crossed the Amazon River from the south, but to date had not yet gone so far as to replace the term for nondomesticated cacao (and remodel it by analogy on walapulu or some other borrowed term for cacao).

This argument leaves open whether the other Tupí-Guaraní languages in the sample also borrowed the word for cacao from LGA. Although deletion of all initial *k*'s seems unlikely, the Parintintin language represents a peculiar departure from the other languages in the sample. Parintintin is from subgroup #6 of Tupí-Guaraní; it is spoken in southwestern Amazonia, close in fact to where Proto-Tupí-Guaraní is believed (using the least-moves hypothesis) to have originated, and it evidently has had little or no LGA influence, for it seems to have been beyond the distribution of the Jesuit missions. Parintintin has the focal generic name *ñumi-* for cacao and many of its relatives (Betts, 1981; Waud

Kracke, personal communication, 2001).[2] Parintintin is also located in the richest area of the genus *Theobroma* in the Amazon basin. At this point, the lexeme *aka* (Araweté, Assurini do Xingu, Tembé, and Wayãpi) or *ako'o* (Guajá, which is arguably closest to the proto-language for this term) may resemble LGA *kakáu* only by coincidence or by borrowing. It is nevertheless intriguing that whereas Ka'apor arguably borrowed *kaka* and extended it to cover non-domesticated cacao species, as discussed above Wayãpi also borrowed walapulu (from a Carib language) for cacao but retained *aka* for the nondomesticated cacao species. If *aka* is closer to the original Proto-Tupí-Guaraní word for cacao than *kaka*, then cacao was most likely not borrowed by Mesoamerican languages from Tupí-Guaraní languages even if *aka* has cognates in Tupian language branches other than Tupí-Guaraní. That is because word-initial epenthesis of a consonant is not likely (Campbell, 1999, p. 33). The only published data now available on another branch of Tupí is from Munduruku, of the Munduruku branch, and the word for cacao appears to be a borrowing, also from LGA, being *kakau* (Strömer, 1932, p. 62; but Crofts and Sheffler (1981, p. 18) indicate *karoba* as the Munduruku term for cacau).

Discussion

In the Ka'apor habitat of today, there are four species of *Theobroma* other than *T. cacao* and *T. speciosum*. These are *T. grandiflorum* (Willd. Ex Spreng.) Schum. called in Ka'apor *k ɨ-p ɨ- hu' ɨ-* and *T. subincanum* Mart. called *k ɨ-p ɨ-'a' ɨ-* for which there is also a synonym,

nuk ɨ-p ɨ- 'ɨ- (Balée, 1994, p. 307). These terms do not appear to be related to the Ka'apor terms for *T. cacao* and *T. speciosum* and they are in a different folk genus. Indeed, the fruit of *k ɨ-p ɨ- hu' ɨ-* , which is widely known in the Amazon region as cupuaçu, is apparently much more esteemed (by the significantly more time that is given to its gathering) by the Ka'apor than are its congeners, cacao and nondomesticated cacao. There is no reason to suppose that this differential appreciation was different in precontact times. The fruit of cupuaçu is eaten as is the fruit of cacao: it is the sweet pulp around the beans in the pod that one eats, but in the case of cupuaçu, this somewhat tart pulp is much more copious. Cupuaçu terms exhibit a tendency to cognate forms also: Guajá *k ɨ-p ɨ-' ɨ-* , Tembé *kupi'a' ɨ- w*, Wayãpi *kap ɨ-ai* (Balée, 1994, p. 307; Grenand, 1989, p. 112) for which one could logically propose the tentative reconstruction of * *k ɨ-p ɨ-'a' ɨ-* β in Proto-Tupí-Guaraní.

[2] The closely related Uru-eu-wau-wau language (also called Yupaú or Tupi-KawahibBJames Welch, personal communication, 2002; Rodrigues & Cabral, 2002), which is also in subgroup #6, denotes a nondomesticated cacao of the forest of central Rondônia (in the southwestern Amazon region of Brazil) as *n ɨ-m ɨ-ta-h ɨ-ma* or *ɨ-m ɨ-ta-h ɨ-ma* (Balée, field notes, 1992). These terms can arguably be glossed as 'smooth cacao.' The initial syllables in these terms (the differences between which may be due to free variation), therefore, are quite similar to Parintintin *ñumi-*.

(Incidentally, Parintintin is once again the odd man out, with *ñumitah* ɨ*-m*, in reference to the cupuaçu fruit only—Betts, 1981, p. 268. But given the underdifferentiation of the large variety of *Theobroma* species in the Parintintin language, it is unlikely if this term is a reflex term). These terms for cupuaçu in Guajá, Ka'apor, Tembé, and Wayãpi have remained phonologically similar because cupuaçu did not become a major export crop, as did cacao. A change in the economic landscape, in which a less-than-salient species, cacao, suddenly surged up incredibly in value and in terms of a monetary valorization system not before known in Amazonia, had the effect of influencing the language(s) most involved in its collection and exportation. Hence, whatever the original word Ka'apor had for *T. cacao* (it may have resembled, for example, *aka'*ɨ or a name containing this form), that term was replaced by a new, borrowed term from LGA, the contact language. In addition, whatever the original Ka'apor term was for wild cacao (*T. speciosum*), and this would have been very close if not identical to *aka'*ɨ (from Proto-Tupí-Guaraní **aka'iβ*—see Table 2.2), that term too was replaced, when the plant was modeled by analogy on *T. cacao*. In other words, Ka'apor *kakaran'* ɨ-, in a broad sense, can be glossed as 'that tree which resembles cacao.' It is possible, indeed, that before the mercantile valorization of cacao and before Ka'apor contact with colonial Luso-Brazilian society, *T. speciosum* was more psychologically salient than *T. cacao*. That is because *T. speciosum* is much more ecologically important and common in old fallow forests (where the Ka'apor once had lived in settled villages between forty and one hundred or more years ago but have since seen a return of forest cover) (Balée, 1994, p. 37). Indeed, *T. cacao* is only occasionally planted in dooryard gardens by the Ka'apor and it is not seen in the high forest or in fallow forests, which is to be expected of a domesticated species. Cacao is relatively uncommon compared to wild cacao and probably this was the case aboriginally in the Xingu and Tocantins basins also.

If *T. speciosum* were more psychologically salient than *T. cacao* before contact, it may be the case that the term for cacao was the marked form in Ka'apor, and wild cacao was unmarked linguistically. In other words, the impact of contact together with landscape modification by the Jesuit mission system, the reordering and transforming of native labor and work priorities, and the sudden high value of *T. cacao* within an imposed, alien system of exchange and valorization, could have not only caused the substitution of the LGA term for the native name of cacao in Ka'apor, that impact may also have brought about a marking reversal with regard to the Ka'apor term for *T. speciosum*. Although this assertion cannot be proven at the present moment, it is clearly a plausible scenario within the context of an historically intricate and significant contact situation.

Conclusions

In summary, it can be hypothesized that in the Ka'apor language, as in some other Amazonian languages such as Quichua, the cacao words (for *Theobroma cacao* and *Theobroma speciosum*) were borrowed, and that this borrowing occurred probably because cacao, as a major export crop, had a profound impact on Indian labor of the Lower Amazon region in the eighteenth century and because that labor was to some extent controlled by Jesuit mission authorities in which LGA (Língua Geral Amazônica) was the contact language. Tupí-Guaraní languages can be ruled out as sources for the Ka'apor word for 'cacao' as well as for the English, Spanish, and Portuguese words for 'cacao'. The evidence here presented of borrowing of the cacao term by Ka'apor further refines comprehension of the Ka'apor past and their relations to other living groups. The evidence suggests that Ka'apor culture and language were influenced by the cacao export economy on the eve of the recognition of a new Amazonian ethnic designation, that of 'caboclo.' The emergence of caboclo culture—together with its entry into the modern world of the eighteenth century as a conceptual, real, named entity—represents the next stage of Amazonian history, after the Ka'apor peaceful experience with Luso-Brazilian society comes to a close, and the antecedents of Ka'apor society extricate themselves from the drogas do sertão trade and the cacao export economy, eventually to become an independent, indigenous society that until 1928 was decidedly hostile to encroachment of the state. Although the Ka'apor were therefore never incorporated into the Amazon peasantry per se, for the origin of caboclos as a distinctive sociocultural system postdates Ka'apor divergence from Luso-Brazilian society, this borrowing by Ka'apor of the term for cacao helps situate the antecedents of the Ka'apor historically in a setting, such as a Jesuit mission, where LGA was the contact language.

The Ka'apor borrowing of the term for cacao is most likely to have occurred farther west than the Tocantins, where the Wayãpi were also located in a Jesuit mission, along the Xingu River. This further strengthens the hypothesized close pairing of Wayãpi and Ka'apor within subgroup #8 of Tupí-Guaraní. Finally, the impact of the cacao export economy shows that a native species in the environment, even a relatively unimportant one, can be renamed in local languages when its historical-ecological setting in the world economy is completely transformed, and when the people speaking those local languages are involved in the labor and technology of that transformation, as was the case with Amazonian cacao. Comprehending the history and uses of cacao and, no doubt, of other highly commercialized species of the past can be most useful for understanding the historical-ecological impact that the expansion of Luso-Brazilian society had on native Amazonian languages and associated ethnobiological vocabularies.

Acknowledgments I thank Beatriz Carretta Corrêa da Silva and Stephen Darwin for constructive comments and suggestions on an earlier draft of this chapter. Judith Maxwell deserves acknowledgement for helping me understand the basic principles of historical linguistics. I am grateful also to all the members of the 2002 Workshop 'Sociedades Caboclas Amazônicas: Modernidade e Invisibilidade' for sharing their invaluable insights into Amazonia and its history.

References

Acuña, C. de. (1963). A new discovery of the great river of the Amazons. trans. Clement Markham. In *Expeditions into the valley of the amazons. Hakluyt Society* (1859, trans. from original of 1641). New York: B. Franklin.

Alden, D. (1976). The significance of cacao production in the Amazon region during the late colonial period: An essay in comparative economic history. *Proceedings of the American Philosophical Society, 120*(2), 103–135.

Aubertin, C. (1996). Heurs et Malheurs des Ressources Naturelles en Amazonie Brésilienne. *Cahier des Sciences Humaines, 32*(1), 29–50.

Azevedo, J. L. (1930). *Os Jesuítas no Grã-Pará*. Coimbra: Imprensa da Universidade.

Baer, W. (1995). *The Brazilian economy: Growth and development*. 4th ed. Westport, Conn: Praeger.

Balée, W. (1984). *The persistence of Ka'apor culture*. Ph.D. dissertation, Columbia University.

Balée, W. (1988). The Ka'apor Indian Wars of Lower Amazonia, ca. 1825–1928. In R. R. Randolph, D. M. Schneider, & M. N. Diaz (Eds.), *Dialectics and gender: Anthropological perspectives*, (pp. 155–169). Boulder, CO: Westview Press.

Balée, W. (1994). *Footprints of the forest: Ka'apor Ethnobotany - The historical ecology of plant utilization by an Amazonian people*. New York: Columbia University Press.

Balée, W. (2000). Antiquity of traditional ethnobiological knowledge in Amazonia: The tupí-guaraní family and time. *Ethnohistory, 47*(2), 399–422.

Balée, W. & Moore, D. (1991). Similarity and variation in plant names in five tupí-guaraní languages (Eastern Amazonia). Bulletin of the Florida Museum of Natural History, *Biological Sciences, 35*(4), 209–262.

Betts, La Vera. (1981). *Dicionário Parintintin-Português, Português-Parintintin*. Brasília: Summer Institute of Linguistics.

Block, D. (1994). *Mission culture on the upper amazon: Native tradition, Jesuit enterprise, and secular policy in moxos, 1660–1880*. Lincoln: University of Nebraska Press.

Boudin, M. H. (1978). *Dicionário de Tupi Moderno: Dialeto Tembé-Ténêtéhar do Alto Rio Gurupi. 2 vols*. São Paulo: Conselho Estadual de Artes e Ciências Humanas.

Bruno, E. S. (1966). *História do Brasil: Geral e Regional. Volume I: Amazônia*. São Paulo: Editora Cultrix.

Campbell, L. (1999). *Historical linguistics: An introduction*. Cambridge, Mass: The MIT Press.

Campbell, L. & Kaufman, T. (1976). A linguistic look at the olmecs. *American Antiquity, 41*(1), 80–89.

Cavalcante, P. B. (1988). *Frutas Comestíveis da Amazônia. 4th ed.* Belém: MCT/CNpq, Museu Paraense Emílio Goeldi.

Cheesman, E. E. (1944). Notes on the nomenclature, classification and possible relationships of Cacao Populations'. *Tropical Agriculture, 21*(8), 144–159.

Cleary, D. (2001). Towards an environmental history of the Amazon: From prehistory to the nineteenth century. *Latin American Research Review, 36*(2), 65–96.

Clement, C. (1999). 1492 and the loss of amazonian crop genetic resources. I. The Relation between domestication and human population decline. *Economic Botany, 53*(2), 188–202.

Coe, S. D., & Coe, M. D. (1966). *The true history of chocolate*. New York: Thames and Hudson.

Conklin, H. C. (1954). *The relation of hanunóo culture to the plant world*. Ph.D. dissertation, Yale University.

Corrêa da Silva, B. C. (1997). *Urubú-Ka'apór: Da Gramática à História: A Trajetória de um Povo*. Masters Thesis, Universidade de Brasília.

Corrêa da Silva, B. C. (2001). 'A Codificação dos Argumentos em Ka'apor: Sincronia e Diacronia'. In: A. S. A. C. Cabral, & A. D. Rodrigues (Eds.), *Línguas Indígenas Brasileiras: Fonologia, Gramática e História*, (pp. 343–351). Atas do I Encontro Internacional do Grupo de Trabalho sobre Línguas Indígenas ANPOLL. Belém, Pará: EDUFPA

Crofts, M., & Sheffler, M. (1981). Dicionário Bilíngue em Português e Mundurukú, Soat a'õ dup wuya'õm pariwat a'om tak. Second Edition. Brasília: Summer Institute of Linguistics.
Cruz, E. (1973). *História do Pará*, vol. 1. Belém: Governo do Estado do Pará.
Cuatrecasas, J. (1964). Cacao and its allies: A taxonomic revision of the genus theobroma. Contributions from the *United States National Herbarium*, (vol. 35, part 6, pp. 378–614). Washington, D.C.: Smithsonian Institution.
Dakin, K., & Wichmann, S. (2000). Cacao and chocolate (A Uto-Aztecan perspective). *Ancient Mesoamerica*, 11, 55–75.
Di Paolo, P.. (1985). *Cabanagem: A Revolução Popular da Amazônia*. Belém, Pará: CEJUP.
Fisher, W. (2000). *Rainforest exchanges: Industry and community on an Amazonian Frontier*. Washington, D.C.: Smithsonian Institution Press.
Gómez-Pompa, A., Flores, J. S. & Fernández M. A. (1990). The sacred cacao groves of the Maya. *Latin American Antiquity, 1*(3), 247–257.
Grenand, F. (1989). *Dictionnaire Wayãpi-Français*. Paris: Peeters/SELAF.
Grenand, P. (1982). Ainsi Parlaient nos Ancêtres. Travaux et Documents de L'ORSTOM, no. 148. Paris: ORSTOM.
Hemming, J. (1987). *Amazon frontier: The defeat of the Brazilian Indians*. London: MacMillan.
Huber, J. (1904). Materiaes para a Flora Amazônica: Notas sobre a Patria e Distribuição Geographica das Arvores Fructiferas do Pará. *Boletim do Museu Goeldi (Museu Paraense), 4*, 392–395.
Hurst, W. J. (2001). The identification of cacao residue in samples of archaeological interest. Paper read at 101st Annual Meeting of the American Anthrpological Association, 24 November, New Orleans, Louisiana.
Jensen, C. (1990). *O Desenvolvimento Histórico da Língua Wayampí*. Campinas: Editora UNICAMP.
Jensen, C. (1999). Tupí-Guaraní. In: R. M. W. Dixon & A. Y. Aikhenvald (Eds.), *The Amazonian Languages*, (pp. 125–163). Cambridge: Cambridge University Press.
Justeson, J. S., Norman, W. N., Campbell, L., & Kaufman, T. (1985). *The foreign impact on lowland mayan languages and script*. Middle American Research Institute, Publication no. 53. New Orleans: Tulane University.
Kaufman, T. (1990). Language history in South America: What we know and how to know more. In: D. L. Payne (Ed.) *Amazonian linguistics: Studies in lowland South American languages*,(pp. 13–73). Austin: University of Texas Press.
Kohn, E. (2002). *Natural engagements and ecological aesthetic among the Ávila runa of Amazonian Ecuador*. Ph.D. dissertation, University of Wisconin-Madison.
Lentz, D. L., & Ramírez-Sosa, C.R. (2002). Cerén plant resources: Abundance and diversity. In P. Sheets (Ed.), *Before the volcano erupted: The ancient Cerén village in central america*, (pp. 33–44). Austin: University of Texas Press.
Leonardi, V. (1999). *Historiadores e os Rios: Natureza e Ruína na Amazônia Brasileira*. Brasília: Fundação Universidade de Brasília.
Maxwell, K. R. (1973). *Conflicts and conspiracies: Brazil and Portugal, 1750–1808*. Cambridge: Cambridge University Press.
Mello, A. A. S. (1966). Genetic affiliation of the language of the Indians Aurê and Aurá. *Opción, 12*(19), 67–81.
Mello, A. A. S. (2002). Evidências Fonológicas e Lexicais para o Sub-Agrupamento Interno Tupi-Guarani. In: A. S. A. C. Cabral, & A. D. Rodrigues (Eds.), *Línguas Indígenas Brasileiras: Fonologia, Gramática e História*, (pp. 338–342). Atas do I Encontro Internacional do Grupo de Trabalho sobre Línguas Indígenas ANPOLL. Belém, Pará: EDUFPA.
Moreira Neto, C. de A. (1988). *Índios da Amazônia: De Maioria a Minoria (1750–1850)*. Petrópolis: Vozes.
Motamayor, J. C., Risterucci, A. M., Laurent, V., Moreno, A., & Lanaud, C. (2000). *The genetic diversity of criollo cacao and its consequence in quality breeding*. Memorias del

primero Congreso Venezolano del Cacao y su Industria. Available: cacao.sian.info.ve/memorias/html/03.html Accessed on 2 June 2002.

Powis, T. G., Valdez Jr., F., Hester, T. R., Hurst, W. J., & Tarka Jr., S. M. (2002). Spouted vessels and cacao use among the preclassic maya. *Latin American Antiquity, 13*(1), 85–106.

Pinkley, H. (1973). *The ethno-ecology of the Kofán Indians*. Ph.D. dissertation, Harvard University.

Purseglove, J. W. (1969). *Tropical crops: Dicotyledons*. New York: J. Wiley and Sons.

Rodrigues, A. D. (1986). *Línguas Brasileiras: Para o Conhecimento das Línguas Indígenas*. São Paulo: Edições Loyola.

Rodrigues, A. D. (1999). Tupí. In R. M. W. Dixon, & A. Y. Aikhenvald (Eds.), *The Amazonian languages*, (pp. 107–124). Cambridge: Cambridge University Press.

Rodrigues, A. D., & Cabral, A. S. A. C. (2002). Revendo a Classificação Interna da Família Tupí-Guaraní. In A. S. A. C. Cabral, & A. D. Rodrigues (Eds.), *Línguas Indígenas Brasileiras: Fonologia, Gramática e História*, (pp. 327–337). Atas do I Encontro Internacional do Grupo de Trabalho sobre Línguas Indígenas ANPOLL. Belém, Pará: EDUFPA.

Schultes, R. E. (1984). Amazonian Cultigens and their Northward and Westward Migration in Pre-Columbian Times. In D. Stone (Ed.), *Pre-columbian plant migration*, (pp. 19–37). Papers of the Peabody Museum of Archaeology and Ethnology, vol. 76. Cambridge, Mass.: Harvard University.

Stone, D. (1984). Pre-columbian migration of Theobroma cacao linnaeus and *Manihot esculenta* crantz from Northern South America into Mesoamerica: A Partially Hypothetical View. In D. Stone (Ed.) *Pre-columbian plant migration*, (pp. 67–83). Papers of the Peabody Museum of Archaeology and Ethnology, vol. 76. Cambridge, Mass.: Harvard University.

Stradelli, E. (1929). Vocabularios da Lingua Geral Portuguez-Nheêngatú e Nheêngatu-Portuguez . *Revista do Instituto Histórico e Geographico Brasileiro*, Tomo 104, vol. 158, 11–768.

Strömer, C. (O.F.M.). (1932). Die Sprache der Mundurukú: Wörterbuch, Grammatik und Texte eines Indianeridioms am oberen Tapajoz, Amazonasgebiet. *Anthropos*, Tomo XI Band. St. Gabriel, Austria.

Taylor, G. (1985). *Apontamentos sobre o Nheengatú Falado no Rio Negro, Brasil. Amerindia* 10:5–24.

Thompson. J. E. S. (1956). Notes on the use of cacao in Middle America. *Notes on Middle American Archaeology and Ethnology, 128*, 95–116.

Wichmann, S. (1999). A conservative look at diffusion involving mixe-zoquean languages. In R. Blench & M. Spriggs (Eds.), *Archaeology and language II: Corrrelating archaeological and linguistic hypotheses*, (pp. 297–323). London: Routledge.

Young, A. M. (1994). *The chocolate tree: A natural history of cacao*. Washington, D.C.: Smithsonian Institution Press.

Chapter 3
Mixed Indians, Caboclos and Curibocas: Historical Analysis of a Process of Miscegenation; Rio Negro (Brazil), 18th and 19th Centuries

Décio de Alencar Guzmán

Abstract The author analyses the process of mixing (*mestiçagem*) in the Rio Negro region during the 18th and 19th Centuries. After presenting the main features of this mestiçagem's components (the Amerindian, the European and the African), the author concentrates on the inter-racial marriage policies prescribed by the Portuguese Crown, as part of a group of projects geared towards the exploitation of human resources in Portuguese America. Guzmán believes that one of the main hindrances to the advance of the studies about the Amazonian *caboclo* societies is the belief that they are independent and self-regulated social systems. Such a conception has prevented a more accurate understanding of such societies as a product of historical transformations.

Keywords Caboclоization · Hybridity · Ethnicity · Colonial ethnic Affiliations · Acculturation · Pombaline laws

Introduction

In 1878, José Veríssimo began his long essay on 'The Indigenous and Mixed Populations of the Amazon' saying: 'Today, America is the huge melting pot of the various races and peoples of the world'. Further on, he remarks: 'the vast Amazon region is a living example of the great fact, which can be seen there in full bloom, if with less variety' (1970, p. 11). Veríssimo's words still ring true today, considering that the issue of miscegenation has assumed important social and cultural dimensions in the area in question.

How did this cultural miscegenation and 'racial interbreeding' come to pass in the Amazon? This is the theme of this present chapter, which will focus specifically on the history of the phenomenon in the Rio Negro region (North-east of the Brazilian Amazon). I will attempt to reflect upon the historical

D.de.A. Guzmán (✉)
Department of History, Federal University of Pará, Pará, Belém, Brazil
e-mail: decioguz@hotmail.com

DOI 10.1007/978-1-4020-9283-1_4, © Springer Science+Business Media B.V. 2009

sources, particularly those for the communities of the Rio Negro Basin and its main tributaries, though without neglecting the broader geographical and historical context in which the issue of 'racial interbreeding', to use Veríssimo's words, took root.

I will highlight and try to articulate different aspects which I consider important to understanding and describing the process of miscegenation in the Amazon: first and foremost, the political and social processes of interaction between cultures and societies in the region; secondly, some of the historical processes of the formation of social representations of Amazonian miscegenation, as well as its importance to the present volume.

Native Communities

In the Rio Negro region, the native communities have historically always been the majority of the local population and they maintained inter-ethnic relations centuries before the arrival of the Europeans. The 'intermixing' between the Indians of the *Tukano* and *Arawak* languages stands out, partly because of the neighborly alliances and inter-marrying practiced by these peoples. Such marriage policies are well-known to ethnologists and other social scientists studying the area, and in fact distinguish the social morphology of this ethnographic province from that of other provinces.

The exogamous and exolinguistic marriages practiced by the *Arawak* and *Tukano* communities, excluding the *Maku*, may well serve to characterize a prime structural element of the history of miscegenations among the native communities of the river basin in question.[1] The exogamous particularities of *Tukano*/*Arawak* inter-marriages can be partly explained by their history of continuous, parallel and sequential migrations for as long as these two peoples have occupied the region. Trade, intermarriage and war relations between the *Tukano* (Uanano and Cubeo) living in regions on the *Arawak*/*Tukano* frontier (the Uaupés, Cuduiary and Querery rivers), for example, gave rise to the more elaborate organization of phratries, with rigid or firmly localized systems of social organization (Chernela, 1988, pp. 35–49; Goldman, 1963).

The issue of 'interethnic contact' has an important contribution to make to the debate on the *visibility* or *invisibility* of caboclo societies in the Rio Negro region. It is in this context, defined by the ethnographies of Eduardo Galvão and, more recently, Ana Gita de Oliveira, that we can see the association between the *ethos* of regional 'caboclo culture' and various indigenous fusions 'visualized' in the sphere of extractivist economic activities, somewhat soured by association with 'white culture'(Galvão, 1979; Oliveira, 1995).

[1] On exogamy in the ethnographic province of Rio Negro, see: Sorensen (1967), Goldman (1963), Jackson (1983), Reichel-Dolmatoff (1971), Silverwood-Cope (1990), Pozzobon (1983), Hugh-Jones (1979), Journet (1995), Wright (1992), and Santos and Barclay (1994–1998).

As such, the regional population of the Rio Negro area is perceived as the outcome of 'change' (read: acculturation) brought about by contact with the 'white man'. The result is a 'pacified' indigenous population, bound to the modalities of work imposed upon it by white civilization and overly dependent on regional commerce. This line of argument largely grounds the ethnographic discourse on our focal region. The 'regional' population along the Rio Negro is therefore 'galvanized' by the negative appraisal the social sciences make of its communities, casting a specter of heuristic opacity across the 'mixed' social relations of the region.

The accumulation of archaeological evidence on the pre-historic social morphology of native communities that could confirm this hypothesis is still in the very early stages. The waves of migration within the indigenous province of the Upper, Middle and Lower Negro River combine a series of contacts between different groups or societies the schema of which has received attention from various ethnologists since the nineteenth century.

Observed by Theodor Koch-Grümberg as far back as 1909, the 'Tukanization of the Aruak' reveals a substantially complex process of cultural interchange and intercourse between the *Arawak* and *Tukano*, which began with the migrations of *Tukano* communities into the North-eastern Amazon, where they mingled with the territory's existing *Arawak* populations (Koch-Grümberg, 1909).

In this regard, Robin Wright's observations on the migratory schemas put forward by Curt Nimuendajú are well known. Wright concludes that Nimuendajú was correct in suggesting the sequence *Maku-Arawak/Tukano -European* as the likely order in which the region was occupied, though he does raise doubts as to the previous local origins of the *Tukano* communities in relation to the *Arawak*. The incipience of more complete and reliable information on the abovementioned data hampers any evaluation on the nature and types of marriage practiced in the native communities before the Europeans reached the area. A definitive conclusion as to the process of interaction or initial historical miscegenation that gave rise to the area's prevailing socio-cultural configuration must therefore be postponed (Nimuendajú, 1982, pp. 168–171; Neves, 1998; Oliver, 1989; Wright, 1992, pp. 256–257).

The Europeans

We know that there was intense movement between the Azores and Maranhão from the seventeenth century onwards, resulting in the latter's occupation and colonization. In 1619, 1622, 1649, 1667 and, again, in 1673, 1674 and 1677, the Crown arranged and financed the transportation of couples from the Azores to Maranhão (Coates, 1998, p. 145).

This practice continued throughout the eighteenth century, mainly due to the urban reform undertaken by the Pombal administration. In 1751, some 502 Madeira Islanders were transported to São José de Macapá (in what is now the state of Amapá) (Mendonça, 1963, p. I:122). Another 486 islanders from the Azores arrived at the same destination the following year (Mendonça, 1963, p. I:35). In a letter to the Marquis of Pombal in 1753, the Governor Francisco Xavier de Mendonça Furtado writes of his intention to distribute the islanders throughout Great Pará (a term which covered the whole Amazon in the eighteenth century). The old towns of Caeté or Souza would receive this contingent as the bedrock for a 'populous country town' to be called Bragança. True to his word, this actually occurred the following year. In the same letter he also mentions two other villages to be settled by Azorean newcomers: one on the Tapajós River and another on the Xingu. Mendonça Furtado also availed of this correspondence to anticipate by some three years the edict of April 4, 1755, which would proclaim on the issue of marriage between Europeans and Indians, and reproduce almost *ipsis litteris* the content of the Governor's letter:

> [. . .] and it would strike me as not only useful, but of most summary importance should His Maj. choose [sic] to decree that not only does marriage between whites and Indian women bring no shame, but that, quite the contrary, it is to be encouraged through the concession of certain privileges, as otherwise I see no way we can populate this [sic] vast State and demonstrate to its natives the honor and regard in which they are held (Mendonça, 1963, pp. I:413–414).[2]

The demographic of the white population in Rio Negro, however, follows other growth parameters. The majority of the region's whites did not derive from 'forced' migrations from the Azorean archipelago or from the Madeira Islands, but rather from European administrative visits or the Territorial Demarcation missions conducted more than twice during the second half of the eighteenth century (Domingues, 2000, pp. 98–105).

Many of the soldiers that escorted these missions stayed on as settlers or residents in the Portuguese colonial villages or were given administrative duties within the bureaucratic apparatus of the main villas and towns along the Negro river basin. In the village of Barcelos – the largest settlement and the administrative seat of the Captaincy of Rio Negro – the white population inventoried in the 1786 census was distributed as follows (Table 3.1):

Table 3.1 Whites and their descendants in the village of Barcelos in 1786

Age	Male	Female	Total
0–7	57	63	120
7–15	15	9	24
15–60	36	39	75
60–90	8	4	12
Total	116	115	231

Source: Mappa de todos os moradores brancos, índios, e pretos escravos existentes na Villa capital de Barcelos, 31 October 1786 (extract), Biblioteca Nacional (Rio de Janeiro), manuscripts 21,1,1, no.7.

[2] On the town of Bragança, see Araújo (1998, pp. 117–22).

Alexandre Rodrigues Ferreira draws our attention to the fact that the total number of free whites in the entire Captaincy of Rio Negro in 1786 was 635 people. He observes that the whites counted in the census mentioned above included the so-called *mamelucos*, the children of Indian women with white men. This fact, which is highly relevant to our study, proves another obstacle we encounter in the historical sources, as does the prohibition of the term 'caboclo' decreed in the marriage law of April 4, 1755, which will be discussed further on (Ferreira, 1983, p. 648).

The majority of this white population formed between 1751 and 1800 largely consisted of members of the Territorial Demarcation Commission, among whom officials, soldiers and laborers, who gradually settled in the towns and hamlets and married the Indian daughters of the colonial village 'headmen' (*principais*) of the Captaincy of Rio Negro (Ferreira, 1983, p. 647).

In this manner the islanders were shifted about countless times in order to populate the Amazon. They were systematically deployed during the Pombaline period to further urban projects considered strategic to the geopolitics of the region, among other interests. With reference to the islanders, Pombal explained that it was necessary to employ them in farming activities in order to ensure the provision of basic foodstuffs. The Marquis exempted such manual labour of the abjection it would normally have held during the eighteenth century. The exercise of 'cultivating the land' would now be no impediment to receiving the 'honors' or 'posts' in the colony which the European settlers so dearly craved (Mendonça, 1963, p. I:29).

The Africans

Throughout the seventeenth century and the first half of the eighteenth there was no systematic or continuous trafficking of African slaves into the Amazonian region.[3] The first real attempt to bring in African captives to work the mines and plantations of Great Pará and Maranhão began with the Maranhão Trade Company, founded in 1682, which stocked the farms of Maranhão state until the company's closure in 1684. When the Maranhão Trade Company ceased operations, the African importation contracts transferred to the Cachéu Company, though with low volumes (Salles, 1988, p. 28). It was only with the creation of the Grand-Pará and Maranhão General Company in 1756 that the 'stock' of African slaves swelled in volume, and continued to do so until the company was shut down in 1778. According to Vicente Salles, the private slave trade was sporadic and did not generate sufficient data to be significant to the overall

[3] To date, there are no monographs or complete studies on the trafficking of black Africans into the State of Grand-Pará and Maranhão during the seventeenth and eighteenth centuries. Most analyses of African slave influx are from studies concerned with other themes and problems of the region's social history. The most important studies are: Marin (1985), Salles (1988 [1971]), Vergolino-Henry and Figueiredo (1990), and Bezerra Neto (2001).

statistics on the slave trade into São Luis and Belém. In fact, the illegal slave trade from Africa or from Brazilian ports only gathered momentum after 1784 (Salles, 1988, pp. 30–43; Freitas, 1819, p. III:453).

The years of the slave trade from Africa to the Amazon were fraught with difficulties when it came to maintaining the continuity and periodicity of the arrivals. However, it was this relative periodicity that peopled many of the colonial villas along the Amazonian Basin and opened up genuine possibilities of miscegenation between the African and native Indian and caboclo. According to the estimates of a contemporary of the Grand-Pará and Maranhão General Company, some 14,000 Africans were introduced to the State between 1755 and 1778 (Ferreira, 1983, p. 648). The historian Manuel Nunes Dias, on the other hand, calculates that the true number could have been as high as 25,365 Africans during this timeframe, mostly from Guinae-Bissau, Cachéu and Angola (Dias, 1970, p. I:465).[4] António Carreira puts the total of slaves shipped in by the Company from these same areas of Africa at 28,835 (Carreira, 1988, p. I:126). There is still no definitive data for the slave trade into Grand-Pará, but when we consider that, of the population of 78,860 in Maranhão at the close of eighteenth century, 36,880 were slaves, while in Grand-Pará, 18,944 of the total population of 80,000 were black African captives and their slave descendents, we can assume that Africans accounted for a sizeable portion of the population (MacLachlan, 1973, pp. 199–230).

Relatively few slaves were introduced to the Captaincy of Rio Negro. On this subject, Alexandre Rodrigues Ferreira wrote: 'As for the blacks, there aren't many [sic], on this river there aren't as many as needed, just like there aren't in the captaincy of Pará' (1983, p. 648). In the census conducted in Barcelos in the year 1786, the slaves are registered as follows (Table 3.2):

Table 3.2 Slaves in the village of Barcelos in 1786

Age	Male	Female	Total
0–7	21	27	48
7–15	15	12	27
15–50	51	41	92
Total	87	80	167

Source: Mappa de todos os moradores brancos, índios, e pretos escravos existentes na Villa capital de Barcelos, 31 October 1786 (extract), Biblioteca Nacional (Rio de Janeiro), manuscripts 21,1,1, no.7.

The total number of Negros and their descendents in the Captaincy in that same year of 1786 reached 274. The population in general, counting whites and Indians, reached 6,642 inhabitants. The low number of Africans in Rio Negro goes some way toward explaining the high incidence of marriages between Europeans and natives throughout the second half of the eighteenth century

[4] Though the author had initially estimated the number of Africans shipped to Great Pará as 14,749, he later conceded that this figure was unrealistically low and corrected it accordingly.

and into the first half of the nineteenth, prior to the influx of migrants from the Northeast in the 1870s (Anderson, 1976, pp. 199–205).

Inter-racial Marriage Policy

From 1750, economic and 'enlightened' Portuguese policy in South America could be gleaned from the orders and innovations of the Secretary of State for War and Foreign Business of the Court of King José I, one Sebastião José de Carvalho e Melo.[5] This policy was part of a package of projects designed to configure the systematic exploration of natural and human resources throughout Portuguese America.

In the broader context of Portuguese state policy, the economic and political optimization of Amazonian resources was the counterpart to similar initiatives in Europe, Africa and Asia. The global dimensions of the Pombaline reforms of the eighteenth century can be seen from the sheer scale of the intercontinental trade carried out by the General Company of Grand-Pará and Maranhão. Between August 1755 and January 1778, the Company – with a commercial fleet of 124 ships with a total capacity of 43,000 tons – built a network of trade posts that enabled it to ship European merchandise, tropical products from the Amazon and African slaves along trade routes that connected Cachéu, the Cape Verde islands, Guinea-Bissau, Angola, London, Antwerp, Rotterdam, Hamburg, Cadiz, Marseilles, Genoa, Pará/Maranhão and the Indian Ocean (Dias, 1964, pp. 115–116).

The measures taken by the king's Minister in relation to economic and political life in the Amazon had simultaneous effects on the region's burgeoning population that only become visible when seen concomitantly. Let us succinctly recall the main Pombaline projects: a) Creation of the General Company of Great Pará and Maranhão with a view to commercializing Amazonian production and introducing African slaves to the region; b) Abolition of indigenous slavery as a means of freeing the local workforce for the formation of able political subjects in effective possession of strategic locations across the vast expanse of South-American rainforest; c) Eradication of the political power wielded by the missionaries in the Indian villages and hamlets; d) population and urbanization of the stretches of territory along the main tributaries of the Amazon River.

In Great Pará and Maranhão, these projects were characterized by *simultaneity* and *indissociability*. The simultaneity can be readily verified by a brief glance at the chronological order in which the projects were implemented (Table 3.3):

---

[5] Better known by the title 'Marquis of Pombal', conferred upon him in 1769 (cf. Maxwell, 1996, p. 2).

Table 3.3 Pombaline Legislation in the eighteenth century

Date	Legislation
17/4/1751	*Decree appointing the Marquis of Pombal's brother, Francisco Xavier de Mendonça Furtado, Governor of the State of Grand-Pará and Maranhão*
3/3/1755	*Royal edict separating the Captaincy of Pará from the Captaincy of São José do Rio Negro.*
4/4/1755	*Seal approving inter-marriage between Europeans and Amerindians.*
6/6/1755	*Royal edict abolishing indigenous slavery in the State of Maranhão. Law establishing the General Company of Grand-Pará and Maranhão.*
7/6/1755	*Seal approving the creation of the General Company of Grand-Pará and Maranhão.*
3/5/1757	*Institution of the "Indian Directorate" (placing missionary hamlets and villages under the administration of Indian chiefs or "directors" appointed by the Governor the State.*
17/8/1758	*Seal approving the "Indian Directorate".*
3/9/1759	*Law expelling the Jesuits (Society of Jesus) from the entire domain of the Portuguese Crown.*

Sources: *Revista do Instituto Histórico e Geográfico Brasileiro*, 61: 59–63; Freitas, 1819, III; Leão, 1796; Senado Federal (2001).

It was against the context of these actions that a policy promoting and privileging systematic inter-marriage between Portuguese men and indigenous women of various ethnicities (particularly those allied with the Crown) took hold in the Amazon region. The policy followed much the same line as the administrative reforms decreed and implemented by Pombal in Portugal (Falcon, 2001; Hespanha, 1993, pp. IV:157–272).

In 1755, a royal decree was approved encouraging Europeans of either sex, from the Homeland or colonies, to marry Amerindians. The edict declared that such marriages would, above all, serve to 'greatly further communication with the Indians' so that the Portuguese dominions could be populated.

The decree also prohibited discrimination against those who took native spouses and lifted all restrictions to their access to public institutions or honors and titles proper to vassals of the Portuguese King. These vassals and their mixed-race offspring would be 'worthy of Royal attention' and be first in line for positions and appointments in their adopted lands for which they were appropriately qualified. While the law gave precedence to expatriate Europeans and *mestiços* in applications for office and professions (employment, honor and dignity), this favoritism was restricted to the villages and hamlets of the colonies (Freitas, 1819, pp. III:419–421).

Nevertheless, what interests us most in this edict is the part that prohibits the use of the term *Caboclos* [sic] with reference to the mestiço children born of these marriages, on the grounds that it was 'injurious' and 'offensive'. 'Caboclo' was therefore struck from the official vocabulary:

> whomsoever is found to have acted to the contrary, regardless of condition or quality, having been duly heard by the Magistrate of the county, shall be sentenced by this same office, with neither right to appeal nor aggravation, to depart the county at a month's

grace and at my discretion; which shall be fulfilled without exception (Freitas, 1819, p. III:420).

Thus we can see that the Portuguese Crown institutionalized the 'invisibility' of the emerging caboclo society as far back as the eighteenth century in the official documentation produced by the authorities of the State of Grand-Pará and Maranhão and throughout Colonial Brazil. The term 'caboclo' was henceforward officially stigmatized.

The historical consequences of this decree are important. It makes it impossible for the historians of today to 'visualize' the figure of the mestiço (caboclo) in the official information sources for the Pombaline period and in relative sources thereafter. Effectively, a cloud descended upon the mestiço population, leaving the Amerindian, European and African in the foreground, isolated one from the other. Therein lies the root of the myth of racial 'purity' in the Amazon. Only cross-reading and referencing of information sources can coax the silent mestiço from this limbo of memory and history.

The encouragement of inter-racial marriages can be found in the 'Indian Directorate' (1757). In this document, the Portuguese Crown declared that, in order to eliminate the 'odious and abominable' gulf between whites and Indians, the directors were to foster the marriage of one with the other, as there was no shame in such matrimony. It also threatened to punish anyone who, once married, should disrespect their husband or wife for being natives - 'finding fault with their Indian condition' (*Directorio*, 1758, pp. 36–37).

Alexandre Rodrigues Ferreira[6] contends that the marriages encouraged in the abovementioned decree from 1755 took place in large numbers across the Captaincy of Rio Negro. He tells of various cases in which whites from the area took Indian wives and obtained the privileges promised in the edict. For example, he relates that on August 10, 1758, the governor Francisco Xavier de Mendonça Furtado appointed João Nobre da Silva Captain Major of Barcelos, the seat of the Captaincy. The captain was chosen above two other candidates for the post for having married the Indian Thereza de Mendonça Mello, daughter of the headman Manoel Gama (Ferreira, 1983, p. 634).

He also mentions the dispensation from military service conceded to soldiers from the Demarcation Commission who married Indian women so they could set up home in the environs of Rio Negro (Ferreira, 1983, p. 633). So numerous were these marriage-related dispensations that they began to undermine the military stability of the Captaincy. To resolve the problem, the Portuguese Crown started authorizing the marriages of soldiers in active service (Ferreira, 1983, p. 635).

[6] Born in Bahia on April 27, 1756. His father had groomed him for ecclesiastical life and in 1768 he took the lower orders. He travelled to Lisbon in 1770 and enrolled on the Law Course at Coimbra that same year. With the reform of the university, he transferred to Philosophy and was appointed director for the Philosophical Mission to the Brazilian territory, begun in 1783. He died in Lisbon in 1815. For more details, see Goeldi (1982).

On the other hand a letter addressed to the Secretary of the Governor of Great Pará dated February 28, 1776, informed of the marriage between a white woman and an Indian in the village of Barcelos.[7] In the eighteenth century, Indian men also took African and African-descendent wives, though these were far less numerous compared with other marriages registered in Great Pará and Maranhão during the same period.

Native Slavery and Miscegenation

Indigenous slavery was a cornerstone of much of the colonial activity in the Amazon from the first European forays into the forest in the sixteenth century up to the nineteenth century. This context of native laboring was the matrix for forms of forced social interaction in the colonial hamlets and villages that would later give rise to the most varied mixed-race marriages. The large contingent of native Indian captives in the colonial precincts – in numbers not yet ascertained by historical demographic studies on the territory of Grand-Pará – occasioned contact between different native ethnicities and, from the mid-eighteenth century, between the Amerindians in general and African slaves.

The log-books of native slaves 'brought down' in the Portuguese canoes around 1739 reveal the presence of various different ethnic groups. These groups were taken from their lands as slaves and forced to work together in Belém and the other main villas of the captaincies throughout the State, particularly Cametá, Vigia, Bragança and Gurupá.[8]

Based on records by the Jesuits Ignacio Szentmartonyi (1749–1755), Manuel Roman (1744) and Achilles Avogadri (1738–1744) – frequent visitors to the area -, the oral histories of groups of *Tukano* (*Desana, Tukano, Makuna*), and the official records of the Portuguese 'slave recovery units', Robin Wright has identified the presence of native slaves from 309 different ethnicities in Rio Negro between 1738 and 1755 (Wright, 1991). This data clearly shows that interaction between various Amerindian groups was not exclusive to the Pombaline period. Since the arrival of the missionaries, these ethnicities had found themselves side-by-side at the catecheses in the mission villages, together constituting a melting pot of languages, forms of social organization and practice, huddled in the same small space and controlled by codes of conduct.[9]

[7] Public records of Pará, Secretariat of the Captaincy of Grand-Pará Collection, Codex 291, Document 317.

[8] 'Book used to register the canoes', (see Fonseca 1739).

[9] On this subject, see 'Regimento de abril de 1680 e leis anexos' and ' Regimento das Missões do Estado do Maranhão e do Pará, de 1 de dezembro de 1686', in: Beozzo (1983, pp. 106–11 and 112–20, respectively).

Conclusions Journeys and Flows – International Connections in Rio Negro

One of the major obstacles to progress in studies on Amazonian caboclo societies is the belief, still present in social scientific studies on the region today, that these societies should be treated as independent, self-reproducing and theoretically self-regulating systems. No tribe or community is or ever has been an island, and the Amazonian world, a totality of interconnected processes or systems, is not and never has been the sum of independent human groups and cultures. So what we often take to be immutable and self-reproducing is not only the result of a constant and complex process of internal and external tensions, but, more often than not, the product of historical transformations.

Throughout the Rio Negro region, the circulation of individuals and families across the frontiers between the various Amazonian nations is another factor that contributed to a general blurring in the social classification of caboclo communities. On the other hand, it is also a factor that displays the wealth of social and cultural practices as the product of the historical traditions of different 'nationalities' interconnected by a way of life and world-view that are very similar in many respects.

From the second half of the nineteenth century, connections between the Amazonian nations under Spanish colonization and the main urban centers on the Brazilian side of the border became increasingly close. The trade posts deep in the forest, often no more than hamlets, villages or even just riverine encampments, constituted a communications network that drove the rubber trade down to the early 20th century.

As such, the indigenous communities on the frontiers maintained close commercial and political ties with the businessmen and local contractors of Manaus and Belém. In addition, the international contacts of the Amazonian North-east (mainly with Europe and the United States) inserted the caboclo communities into a global context of connections and changes that occurred at an accelerated pace from 1870 up to the 1920s (Stanfield, 1998, pp. 115–129).

This essay has gone into studying structural aspects of the formation of mestiço Amazonian society from the second half of the eighteenth century. This was the historical period of most accentuated social and demographic change in Amazonia as a whole and in the Rio Negro Basin in particular. The various aspects presented in this text (marriage, slavery, urbanization, etc.) demonstrate the variety of politico-economic arrangements implemented in the Rio Negro zone, producing a type of society there that was very different from Amazonian society in general.

Perhaps the main interest in seeking to elucidate the particularity of the process of miscegenation in the Northeast Amazon resides in the fact that the social systems that took hold there do not present us with any causal historical connections, contrary to what the region's historiography and ethnography would have us believe, always proposing their conclusions in terms of evolutionary linearity.

Rather, the interest would seem to lie in observing the miscegenation of this 'regional population' in terms of the variability and different combinations of the elements of its history.

Besides this, as mentioned earlier, the human societies that inhabit the open Amazonian borderlands between Brazil and the Andean nations are perhaps best described as a continuum rather than brusque rupture against which mestiço societies on all sides clashed. Perhaps the historical dynamic of these societies, at least in the Amazonian border regions, is intimately linked to the ebb and flow of 'caboclo' populations that have circulated and mingled down through the centuries, separated only by the initiatives and ideologies of their nation states.

References

Anderson, R. L. (1976). *Following Curupira: Colonization and migration in Pará, 1758 to 1930 as a study in settlement of the humid tropics*. Ph.D. dissertation, Department of History, University of California, Davis.

Araújo, R. M. de. (1998). *As cidades da Amazônia no século XVIII: Belém, Macapá e Mazagão*. Porto: FAUP.

Beozzo, J. O. (1983). *Leis e regimentos das missões: Política indigenista no Brasil*. São Paulo: Loyola.

Bezerra Neto, J. M. (2001). *Escravidão negra no Grão-Pará (sécs. XVII–XIX)*. Belém: Paka-Tatu.

Carreira, A. (1988). *A Companhia Geral do Grão-Pará e Maranhão*. São Paulo: Companhia Editora Nacional.

Chernela, J. (1988). Righting history in the Northwest Amazon: Myth, structure, and history in an arapaço narrative. In: J. D. Hill (Ed.). *Rethinking history and myth: Indigenous South American perspectives on the past* (pp. 35–49). Urbana: University of Illinois Press,.

Coates, T. J. (1998). *Degredados e órfãs: Colonização dirigida pela Coroa no Império Português, 1550–1755*. Lisboa: Comissão Nacional para as Comemorações dos Descobrimentos Portugueses.

Dias, M. N. (1964). A tonelagem da frota da Companhia Geral do Grão-Pará e Maranhão (1755–1788). *Revista de História da USP*, *28*, 13–40.

Dias, M. N. (1970). *Fomento e mercantilismo: A Companhia Geral do Grão Pará e Maranhão (1755–1778)*. Belém: UFPA.

Directorio, que se deve observar nas povoaçoens dos Indios do Pará, e Maranhão em quanto [sic] Sua Magestade não mandar o contrario. Lisboa: Na officina de Miguel Rodrigues, 1758.

Domingues, A. (2000). *Quando os índios eram vassalos: Colonização e relações de poder no Norte do Brasil na segunda metade do século XVIII*. Lisboa: Comissão Nacional para as Comemorações dos Descobrimentos Portugueses.

Falcon, F. C. (2001). Pombal e o Brasil. In: J. Tengarrinha (Ed.). *História de Portugal* (2nd ed., pp. 227–243). Bauru/São Paulo/Lisboa: Edusc/Unesp/Instituto Camões.,

Ferreira, A. R. (1983). *Viagem filosófica ao Rio Negro*. Belém: MPEG.

Freitas, J. I. de. (1819). *Colleção chronologica de leis extravagantes, posteriores à nova compilação das ordenações do Reino, publicadas em 1603*. Coimbra: Real Imprensa da Universidade.

Fonseca, J. G. da. (1739). *Livro que há de servir para o registro das canoas que se despacharem para o sertão ao cacao, e ás pessas, e das que voltarem com escravos e nelle haõ de assinar termo os cabos das mesmas canoas. Teve principio no anno de 1739. Por ordem do Illmo. e Exmo.*

Senhor João de Abreu de Castelbranco Governador e Capm. Gnl. do Estado. Sendo secretario delle José Glz. da Fonseca. Belém: Arquivo Público do Pará.
Galvão, E. (1979). *Encontro de sociedades: Índios e brancos no Brasil.* Rio de Janeiro: Paz e Terra.
Goeldi, E. A. (1982). *Alexandre Rodrigues Ferreira.* Brasília: Editora da Universidade de Brasília.
Goldman, I. (1963). *The Cubeo-Indians of the Northwest Amazon.* Urbana: University of Illinois Press.
Hespanha, A. M. (Ed.). (1993). *História de Portugal: O Antigo Regime (1620–1807).* Lisboa: Editorial Estampa.
Hugh-Jones, C. (1979). *From the Milk River: Spatial and temporal processes in Northwest Amazonia.* Cambridge: Cambridge University Press.
Jackson, J. (1983). *The fish people, lingüistic exogamy and Tukano identity in Northwest Amazon.* Cambridge: Cambridge University Press.
Journet, N. (1995). *La paix des jardins: Structures sociales des indiens curripaco du haut Rio Negro (Colombie).* Paris: Institut d'Ethnologie/Musée de l'Homme.
Koch-Grümberg, T. (1909). *Zwei Jahre unter den Indianern: Reisen in Nordwest-Brasilien 1903–1905.* Berlim: Wasmuth Verlag.
Lapa, J. R. do A. (1973). *Economia colonial.* São Paulo: Perspectiva.
Maclachlan, C. (1973). The Indian labor structure in the Portuguese Amazon, 1700–1800. In: D. Alden (Ed.). *Colonial roots of modern Brazil.* Berkeley: University of California Press.
Marin, R. E. A. (1985). *Du travail esclave au travail libre: Le Pará (Brésil) sous le régime colonial et sous l'Empire (XVIIe–XIXe siècles).* Thèse de Doctorat en Histoire – École des Hautes Études en Sciences Sociales, Paris.
Maxwell, K. (1996). *Marquês de Pombal: Paradoxo do Iluminismo.* Rio de Janeiro: Paz e Terra.
Mendonça, M. C. de. (1963). *A Amazônia na Era Pombalina.* Rio de Janeiro: IHGB.
Neves, E. G. (1998). *Paths in dark waters: Archaeology as indigenous history in the Upper Rio Negro Basin, Northwest Amazon.* Ph.D. dissertation, Department of Anthropology, Indiana University, Bloomington.
Nimuendajú, C. (1982). Reconhecimento dos rios Içana, Ayarí e Uaupés [1927]. In: *Textos indigenistas: Relatórios monografias, cartas* (pp. 168–171). São Paulo: Loyola,
Oliveira, A. G. de. (1995). *O mundo transformado: Um estudo da 'cultura de fronteira' no Alto Rio Negro.* Belém: Museu Paraense Emílio Goeldi.
Oliver, J. R. (1989). *The archaeological, linguistic and ethnohistorical evidence for the expansion of Arawakan into Northwest Venezuela and Northeastern Colombia.* Ph.D. dissertation, Department of Anthropology, University of Illinois, Urbana.
Pozzobon, J. (1983). *Isolamento e endogamia: Observações sobre a organização social dos índios Maku.* Dissertação de Mestrado, Antropologia, – Universidade Federal do Rio Grande do Sul, Porto Alegre.
Reichel-Dolmatoff, G. (1971). *Amazonian cosmos: The sexual and religious symbolism of the Tukano Indians.* Chicago: University of Chicago Press.
Salles, V. (1988). *O negro no Pará: Sob o regime da escravidão.* Brasília/Belém: MEC/Secult/FCPTN.
Santos, F., & Barclay, F. (Comp.). (1994–1998). *Guía etnográfica de la Alta Amazonía.* Quito/Balboa: Flacso/IFEA/Smithsonian Tropical Research Institute/ Ediciones Abya-Yala.
Senado Federal. (2001). *500 anos de legislação brasileira.* Brasília: Senado Federal. (CDROM).
Silverwood-Cope, P. L. (1990). *Os Makú: Povo caçador do noroeste da Amazônia.* Brasília: Editora da Universidade de Brasília.
Sorensen, A. P. (1967). Multilingualism in the Northwest Amazon. *American Anthropologist, 69*, 670–684.
Stanfield, M. E. (1998). *Red rubbser, bleeding trees: Violence, slavery and empire in Northwest Amazonia, 1850–1933.* Albuquerque: University of New Mexico Press.

Vergolino-Henry, A., & Figueiredo, N. (1990). *A presença africana na Amazônia Colonial: Uma notícia histórica*. Belém: Arquivo Público do Pará.

Wright, R. M. (1991). Indian slavery in the Northwest Amazon. *Boletim do Museu Paraense Emílio Goeldi, Antropologia, 7*, 149–179.

Wright, R. M. (1992). História indígena do Noroeste da Amazônia: Hipóteses, questões e perspectivas. In: M. C. da Cunha (Ed.). *História dos índios no Brasil* (pp. 256–257). São Paulo: Companhia das Letras.

Chapter 4
'Sempre Ajeitando' (Always Adjusting): An Amazonian Way of Being in Time[1]

Mark Harris

Abstract This article argues that the modernity of *caboclo* societies is characterised mainly by its constant renovation of the past in the present, a strategy that has given them great reproductive (social and biological) success and that was critical for their adaptation to unstable economic and political conditions and to a scenario of socio-cultural collapse. Resilience and flexibility are, for the author, the riverine populations' main features. In the text, Harris dialogues with two previous forms of referring to the genesis of these populations: the "caboclization" process by Eugene Parker, elaborated in the mid-1980's, and the mercantilist theory, formalised by Stephen Nugent in the beginning of the 1990s. For him, as one imposes abstract categories and concepts with the aim of building collective entities such as caboclo "culture" or "identity", one misses out what is richest in the analysis object: the heterogeneity, the ambivalence, the ideology of "mixture" and the "opening" before the unknown, which emerges with the analysis of specific biographies in their respective socio-economic contexts.

Keywords History · Historicity · Caboclo identity · Religion · Shamanism · Kinship

---

M. Harris (✉)
Department of Social Anthropology, University of St Andrews, St Andrews, Fife, Scotland
e-mail: mh25@st-andrews.ac.uk

[1] I would like to thank the Economic and Social Research Council, the British Academy and the University of St Andrews for financial support in connection with the research for this paper. I am most grateful to Tristan Platt for his comments on an earlier version, even though this rewrite will not satisfy his demanding and often brilliant suggestions. The commentaries by Henyo Barretto and Eduardo Brondizio, discussions with Décio Guzmán and Aldrin Figueiredo, the editorial suggestions of Rui Murrieta and Cristina Adams were most very helpful in rewriting this paper and correcting factual errors. Most of all I want to thank the organisers and funders of the conference in São Paulo for making possible such a stimulating and enjoyable event.

C. Adams et al. (eds.), *Amazon Peasant Societies in a Changing Environment*,
DOI 10.1007/978-1-4020-9283-1_5, © Springer Science+Business Media B.V. 2009

Caboclos and History

Caboclos, like many other peoples throughout history, often contribute to the demise of their own culture'. (Parker, 1985, p. xli). '*Caboclos have neither identity nor supporters. They deserve both.* (Parker, 1985, p. xliv)

It is difficult to generalize about caboclos. Cleary (1993, p. 335)

For Eugene Parker the caboclo is an outcome of the colonization of the Amazon by the Portuguese invaders. It is a category of people who share similar cultural patterns, such as the way they exploit environmental resources and their beliefs in an enchanted environment; and lives, or lived, in predominantly rural communities of kinspeople. Caboclo culture and society emerged some time before the Cabanagem rebellion (c. 1835–1840) in a process called by Parker 'caboclization'. At the time of the publication of his edited volume (1985) Parker argued there was a breaking up of this traditional way of life following the 'second conquest' of the Amazon by the Brazilian military government in the 1960s and 1970s. This resulted in the decaboclization of the rural population as they migrated to towns and abandoned the adaptive strategies characteristic of caboclos in the past.

Parker's achievement is to have developed an historical framework for contextualizing caboclo lives, a project initiated by Eric Ross (1978), Eduardo Galvão (1979) and Stephen Nugent (1981). Nevertheless, this took place from the perspective of a cultural ecological framework (introduced through Charles Wagley and developed by Emilio Moran) which could not fully incorporate complex political and economic transformations (Harris, 2000; Nugent, 1981). Thus the deforestation encouraged by road building and logging, and land clearing for cattle pasture was seen in apocalyptic terms, as a second conquest, rather than as another key moment in a series of interventions, and accompanying resistance, whose effects were not entirely predictable and, given the heterogeneity of the region, unlikely to have homogenous consequences (Cleary, 1993, p. 336). In this chapter I shall explore an Amazonian way of being in time which is not about gain and loss of cultural traits. I argue there is no caboclo culture, as though it had an ontological reality as a web of beliefs and procedures with an immobile legacy. Similarly, there can be no caboclo identity, in the sense of an ethnic boundary separating the caboclo way of life from another (Lima-Ayres, 1992, 1999). Both these positions create false problems. How can we understand the history of riverine peasantries (i.e. caboclos) without falling into the trap of reifying culture?

The answer I offer here is that aspects of Brazilian Amazonian history have produced a mode of being which has become characteristic of the communities living on the floodplains. Two aspects of an historical approach are examined: (1) economic history and relations of kinship (2) the role of popular Catholicism in various ideological transformations. The intention is to arrive at some general remarks on the particularities of modernity in Amazonia which began with the conquest of the Americas. This framing of the study raises many important questions.

The formation of Amazonian peasant societies sometime between the expulsion of the missionaries (1757) and the Cabanagem (1835–1840) makes them invented by and dependent on colonizing and modern forces.[2] But at the same time, they are distinctly local (relationally and environmentally) and independent (owning key economic resources). Can we really separate the external and internal so clearly with respect to these 'original moderns'? What is the proper context in which they should be studied? Amazonianist? Portuguese and Iberian? I pose these questions to introduce the multi-layered character of the material. There are neither single constituencies nor answers to these questions of the historical identity of the heterogeneous peasantries of Amazonia. Moreover, it is analytically problematic to talk of syncretism or hybridity (two discourses concerning mixing) without a precise understanding of the transformative process of conjuncture of specific traces.[3]

Deborah Lima has provided an authoritative ethnographic description of floodplain peasants based on her work on the Mamirauá reserve upriver of Tefé (Amazonas State). In her highly suggestive article with Alencar it is argued that '*ser vargeiro significa viver o presente*' (to be a floodplain resident means to live in the present Lima & Alencar, 2001, p. 44). This presentism, or mastery of the present, includes an understanding of the past as discontinuous with the present, and a lack of group identity consolidated around social memories or a myth of origin. These insights from Lima and Alencar are based in long-term fieldwork and fit well with the cultural ecological view of the caboclo seen above. Given the emphasis on adaptive processes it is not surprising that again the themes of flexibility and resilience reoccur.[4] Nevertheless Lima and

---

[2] This understanding of the reconstitution of the riverine peasantries follows Mintz (1989) in the Caribbean: they come from no pre-colonial heritage, they have no glorious past to invoke, and no shared folk memory of a common origin. More positively, they originated as a direct result of the colonial economy, formed in the gaps of dominant agrarian institutions for the provision of cheap products, and controlled by debt bondage. For the most part they remained anonymous and amorphous until the beginning of the 19th century.

[3] In order to know how local Catholicism, for example, has come to be what it is we need to know under what conditions, political and ideological, did certain discourses mix, or else converge or repel, and transform in the process. This perspective is well developed in the study of Andean histories and societies e.g. Dean (1999) and Wachtel (1990).

[4] Moran (1974, p. 136) argues '[the *caboclo* is] the most important human adaptive system found in the region'. The *caboclo* is an 'adaptive system', because s/he has successfully endured in both harsh environmental and disadvantageous historical conditions. Thus the two themes of adaptation to the environment and accommodation to external economic demands are inseparable in the work of Moran and Parker. Both environmental adaptation and economic accommodation figure as determinants of social organisation. *Ribeirinhos* are supposed to accommodate to market conditions in a similar way to their adaptation to environmental pressures. This flexibility is said to be a function of the Amazonian peasantry, being a result of the colonisation of the Amazon valley by Europeans and the consequent destruction of aboriginal societies starting from the seventeenth century. According to their argument, this new society was a product not of the people who lived it but the external conditions which generated it.

Alencar offer a way out of the paradox of the cultural ecological view, which understands caboclo culture to be both a blueprint for using resources and an adaptive system (Harris, 2000). Their study draws out the *vargeiros'* narratives of the changes which confront them on a daily basis and how they perceive them, revealing their privileging of the immediate here and now. They reveal that change is not necessarily driven by external processes or pressures (e.g. ecological, economic or political). Instead the *vargeiro* narratives indicate their own cultural attachment to reinvention and living for the moment. In other words, the floodplain orientation is produced in the attention to the ever changing present, from an openness to experience rather than a template imposed upon the flow of life.[5]

I want to open this discussion with a brief summary of one man's life history. The man's son gave this account to me during the summer of 2001. In 1911, Abraham left Morocco for Brazil. He was 13 and traveled alone. His aim was to make his fortune there and return with his riches to his family. He made his way to the Lower Amazon, eventually settling down in Óbidos. There he married a woman from the neighboring municipality of Oriximiná, and established a trading business and later acquired a farm on the Sapucaia River (a tributary of the Trombetas River), where they went to live. His family carried out a diversity of economic activities, including raising cattle, marketing their meat, and cutting and selling firewood to steam boats. With his children in need of schooling, the family moved from the farm to the town of Oriximiná. From there he developed a trading enterprise, basically a kind of mini-market, *Casa Israelita*, selling anything from shoes to popular remedies. There the two children received an education from the town's judge, taking exams in Belém and qualifying in medicine and law. The family then dispersed to different cities. The doctor moved to Alenquer, practicing there and in other nearby cities. The lawyer became a professor at the Federal University of Pará and later worked as the state's secretary for education. Abraham died in 1974 in Alenquer, where he had moved with his son and owned an earthenware factory. A few years later his bones were taken to Israel to be entombed next to his family, who used to live in Belsheva, Negev. Nowadays his grandchildren have few direct links with the Amazon, except that one of them is a social anthropologist, based in the Northeast of Brazil, who studies the Amerindian peoples who live in the upper Rio Negro, Northwest Amazonia.

These are only the bare facts of a fascinating story. The life of Abraham typifies so many of the processes that have enveloped the Amazon in global socio-economic transformations. However, Abraham's life of migration and economic success is also unique – only he could have managed his affairs in the way he did, depending on his personal circumstances and motivation. Abraham's biography is both extraordinary and ordinary, like and unlike other biographies

[5] I want to acknowledge the work of Cleary (1993, 1998) which has drawn out the complexity and particularity of the historical features of Amazonia, especially the importance of the informal economy.

of the people who have made and continue to make up Amazonia's population. This brief example can throw much light upon the understanding of the region's history, connecting the particular with the general.

Abraham was part of a larger process of Jewish migration from Morocco and Egypt at the beginning of the twentieth century and settled in towns on the Amazon River, such as Santarém, Óbidos and Monte Alegre (Nugent in this volume). These men also married local women from Catholic families. They also set up successful commercial enterprises, some of which continue to this day. While the men's Jewish origins remain important to their families (e.g. visits to Israel), the children were raised in a predominantly Catholic context. The children grew up in a culturally open environment. Ideas of dolphins' transformative capacities, if we take this one example of Amazonian culture, would have been known by newcomers and learnt by their children (whether they really thought dolphins can turn into humans is irrelevant and indeed, the same goes for 'caboclos'). They would have inhabited the same popular culture, even if they did not share the same history. We are dealing then with a form of life that is more permeable and malleable than Parker implies. It is neither confined to a single group, nor is that population homogenous.

In addition to the North African Jewish migrations, there have been other well-documented and significant movements to the Amazon (e.g. North-eastern Brazilian, Japanese, Italian) since the independence of Brazil. These people too have learnt about this folk culture of the Amazon. Some of them, for example Italian migrants from the beginning of the twentieth century, live in floodplain areas and pursue diverse economic activities. They too tell stories of dolphins (*botos*) and large snakes (*cobra grandes*). Are they caboclos? Can one be more or less caboclo? These questions reveal the difficulty of thinking in such terms, of imposing abstract concepts, such as culture, on real life processes and people. The answers effectively reify Amazonia as a cultural object (Nugent, 1997). What I find more rewarding is the exploration of the link between these biographies and their socio-economic context. This means pushing to one side matters concerning what would constitute caboclo culture and identity. These 'emplaced narratives' can then be used to give historical content to the peculiar cosmopolitan and modern character of Brazilian Amazonia.

Thus in Abraham's life story we can view the personal travelling of impersonal forces. It reveals life in movement, from international migration to the union of old and new worlds, Moroccan and Brazilian, Jewish and Catholic, the desire for 'a better life', in terms of both economic and educational opportunities. So the life history is remarkable on a personal level, but unremarkable in Amazonian terms, since, in outline, it is shared by many others. It is to the modernity of the Amazon region that these migrants' lives speak. What is specific to the region is the way a predominantly river based way of life envelopes these larger processes. The river as both a material network and symbolic presence shapes cultural and political possibilities. There is a loosely based form of life deeply connected to the rivers which was developed by caboclos in the ruins of colonial conquest. Following Lima and Alencar then we need to put the river environment at the core of any discussion of caboclos (*ribeirinhos* and vargeiros) and that means understanding their historical mobility.

In the rest of this article I shall avoid using the term caboclo for two reasons. First, it is not one of self-ascription: it is a local term of abuse. Secondly, its usage would involve an understanding of its historical discourses of national and regional identity, tasks already attended to (Lima-Ayres, 1992, 1999). Instead I will use ribeirinhos often referring to them collectively as a 'they', a linguistic obligation I know to be an ethnographic travesty. This means, at times, I assume a level of generality between various riverine areas of the Amazon (Tefé, Óbidos, Santarém, Gurupá and Vigia), since in this paper I have opted to make comparisons than select differences. I thus ask the reader to bear in mind that there are both historical commonalities and significant local ecological and ethnographic variations between these populations.

Modernity and Invisibility

Galvão's outline (1979) of the encounter between tribal and national societies stands as one of the first, and still one of the finest, attempts to engage analytically with the historical conditions which led to the emergence of the riverine populations of the Brazilian Amazon. Somewhat belying his theory of the gradual formation of the Amazonian peasantries, Galvão (1979) notes the unevenness of the economic history of Amazonia. Drawing attention to the boom-bust nature of the economic cycles in the Amazon, he writes there is often a resurgence of indigenous techniques of adaptation during the 'bust' periods. This involves a return to a baseline which prevents tribal units from total disintegration. Furthermore, there are significant regional differences which affect the development of locally oriented interests.

It is Nugent (1993) who has most clearly written about these contradictions in relation to the role of merchant capital in the region. He examines the important relationship between the historical viability of the societies and economies of Amazonia and the degree of penetration of international capital. He argues in times of low external market demands local communities develop internally. Conversely, when there is a high demand for Amazonian products, and concomitant control of labor through debts and prices of products, these same people become more atomized and oriented to the outside. This theory of contraction and expansion of local societies is quite different from the caboclization account. The cultural ecological view understands ribeirinho society and culture to be formed in a linear sequence, which reaches a climax ('solidification' is Parker's word), leading inexorably to its demise. This kind of perception of history is too neat, as Eric Wolf (1982) has shown: ribeirinho histories should be resituated across and in relation to these fluctuations. Following Nugent, and recent ethnographic studies, such as Lima-Ayres (1992), Heraldo Maués

(1995) and Acevedo and de Castro (1998) history, from the ribeirinhos point of view, has been a series of upheavals, each of which have left their imprint, often changing radically the social context. Nevertheless, it is often the same families and their descendants who continue working in the new social and economic conditions. The 'caboclo complex', as Nugent (1993, p. 205) calls the various peasant economies and histories in the Amazon (riverine, Northeastern and Japanese for example), is a repertoire of petty commodity forms which has emerged in relation to the character of regional development, the most prominent feature of which has been the role of merchant capital (see discussion below).

The boom-bust view of the economic history of the Amazon is also not without its limitations. Historians (Gentil, 1988; Santos, 1980; Simonssen, 1954; Weinstein, 1983; WinklerPrins, 2005) have documented the forms of production and distribution relating to a few high profile export commodities, for example rubber, fish, cacao and jute. Despite this history of Amazonia, we know little about the day to day lives of those people actually doing the work of producing and distributing these commodities. This includes aspects such as the full range of economic activities and how their organization changes over time with market cycles; the organization of labor; the specific histories of regions and communities, through migration, marriage practices, land claims; and the shaping of material life by the Amazonian cultural imaginations and memories. To give one example: the history of the second half of the nineteenth century is represented as one dominated by the collection of rubber and its incumbent interests. Yet in the riverine areas of the Lower Amazon, there was much less production of rubber than in Acre. Still the former region saw significant commercial development and urban growth. What were the people who lived there doing? Why did the money end up in these riverside towns? Who brought it there?

These kinds of questions point in the direction of what Ferdinand Braudel has called the shadowy zone, lying underneath the market economy, representing the elementary activity of daily life (1981, p. 23). This is the invisible economy, which is not recorded, in official reports or statistics. But it is present in oral narratives (Acevedo & de Castro, 1998; Harris, 2000; Lima & Alencar, 2001) of the past told and inscribed on the landscape (for example, deforestation to provide firewood to passing steamers, Raffles, 2001 for a natural history of the creation of Amazonian landscapes). This vast mass of undocumented and invisible activity lies outside the official, or export-commodity oriented, understanding of the Amazonian economy and history, requiring scholars to reexamine the nature of the riverine population's accommodation to market conditions (Larson & Harris, 1995 on Indian market participation in the Andes).

This idea can be taken forward to reveal actual ribeirinho economy and society. Nugent (1993) has argued the peasant economy is formed by merchant capital, that is capital which produces profit only from serial increase in the price of a product as it is traded, rather than from labor (as in industrial capital, which develops from merchant capital and the drive to increase profit). As such the peasant economy displays the contradictory features of a transitional system. Economic production is oriented towards the market, but prevalent relations and organization of production are not capitalist, since there is a clear separation between worker and boss and no single owner of resources. The system continually appears as though it is moving from part-capitalist to fully formed capitalism, as its merchant capital roots give way to a more profitable form of production. In fact it never does, because the relations of production are never transformed, since they remain peasant in character. This is because labor is never fully incorporated, only the products of labor and by debt slavery. The repression of these relations arises from their being incorporated into the global economy through demands from traders for exchangeable products and not for labor. Merchant capital is degenerative for Amazonian development, but it is instrumental in the flourishing of kin-based peasantries. There is then a basic contradiction at the heart of modern Amazonia: it is capitalist in name and in the drive for profit and expansion, but non-capitalist in its character and relations and in notions of ownership (Cleary, 1993; Harris, 2000).

Precisely this dichotomy is also found in the relationship between Church based Catholicism and its regional and popular religious expression (see discussion below). Both non-capitalist relations and popular Catholicism are local refractions of their larger counterparts, and are therefore contingent on them; this does not mean they do not come to develop their discourses.

The riverine populations seem to be doing more than accommodating the prevailing demands. These people are able not just to accommodate the fluctuating markets, but also reorganize and reproduce in the new conditions in which they find themselves each time. As such they have developed a capacity to embrace change at each new stage without it leading to the demise of their current way of life. On the contrary, their economic openness, that is their ability to deal with rapid changes, serves their reproductive potential extremely well. For this local peasant economy is resilient enough to expand in times of relative market stagnation. The key to this success is the organization and control of labor and resources through relations of kinship (Lima-Ayres, 1992; Harris, 2000). This achievement means the stress of their lives is in the mastery of the present, rather than the past and holding onto some assumed heritage. I am suggesting that the lack of information on the invisible economy led to the underappreciation of the historical agency with which caboclos have acted.

It is perhaps because peasant productive relations are based around kinship, that is they are locally generated, they are invisible (or even uninteresting to all but anthropologists). I do not have the space to detail the material and moral importance of kinship to social reproduction. Suffice it to say that since the rural communities and villages along the rivers are formed by bonds of kinship they are genuinely local forms. Kin relations and marriages allow for continuous control of and access to nearby resources, such as land, lakes and streams. Proximity and ongoing usage of land and water confers usufruct rights. Typically, riverine communities have dense networks of kinspeople, organized in clusters of families (Furtado, 1993; Harris, 2000; Lima's and Futemma's chapters in this volume). Labor is both organized hierarchically, with parents controlling their children's labor for as long as possible, and through horizontal relations between peers (cousins, co-parents and neighbors; Murrieta, 1998, 2001 and in this volume). In addition, there are a number of marriages between cousins (from first and beyond),[6] which help to renew and continue the previous generation's relations (Lima-Ayres, 1992; Harris, 2000).

This section has presented a general argument. Elsewhere (Harris, 2000, 2005) I have presented in historic and ethnographic detail the evolution of a floodplain area of the Lower Amazon. This economic history, from the ribeirinhos point of view, can neither be simply characterized as one of boom and bust, nor one of caboclization, of the solidification of a regional culture. The boom-bust model ignores the dynamism of the local forms of reproduction. The linear evolution model falsely understands the change from tribe to peasant to proletarian as an historical and environmental inevitability. All social and economic activity is functionally adaptive, so there is no room for contradictions in the system. I have tried to show some of the problems in both these models of Amazonian development and to indicate a way forward. The invisibility of local dynamism is as much due to its unremarkability as to its basis in kinship (Futemma, Murrieta and Brondizio's chapters). The modernity of the system goes hand in hand with the invisibility of peasant societies. The contradictions have given rise to an economy alternating between stagnancy and activity, transformation and maintenance; or in the words of my informants, *sempre ajeitando*.

Popular Religion in Amazonia

I shall now develop the argument on the historical participation of ribeirinhos by considering popular religion. I shall explore the link between the local dynamism, discussed above, and what I shall call their 'condition', which is to embrace social and economic change and influences external to their immediate social universes. They do not reject or resist these influences through overt political action, though they may be more covertly in everyday forms of opposition. From fulfilling a request from a boss for the supply of more catfish to the taking out a loan from a bank to finance jute cultivation, from the fascination with *telenovelas* (soap operas) to pasting mail order catalogue pages on walls in houses, from accepting gifts in exchange for votes to the distant reverence given to white foreign priests with strange accents, the outside is observed and acted upon with a special and keen interest. In accepting these powerful and seductive forces and in their attempts to deal with them, there is

[6] First cousin endogamy is widespread among European peasant communities (for example O'Neill, 1987; Segalan, 1986), as well as close–knit capitalist élites.

moreover a process of negotiation. In few areas of social life is this more evident than in the nature of Catholicism and popular religiosity in general.

By the time the missionaries were all but expelled (1757), Catholicism had been imposed on the contacted Amerindians – at least nominally (Hemming, 1978). Not only was Catholicism the religion of the conqueror, the Church (rather than the missionary orders) was joined to the state in a special relationship. Nevertheless, until the proclamation of the republic in 1889, the Church in the Amazon was relatively weak (Maués, 1995, p. 87). Only a few priests covered huge areas, and most would have concerned themselves with the urban-based elite. It is curious that the institutional weakness of the Church should not have led to the erosion of Catholicism as a powerful force in everyday life. On the contrary, popular religiosity had, and continues to have, a strong presence in all cultural dispositions. This makes it difficult to separate 'religion' from other areas of social life. Despite all the migrations and the search for better lives, and the environmental destruction, the process of disenchantment is unevenly distributed (though Slater, 1994 argues it is widespread in urban centers and the rise of Protestantism must be taken into account, see below).

Furthermore, it is important to distinguish between religion as an institution (Church) and as experience and knowledge (popular beliefs and practices). Most of the time these twin aspects have been at odds, in that the Catholicism of the priests and missionaries has had a different orientation to that of the lay population. They appear superficially to be in the same discursive universe and they obviously depend on each other, but closer investigation reveals important differences (for example Maués, 1995 on saints; and Cannell, 1999 on concepts such as pity, death rituals, and altruism in the lowland Philippines). Arguably, the differences are now coming closer together, as the Church has strengthened, following the period of romanisation and strong ecclesiastical control of popular religiosity from end of nineteenth century and the return of missionary orders in 1909 (Maués, 1995 for an excellent discussion of this process).

Thus while the Catholicism(s) of missionaries, church priests and settlers caught on relatively quickly, it came at a cost to official doctrines. Examples from outside the Amazon, such as candomblé and the appropriation of saints as spirits, are well known. A more subtle and less defined process of modification also occurred in Amazonia, (it could have been less clear because it was not associated with an oppressed identity, i.e. enlsaved Africans). What interests me is the 'local religion' that developed in the post-missionary period and the beginning of twentieth century corresponding to the river based form of life mentioned earlier.[7]

[7] I should properly start with the missionary period and juxtapose this with the Catholicism introduced by colonists who were not linked with any religious order, i.e. the Portuguese and other Europeans who were encouraged to marry indigenous women. Whether these add up to a great and little tradition is another matter but their difference is important. There are some excellent analyses to consult here; on missions see Sweet (1974), and Block (1984) and on Portuguese Catholicism see Sanchis (1992).

In this I am dependent on the work by Galvão (1955) and Maués (1995), amongst others, who have studied popular religion.

The Catholicism that grew in this period is significant to understanding the condition of ribeirinho identity as a whole, complementing the previous section. In making these assertions, I assume that there is some determining relationship between the mercantile economic system and patterns of relations, ideas and practices. Logically, it is possible to follow this with another assumption: there is some coherence in the ideology which supports the economic system, a recurrence of similar patterns in different social contexts and institutions. It is precisely the finding of coherence in religious beliefs and practices that has motivated the studies by Galvão (1955) and Maués (1995). The problem however lies with the analysis of the myriad of historical influences, some which survive by force of institutional power and some which survive other ways, such as long established bodily habits. Again, my aim in raising these matters is to complicate the presentation of the linear model of caboclization. This is rather well expressed in the following quotation taken by Heraldo Maués' study.

Another concern of the vicar [of Vigia in the State of Pará was ... in controlling the saint Nazaré festival. His term as vicar in Vigia ended in the same year as the second Vatican Council started, out of which came a new orientation to parish work, a renewed theology and political practice. The phase of romanisation [centralisation of Church creeds] had ended. Nevertheless the case of Vigia, like the Salgado region, demonstrates that its effects were going to last, like the deep structural elements, which arose in the imperial period, and even in times before that. Maués (1995, p. 85)

And I would add that the same sense of 'structural continuities' applies to the whole area of the primary colonizing efforts of the Portuguese, that is along the main trunk of the Amazon and its major tributaries. The different elements of this quotation reveal some of the main themes in analyzing the popular Catholicism of the Amazon. There is significant control, not just in doctrinal matters but also in social affairs, such as community organization, abortion, and so on (what Maués, 1995 calls the Church's *desejo de totalidade*). In reaction, the lay population may refuse to comply with these attempts, as seen above in the failure to discipline the *festa* de Nazaré. This may take the form of resistance to, or subversion of, the Church. What are the influences on popular religious experience? Maués characterizes these influences as transforming structures, which almost lie like geological strata on top of one another, but leave important traces in contemporary worldviews. What are these traces and how are they still present? Is syncretism the best concept to understand them?

These are difficult, but important, questions. What was the influence of medieval popular Catholicism brought by the first settlers? Missionaries were fully active from the mid to late seventeenth century by which time Catholicism had already established itself.[8] For example, it was common in medieval

[8] Bettendorf was in Santarém in 1657.

Europe to use a stalk from certain plants to beat a possessed person (Ginzburg, 1983 for example). In popular lore, the stalk would scare off the demon in the body. The same practice is still in use today in the Amazon (Harris, 2000 and Alencar personal communication in the region above Tefé).[9] Another example is the devotion to saints, and their festivals; though there are significant differences between contemporary old and new world accounts (e.g. Pina-Cabral, 1983 in Portugal, and Christian, 1989 in Spain). Furthermore, the predominantly domestic context for religious practice could be an inheritance from Iberian Europe. But it could also be a reaction to the disapproval of the Church of popular practices, pushing it out of sight into private spaces. Certainly this is what seems likely to have happened in the case of shamanic healing, which raises questions about the origin of *pajelança*. Maués (1995, p. 33) sees influences from indigenous Amerindians, as well as from *catimbó* in the Northeast and from Kardecist spiritualist beliefs (1995, p. 250).

Of course, speculation about origins is just that. Nevertheless it does raise interesting issues concerning the nature of knowledge and its reproduction, (e.g. are some cultural practices more resistant to change than others? Does such transformation depend on external intervention or on cognitive psychological factors?) which lie beyond this paper. Such continuity should not surprise us though given the relative success of the social reproduction of the riverine populations (again urban and rural), that is their ability to change, to incorporate the outside, without drastic reorganization. *Plus ça change...*

Irresistible Catholicism?

I want now to bring the discussion down to an ethnographic level in order to highlight some of these problems. In 2001 I returned to the site of my doctoral fieldwork (1992–1994), a floodplain village near Óbidos (Pará), after an absence of more than seven years. There had been many changes, including erosion of the *restinga* (levee) and some river islands. The electricity generator was no longer working and, more importantly, the school was only taking pupils up to the third grade (*terceira série*) of primary education (under the previous teacher it had gone up to fifth grade). Thus a few young people had moved to Óbidos or Oriximiná to continue their studies, staying with relatives or friends. In one case a whole family had moved to educate their four children, but kept their land and house, in the expectation of their return. However, four other families had also moved out. And it took a while for me to get an explanation that made sense

[9] Beating with plants is common in the Andes (T. Platt, personal communication, who has challenged my assumption of diffusion from Europe). In the Andes the concept of 'convergence' has been used, where similar ideas are found in both mediaeval European and pre–Colombian Andean civilizations, one becoming 'superimposed' on another with the European invasion. Then comes the interesting question: are people only able, generally, to recognise 'otherness' insofar as they can find a similarity to something they already understand?

since they did not move for education. Eventually I learnt that they had converted to an *Assembleia de Deus* Church (in 1998) and had been forced to leave. The converted Protestants (eight adults in a dense network of houses) had tried to live harmoniously with their Catholic neighbors. We should not forget that these people are all closely related. Eventually the underlying tension became too much. The *crentes* had asked for changes to be made to the school, such as not singing the national anthem (because it contained Catholic references), and had refrained from the normal leisure time activities of drinking, smoking and so forth. They would not eat the same kinds of fish and dressed differently, would not work on Saturday. All these differences caused a rift, which became more and more irreconcilable. Thus the families left and joined the small protestant communities in nearby towns.

When I asked questions about the rift and its causes, I was met with some strong answers. My concern here is with the Catholic perception not the reasons for conversion by those Protestants. Many people replied that '*crentes não segue um caminho bom*' or that '*só tem uma religião, que é fé em Deus*' or '*só tem um Deus*'. There was agreement that Protestants do not pray properly, because they do not do the sign of the cross; they do not have icons, indeed their churches are bare, so there is no beauty; and they were mad to give up the good life, comprising of drinking, going to festas, and so on. Some said the newly converted protestants seemed the same, they had not changed in their personality, so why did they bother loosing so much when they gained little or nothing. The strongest condemnation came from the *pajé*, a sixty five year old man. He told me he refused to cure Protestants, if they came to him (which is unlikely). He accused them of being a *seita ruim*, implying they are a 'satanic cult', because they do not pray and have icons to direct their prayers. These are strong denunciations, given their openness to external influences.

However, the challenge of Protestantism to the Catholic worldview of floodplain dwellers is threatening. Like large cattle owners taking over valuable land on the floodplain or large fishing boats taking away the means of trade and subsistence, Protestantism threatens the whole existence and historical momentum of rural livelihoods. This is no exaggeration. Conversion demands a radical break from the present. It specifies a series of personal life and bodily comportment changes, such as what foods can be eaten, what clothes can be worn, no drinking or smoking or adultery, avoidance of fighting and Catholic festas. Participation in most of these activities are central to everyday life, as perceived by Catholics. What is also shed in conversion is the range of beliefs associated with the enchanted environment. Protestants regard *boto* stories, and other beings (cobra grande etc.) the power of the pajé, *panema* (bad luck in hunting and fishing) and so on as superstitions to be discarded in their search for true knowledge. All these are integral to the Catholic worldview. Moreover, the fact that these few people were prepared to leave their close kinsfolk and communities for the sake of their new beliefs suggests that Protestantism is highly individualistic (Lehmann, 1996; Burdick, 1998 for elsewhere in Brazil). Again this individualism is at odds with the collectivist orientation of Catholic ideology (Lima-Ayres, 1992; Maués 1995; Wagley, 1976).

Protestantism in these densely organized villages is not simply a doctrinal conversion; it is a refutation of a whole way of life.[10] It involves a rejection of the past self. Converts should come to inhabit a different discursive universe. Thus the change should be a complete cultural one, where a person may even come to acquire new sets of kin in an attempt to rebuild networks. In the same way, for Catholics the link between popular religion and everyday practice has, until recently, been comprehensive. Thus one of the most common acts of self-ascription by floodplain people is *somos católicos* (we are catholics). They are not offering their precise religious identification, but a statement of their being in the world, their total presence as human beings. They are also invoking their enjoyment of many kinds of fish and food and drink (notwithstanding the complex system of avoidance and taboo, Maués & Maués, 1978), their use of shamans and herbal medicine, their stories of encounters with *encantados*, *visagens* (enchanted beings and ghosts) and evil spirits, the accusations of witchcraft; and the pity (*misericórdia*) they pray bosses, saints and God have for their condition, and the desire to be blessed in their humility and poverty. As such somos católicos is a positive affirmation of how they perceive themselves and their connection to an environment. Many times I heard these same people say to become a protestant is to '*se meter na religião*' (to get involved in religion). This implies Catholicism is seen almost as a statement of fact, of common sense knowledge and observation. And not a matter of belief. Clearly this self-identification is incredibly important and goes beyond any religious affiliation. Somos católicos then is an umbrella, all inclusive, term, and almost a denial of (perhaps indifference to) identity and difference (Maués, 1995, pp. 168–169 for a different but complementary reading of the contrast between Catholicism and Protestantism).

So while I have argued the riverine populations have a cultural openness to external influences, for the majority Protestantism is a change too far. The liberal, tolerant, fatalistic and communitarian quality of popular Catholicism in Amazonia cannot cope with this particular challenge. Furthermore, this is a grassroots reaction in the Óbidos region. Such anti-*crente* messages are not being divulged by the priests and the active lay population of the Church. I want to stress that the numbers of converts is relatively small, but they are growing. The Protestants are tireless in their efforts. IBGE did not publish statistics on religious affiliation for each municipality for 1991. As a whole, Pará had a population of 4,949,199 people, of which 4,224,900 (85%) were Catholics and 536,725 (9%) were Protestants. In Brazil the split is now 74% Catholics and 16% Protestants (Folha de São Paulo, May 2002).

[10] Elsewhere I have argued this conversion is not quite the moment of rupture my informants, from both sides, claim. Instead there are significant continuities from one to the other (Wachtel, 1990 for an Andean reading of the conversion to Pentecostalism and its fusion with Catholicism).

It is important to note that three shamans I got to know in the region made some of the most acute observations and strongest criticisms about the differences between Protestants and Catholics. The reasons for this could well be personal. However there are historical factors and cosmological designs that I would like to argue make the shaman particularly interested in the 'success' of Catholicism. I now want to look at this connection between the Catholicism of ribeirinhos and the work of the pajé. The link between these activities and beliefs will provide further evidence of the transformation of external impositions or influences.

According to Hemming (1978) and Maués (1995), the missionaries persecuted the indigenous Amerindians pajés, more than anyone else. The pajé was seen as the embodiment of evil and trickery, and able to command the attention of his fellows. Without the pajé, the Europeans hoped the Indians would turn to Christianity for their needs and salvation. Given this history, it is not surprising this same European fear of shamans, and their suppression, was repeated during the rubber boom in an attempt to control Indian labor (Taussig, 1987). Shamans were also persecuted by the Inquisition. As I have said, in the post-missionary period, and the consequent absence of strong state or church institutions, there was a development of local interests. Slavery was still active in this time, particularly in the Lower Amazon and the village Directorates was desperately trying to promote colonist settlements, agriculture and control labor (Hecht & Cockburn, 1989). During this period (1755–1798), the elements of an ideological reorganization were in place, based around a deeply hierarchically organized and racially stratified society. Shamnism and a whole complex of associated beliefs, such as encantados, *espirítos maus* (evil spirits) and so on, which survived the previous period, then became reintegrated into popular Catholicism. Indeed, perhaps the only way for shamans to continue their work was to seek the links between their work and popular Catholicism. Although this is a speculative claim that requires further research, I suspect that shamanism, as popular religion did (in form of devotion to saints), flourished in the late eighteenth and nineteenth centuries.

Why? In answering this question, I want to work backwards from Maués' brilliant analysis of religion in Vigia and my own ethnographic research. Maués suggests there are many sources for the religious beliefs and practices in Amazonia, which have already been mentioned above. Nevertheless, he shows these themes are integrated in a cosmological model of the universe from a *ribeirinho* point of view. He emphasizes that there are inconsistencies in this model, because there is a degree of heterogeneity across informants' accounts. Moreover, he says that the integration of the multiple influences in a cognitive map is 'incapable of being explicitly verbalized, in its totality' (1995, p. 254). This fits with other anthropological studies of ideology in that they are not fully available to consciousness (Bloch, 1989 for example).

Maués writes this map is a predominantly Christian one, despite its incorporation of other non-Christian elements, such as beliefs about enchanted beings and their presence in the environment. This rather inconsistent cosmology is nevertheless distinctly Amazonian (1995, p. 256). For example, hell and purgatory have been replaced by a conception of the river as the deep (*o fundo*) where spirits, good and bad, dwell. The river is not in itself a bad place, but inherently ambiguous (compare the river with the mines and the underground in the Andes). The deep contrasts with heaven and the sky which are occupied by God and the saints. Spirits are known in the human world by either their malevolence, or through their benevolence in curing sessions when they are called upon by shamans. Saints occupy a materiality in their icons (*imagens*), which bear a likeness to the saint. They are invoked in prayer and celebrated in festivals, offering thanks to a saint. Maués finds a homology between saints and enchanted beings and their powers. Both saints and encantados were human beings; saints live in a divine place (supernature), spirits in natural spaces (nature); both can be called upon with prayer, saints by individuals or a collectivity, and spirits by a shaman. Both are ambiguous, since they can do good and bad; saints punish disrespect and the non-fulfillment of promises. These correspondences suggest a degree of coherence in the various ideological influences of the religion of the Amazonian interior. Furthermore it is a clear indication of the separation of the Catholic Church from popular culture. Indeed, Maués' argument is that the key theme in the study of Amazonian religion is the degree of ecclesiastical control on lay beliefs and practices.

The point is that the church felt there was something to be controlled. The period of strong ecclesiastical control started at the end of the nineteenth century, with the proclamation of the republic and the end of the link between the imperial state and the church (*padroado*). This heralded a new phase of the church in Brazil and the beginning of romanisation, where bishops attempted to standardize the proliferation of saints and their cults and forms of worship.

Historically we can assume, on the other side of the fence, that ideas concerning enchanted beings, visiting pajés, and devotion to saints and saints' festas in the Amazon had a pertinence to the lived experiences of riverine populations. I would like to add to Maués's study a brief analysis which also demonstrates the development of local religious expression, as well as the integration of shamanism within Catholicism. These data are based in my fieldwork in the Óbidos region.

Pajés are sought for series of reasons. Their powers can cure many types of disease, particularly ones with fever and general weakness of the body. They call upon their own personal spirits to detect the affliction which also involves the use of *tauari* cigars to draw the evil spirits out of the client's body. The illnesses can come from a variety of sources. They can be natural, in which case the pajé will make a diagnosis and recommend some home remedies and possibly some drugs only available from the pharmacy. Alternatively the affliction could be caused by an evil spirit, which may have been sent by a witch, or the person may have abused, perhaps unwittingly, an *encantado*. The pajé then is one of the first ports of call for anyone who is with a condition that cannot be easily treated or where special knowledge is needed.

Socially, the pajé has no privileged position. He is respected only as an individual with powers, with which he is born (*dom de nascença*) to see and control spirits, to travel to the city in the river and underneath land (*o encante*). He or she is constantly under surveillance however. The quality of their curing, their knowledge, whether they are witches are examples of some of the points of discussion. Needless to add, there are continual disagreements over judgments. Occasionally pajés come into conflict with the medical establishment, but most in the town of Óbidos said they had a good relationship with doctors. The priests (Brazilian) and Franciscan Friars (German and American) I talked with in Óbidos said they had a policy (since Vatican II and Medellin) of 'valuing local culture' of which the pajé was a part.

As I have said, for rural dwelling people and recent migrants to Óbidos, there was no contradiction between Catholicism and pajelança. Furthermore, the pajé regards, him or herself, as a Catholic. Maués reports the same for the people of Vigia (1995, p. 237). In fact some of his informants sought justifications for shamanic practice in the bible. However the connections go further than this and this is what I want to turn to.

In conversations with three pajés in the Óbidos area, I talked about the nature of their work. They described *pajelar*, the verb, as one of misericórdia, of mercy and compassion towards one's fellow human beings. Their ability to cure is given from birth (though they denied it was from God) and they do not learn anything from books or others. Since their power is a gift obtained without effort, they cannot charge for their services of diagnosis and treatment. Their work is a form of charity, of love, for their fellow sufferers. This understanding is clearly linked to the concept of misericórdia as the character of God's love, and which is repeated in mass. God's love is unconditional, pitiful and forgiving. Thus the conceptual armory of Catholicism has become translated to another context. The analogy is more complicated than this, since God is seen as unambiguously good, despite working in mysterious ways; whereas the pajé can harm as well as heal. When I suggested this link to informants, one responded that God too would have to know good and bad, since he is omniscient.

At the same time, an equivalence can also be drawn between a saint and pajé. A saint can be prayed to for specific requests such as help for recovery from an illness. Both have powers to perform cures and to offer compassion and pity. Though no one uses the term '*milagres*' (miracles) to talk about the pajé's gifts to heal. Both travel in spirit, either from the sky to an icon or to the underwater city and back. Again, there is not a full set of correspondences, because the relationship between saint and person is reciprocal and often intimate. Pajés receive nothing for their work, except a respectful following. They also have the power to bless, *benzer*, a person. The blessing involves receiving the protection of the spirits of the pajé and a diagnosis of any ailments. The pajé uses his cigar to blow smoke over a person as he recites his prayers (*rezas*), which invoke his *mestres* (masters), *os encantados*. In pattern, this is similar to a priest who blesses an icon or a person, who uses water to bless, and murmurs prayers as he stands over the anointee. These convergences are striking and are suggestive of a process of creative imitation of the other, the outside, that which is different. However, in this process of imitation they hold onto some previous dispositions, creating something new and original (Taussig, 1993).

These analogies, while being briefly stated, are too striking to be ignored: the use of prayer (magical words) by pajés and priests to communicate with a divine or enchanted order, the concept of pity stretching across Catholicism and pajelar, and the importance of blessing, of receiving a sacred favor by either a divine being or by enchanted ones. There are more connections that can also be made (as well as the others Maués gives), but I have made my main points. When I stated above that local Catholicism flourished in the post-missionary I had in mind the pajé as a central figure in this development. In the absence of a priest to bless and offer spiritual favors and protection, it is possible to speculate that the pajé opened up a new position for himself. He would have borrowed elements from Catholicism, and absorbed them within his own local context. The pajé took care of the sacred, which, in keeping with Catholicism, was kept separate from the profane, needing an intermediary to reach. But he has no social privileges and is as much a part of communal daily life as everyone else. The pseudo-priest's power also explains why the Church has intermittently persecuted the shaman.[11]

The main work of the pajé is to cure afflictions. Thus he is not only a kind of priest, he is a kind of physician. His diagnostic ability and lists of recommended medication are evidence of an extensive medical knowledge, both pharmaceutical and herbal. Once again this process can be seen in terms of the pajé creating a role for himself in the long-term absence of health institutions available to poor people.

In recent work on shamanism, a series of authors have noted the link between the state and shamanism (e.g. Humphrey & Thomas 1994; Taussig, 1987). Far from being an outdated and conservative social force, shamanism has been shown to be highly flexible, both in terms of its knowledge of the world and its adaptation to new forces. Thus in a film about shamanism amongst the Yaminahua in the Peruvian Amazon (made by Reid & Townsley, 1993) a shaman is filmed singing new songs about the power of boat engines, which has recently been introduced to the area. A visit by the shaman and his apprentice to a city cinema is also filmed. The experience of watching a film is likened to the hallucinatory voyage of a shaman to the spirit world. This visual ethnography expertly reveals the way in which the shaman is intensely aware of the power of the whites, and their technology, incorporating these potent outside forces within the Yaminahua shaman's knowledge. In the process something new is created. Similarly, Taussig shows that the power of the shaman is as much a myth as a reality. The fear of sorcery by the rubber tapper barons in the

[11] I have found virtually no historical material on the *pajé* (both as office and practices). But I am aware there has recently been a study by Aldrin Figueiredo, which I have not seen. It would be fascinating to do a distribution map of usage of the word pajé, as well as other Tupi terms, in Brazil and beyond.

nineteenth century Putumayo region of Peru created the fiction of 'shamanism' as some kind of elemental force that threatened the very existence of the Europeans. He writes that 'Sorcery and so called shamanism... present modes of always locally built experience and image formation in which such social knowledge is constitutive.' (Taussig, 1987, p. 463).

What is apparent in these examples is that there is no single identifiable set of characteristics which constitute shamanism as an authentic category of social analysis. It is forever modifying in relation to wider circumstances. Indeed, following Lévi-Strauss' analysis is also relevant, since he argues the American shaman is only successful because of the group's acceptance of his curing, not the other way round. Thus the shaman is highly skilled in his perception of other people and changing his behavior. In the ribeirinho case, shamanism has adapted to the prevailing historical circumstances. Thus shamanic activities have been shaped by the absence of medical care for the poor, a strong hierarchical social and economic order based on patron client relations, a weak church institution (until twentieth century) and a Catholicism which was oriented to saints, festivals and stressed tolerance and the collectivity.

The question I asked previously was why did a popular Catholicism flourish in the nineteenth century. It was not so much Catholicism that strengthened in this time as a new set of folk practices and beliefs. It was Catholicism in name, since this was the acceptable face a folk view could present to the world, as well as to its uncertain self. Furthermore, this self was heterogeneous, since there were local variations and uneven developments of markets and patron-client relations in the region. Bastide (1978) argued that the slaves, who elaborated the African religions in Brazil, had been thrown into a new social framework and had to find new ways of understanding their experiences. Similarly ribeirinhos of the Amazon, living along the waterways, created a new religion from available material and imaginative resources. This religion too had to collude with the prevailing ideology, which justified exploitation and prejudice on basis of race and the legitimacy of patrons.

In presenting this ethnographic evidence of what the auto-identification of being a Catholic means and the conceptual overlap between *ribeirinho* shamanism and Catholicism, I have emphasized, following Maués (1995), the historical difference between official and popular religious expression; antagonistic, but also intimately linked. On the surface the differences appear subtle but analytically there are significant contrasts as to meanings and references. To anthropologists of religion this would come as no surprise.

What is significant about the Amazonian case is the way in which a fresh social category of people, not necessarily homogenous, and within limited conditions, converted an imposed Catholicism and combined it other traditions, such as popular medieval beliefs and indigenous shamanism and reinvented them. Thus they have attended to, and in some cases accepted, the manifold influences with which they have been presented over time. Thus there has been an active and creative process of searching on behalf of the riverine populations in constituting not just their material lives (as seen above), but in ideological matters as well. Thus attention is constantly being drawn to the present, to what is at hand. This does not mean there is a complete rupture with the past at each moment, but that continuity inheres in embodied practices, rather than in consciously producible meanings.[12]

A Way of Being in Time

How does this popular religious expression relate to the mercantile based economic system? In order to answer this question by way of conclusion, I shall return to the subtitle of the paper which is derived from an article by the social anthropologist Maurice Bloch (1998). He compares different types of memory and how they relate to notions of kinship and the person held by different societies. In particular he is interested in how these concepts affect the way an individual relates to history. He distinguishes between two ends of a theoretical pole. At one end is Aristotlean view, who saw individuals as in history and shaped by their learning experiences; what we may now call a processual approach. He contrasts this with Plato's perspective and his innatist vision of human qualities, thus all history is external to what a person is, unable to change him or her. In turn he shows how these two positions match onto different cultural ways of being in history, one in the lowland Catholic Philippines and the other amongst the Sadah in Yemen. Bloch examines how there can be different senses of the past (even if we begin from cognitive universalism) and, contrary to the commonplace assumption, change is not always activated by external events, and nor is the force of continuity propelled by tradition. In this paper I have attempted to show how in Amazonia this opposition of essentialism versus hybridism is too simplistic, for it can lazily appeal to the division between indigenous Amerindians and mixed blood ribeirinhos. The latter are not in any simple sense hybrid products of external forces, imitating more powerful outsiders; indeed they appear adroit at combining essence (length of residence to claim land rights, or tradition) with historical transformation (reproductive success and more generally modernity). Instead there are continuities and convergences, as well as discontinuities and resistance.

The riverine peasants of Amazonia bear some similarities to the lowlanders of the Philippines. Both groups say they are poor and exploited by the rich and

[12] At the same time we need to put this ideology in the context of the different political conjunctures in history which affect how this remaking actually works out. I am aware that I have offered little historical material to support this argument. Part of the difficulty here is the absence of local records (i.e. municipal) to develop a thesis of how it relates to local political processes in successive historical conjunctures. The degree of accommodation versus resistance might well vary according to political conjuncture. Creative adaptation is one option, and this may be legitimising of Church and State, but it may also switch to resistance. As it stands my argument rests on the ideology of mixing, with little support from what might really be going on. However, I would not be making it if I did not think there was no confluence.

traders. Neither has a pre-colonial nor triumphant past to appeal to. They both feel themselves to be far from the centers of power, to be exposed to the vagaries of outsiders, and decisions, which are never taken for their benefit. The development of ecologically based identities in Amazonia (e.g. ribeirinhos and vargeiros) around resources, such as water and land, signifies a strong element of an attachment to the here and now of an environment. Amongst ribeirinhos there is a value put on the ability to negotiate successfully the conditions of the present (sempre ajeitando, or as Lima & Alencar, 2001, p. 46 put it: '*a busca constante por melhores condições de vida*' (the constant search for better conditions of life)). Little interest is shown in conserving the past, either materially or ideologically. What people do now will bring about new possibilities, which are seen to dissolve the past. At the same time, in the twentieth century they have allowed (in the sense they do not resist) their lives to be transformed by powerful outsiders, whether they be priests, bosses, large cattle owners, traders or saints or even spirits. The floodplain is constantly changing and the river is forever altering land formations. If people tell stories about the past, it is often related to the present, to a contemporary and personal experience. Their identity is therefore a product of what they are in the present, and is in contrast to what they were in the recent past (Lima & Alencar, 2001, who also stress that the past can be seen positively, as well as negatively). Thus their way of being in time can be compared to riding on the crest of a wave. To conserve this identity, to draw a boundary around it, would run counter to its nature. The non–reifiable aspects of ribeirinhos, such as immigration, economic cycles and environmental flux, conspire to give continuity to the momentum of the wave.

This Amazonian riverine way of being in history is connected to the structural-economic conditions and ideological dispositions discussed in this paper. It is brought about by the systematic features of the region's modernity. The demands placed on the riverine peasantry have oriented it to the present, to fulfilling the demands rapidly (e.g. supply of a product). The ideological basis for this state of affairs derives from a Catholicism, which stresses hierarchy, tolerance and allegiance to saints; as well as from a colonial and post-colonial policy which promoted incorporation through miscegenation. The world is there to be taken on and negotiated with. The openness to external influences and consequent modification of them places a value on being and doing in the current flow of life and represents their ambivalent response to the modernity whose westerly winds carried it into the world system. The next step in demonstrating this argument is to examine how this way of being has been constructed, and in response to what political pressures.

References

Acevedo, R., & de Castro, E. (1998). *Negros do Trombetas: Guardiães de Matas e Rios* (2nd ed.). Belém: UFPa, NAEA.

Bastide, R. (1978). *African religions of Brazil*. Baltimore: John Hopkins University Press.

Block, D. (1984). *Mission culture on the upper Amazon*. Lincoln: University of Nebraska Press.

Bloch, M. (1989). Cognition and ideology. In: *Ritual, history and power*. London: Athlone.
Bloch, M. (1998). Internal and external memory: Different ways of being in history. In: *How we think they think*. Boulder: Westview.
Braudel, F. (1981). *The structures of everyday life: Capitalism and civilisation 15th to 18th century* (Vol. 1). London: Collins.
Burdick, J. (1998). *Blessed Anastácia: Women, race and popular christianity in Brazil*. London: Routledge.
Cannell, F. (1999). *Power and intimacy in the christian Philippines*. Cambridge: Cambridge University Press.
Christian, W. (1989). *Person and God in a Spanish valley*. Princeton: Princeton University Press.
Cleary, D. (1993). After the frontier: Problems with political economy in the modern Brazilian Amazon. *Journal of Latin American Studies, 25*, 331–349.
Cleary, D. (1998). 'Lost altogether to the civilised world': Race and the cabanagem in Northern Brazil 1750–1850. *Comparative Studies in Society and History, 40*, 109–135.
Dean, C. (1999). *Inka bodies and the body of Christ*. Durham: Duke University Press.
Folha de São Paulo. (2002). *A encruzilhada da Fé*. Sunday 19th May, São Paulo (Mais supplement).
Furtado, L. G. (1993). *Pescadores do Rio Amazonas: Um Estudo Antropológico da Pesca Ribeirinha numa Area Amazônica*. Belém: MPEG, CNPq.
Galvão, E. (1955). *Santos e Visagens: Uma Estudo da Vida Religiosa de Itá*. São Paulo: Editora Nacional.
Galvão, E. (1979). The encounter of tribal and national societies in the Brazilian Amazon. In: M. Margolis & W. Carter (Eds.). *Brazil: Anthropological perspectives, essays in honour of Charles Wagley*. New York: Columbia University Press.
Gentil, J. (1988). A juta na agricultura de várzea da área de Santarém, Medio Amazonas. *Boletim do Museu Paraense Emílio Goeldi, Antropologia, 4*, 118–199.
Ginzburg, C. (1983). *The night battles*. Baltimore: John Hopkins University Press.
Harris, M. (2000). *Life on the Amazon*. Oxford: Oxford University Press.
Harris, M. (2005). 'Riding a wave: Embodied skills and colonial history on the floodplain of the Amazon', *Ethnos*, D. Boyer (Ed.), special issue on 'Revisiting the Anthropology of Knowledge', *70*, 197–219.
Hecht, S., & Cockburn, A. (1989). *The fate of the forest*. Penguin: Harmondsworth.
Hemming, J. (1978). *Red gold: The conquest of the Brazilian Indians*. London: Macmillan.
Humphrey, C., & Thomas, N. (Eds). (1994). *Shamanism, history and the state*. Ann Arbour: University of Michigan Press.
Larson, B., & Harris, O. (Eds.). (1995). *Ethnicity, markets and migration in the Andes: At the crossroads of history and anthropology*. Durham: Duke University Press.
Lehmann, D. (1996). *Struggle for the spirit*. Cambridge: Polity.
Lima, D., & Alencar, E. (2001). A lembrança da História: Memória social, ambiente e identidade na várzea do médio Solimões. *Lusotopie, 2001*, 27–48.
Lima-Ayres, D. (1992). *The social category caboclo: History, social organization and outsiders social classification of the rural population of an Amazonian region*. Ph.D thesis, University of Cambridge.
Lima-Ayres, D. (1999). A construção histórica do termo caboclo. Sobre estruturas e representações sociais no meio rural Amazônico. *Novos Cadernos do Naea, 2*, 5–32.
Maués, H. (1995). *Padres, Pajés, Festas e Santos: Catolicismo Popular e Controle Eclesiastico*. Belém: Cejup.
Maués, R. H., & Maués, A. (1978). O modelo da 'reima': Representações alimentares em uma comunidade amazônica, *Annuário Antropologico*, 77, 120–147.
Mintz, S. (1989). *Caribbean transformations*. New York: Columbia University Press.
Moran, E. (1974). The adaptive system of the Amazonian *caboclo*. In: C. Wagley (Ed.), *Man in the Amazon* (pp. 136–159). Gainsville: University of Florida Press.

Murrieta, R. S. S. (1998). O dilemma do papa-chibé: Consumo alimentar, nutrição e práticas de intervenção na Ilha de Ituqui, Baixo Amazonas, Pará. *Revista de Antropologia, São Paulo, 41*, 97–150.
Murrieta, R. S. S. (2001). Dialética do sabor: Alimentação, ecologia e vida cotidiana em comunidades ribeirinhas da Ilha de Ituqui, Baixo Amazonas, Pará. *Revista de Antropologia, São Paulo, 44*, 39–79.
Nugent, S. (1981). Amazonia: Ecosystem and social system. *Man, 16*, 62–74.
Nugent, S. (1993). *Amazonian caboclo society: An essay on invisibility and peasant economy*. Oxford: Berg
Nugent, S. (1997). The coordinates of identity in Amazonia: At play in the field of culture. *Critique of Anthropology, 17*, 33–51.
O'Neill, B. J. (1987). *Social inequality in a Portuguese hamlet: Land, late marriage and bastardy, 1870–1978*. Cambridge: Cambridge University Press.
Parker, E. (1985). Caboclization: The transformation of the Amerindian in Amazonia 1615–1800. In: E. Parker (Ed.), *The Amazon caboclo: Historical and contemporary perspectives* (pp. 1–50). Studies in Third World Societies, 32, Williamsburg: College of William and Mary.
Pina-Cabral, J. (1983). *Sons of Adam, daughters of Eve*. Oxford: Oxford University Press.
Raffles, H. (2001). *In Amazonia: A natural history*. Princeton: Princeton University Press.
Reid, H., & Townsley, G. (1993). *The shaman and his apprentice*. London: BBC Films.
Ross, E. (1978). The evolution of the Amazon Peasantry. *Journal of Latin American Studies, 10*(2), 193–218.
Sanchis, P. (1992). *Arraial: Festa de um povo. As Romarias Portugueses*. Lisboa: Don Quixote.
Santos, R. (1980). *História Economica da Amazônia*. São Paulo: T.A. Quieroz.
Segalan, M. (1986). *Historical anthropology of the family*. Cambridge: Cambridge University Press.
Simonssen, R. (1954). *A história econômica do Brasil*. São Paulo: Editora Nacional.
Slater, C. (1994). *Dance of the dolphin: Transformation and disenchantment in the Amazonian imagination*. Chicago: University of Chicago Press.
Sweet, D. (1974). *A rich realm of nature destroyed: The middle Amazon valley, 1640–1750*. Unpublished Ph.D thesis, University of Wisconsin, Madison.
Taussig, M. (1987). *Shamanism, colonialism and the wild man*. Chicago: Chicago University Press.
Taussig, M. (1993). *Mimesis and alterity: A particular history of the senses*. London: Routledge.
Wachtel, N. (1990). *Le retour des Ancêtres: Les Indiens Urus de Bolivie, Xxe-XVIe siècle: Essai de Histoire Regressive*. Paris: Gallimard.
Wagley, C. (1976, [1953]). *Amazon town: A study of man in the tropics*. New York: Macmillan.
Weinstein, B. (1983). *The Amazon rubber boom 1850–1920*. Stanford: Stanford University Press.
WinklerPrins, A. (2005). Jute cultivation in the Lower Amazon, 1940–1990: An ethnographic account from Santarém, Pará, Brazil. *Journal of Historical Geography, 32*, 818–838.
Wolf, E. (1982). *Europe and the people without history*. Berkeley: University of California Press.

Part II
Sustainability and Development Policies

Chapter 5
Traditional Peoples: Introduction to the Political Ecology Critique of a Notion

Henyo Barretto Filho

Abstract This chapter discusses the polemical issue of the permanence or not of "traditional peoples" – a classification that includes the various *caboclo* categories – living within protected areas. The author presents and discusses the problems posed by the attempt to define "traditional peoples" in Brazil. In the specific case of the Amazon, the Amerindian and caboclo populations are normally classified in this category because, on the one hand, their cultivation practices do not prevent the regenerative system of the humid tropical forest from working, and, on the other, because the impact caused by these groups' economic activities does not exceed the impact caused by small scale natural disturbances. Barretto Filho vehemently questions such claims, recalling that a great part of the Amazonian Forest today can be seen as a vast "cultural forest", to use a term introduced into ecological anthropology by William Balée at the end of the 1980s. As we declare these groups "traditional" we run the risk of rendering such people hostages to an a historical definition excluding them from social change.

Keywords Traditional peoples · Political labeling · Identity · Development · Conservation · Conservation units

Introduction

Within the debate on the supposed problem of the presence of human groups inside protected areas – defined as instruments for the in situ conservation of biodiversity – a notion was forged that has been freely applied in reference to historically specific and distinct social groups. 'Traditional people' has become an umbrella category (Esterci, 2001) in Brazil for a number of social groups, excluding Amerindians and remnant pockets from *quilombo* maroon colonies (cf. Lima, 2001b), that, despite being culturally distinct within the wider fabric

H.B. Filho (✉)
International Institute of Education (IEB), Brasilia, Brazil
e-mail: henyo@iieb.org.br; henyo@terra.com.br

C. Adams et al. (eds.), *Amazon Peasant Societies in a Changing Environment*,
DOI 10.1007/978-1-4020-9283-1_6, © Springer Science+Business Media B.V. 2009

of national society and characterized by culturally and historically specific forms of land occupation and appropriation of natural resources, are denied differentiated legal treatment[1] that recognizes their right to occupy their traditional homelands. The notion is ambivalent because, whilst serving as a negatively-charged run-off category for everything that is not Indian, not *quilombolo*, not rubber-tapper (as Lima & Rola, 2001, suggest), it actually subsumes all of those categories – and others besides – whose cultural distinctiveness is expressed in terms of specific territorialities.

In this article I aim to trace the genesis of the notion of 'traditional peoples' and draw up a sociological critique of its use in characterizing conflicts involving social groups residents in or around protected areas, especially those established in the Brazilian Amazon. Among other factors, the focus on the Amazon is justified by the fact that the notion has been systematically applied in the sphere of conservation and 'sustainable development' policies for the region since at least 1988 in an attempt to identify and bring to public knowledge the so-called 'historical' peasant societies generated by the colonial incorporation of the Amazonian region – i.e., the caboclos. As Lima points out, caboclo is a complex classification, combining dimensions of race, geography and class, used as the common sense stock representation – though not exclusively, as we shall see later – of the rural Amazonian population in the gallery of regional Brazilian archetypes. This type of social formation of the region's peasantry, marked by its own particular form of social organization, natural resource usage and land occupation (Lima, 2001b), or, indeed, by a given socio-cultural model of environmental adaptation (Arruda, 1997, p. 353), has been plastered over with the notion of 'traditional people' in contemporary times, recently converted into a legal category and institutional *démarche* for handling the region's social groups. Things being so, a present-day anthropology of 'caboclo social systems' (Nugent, 1993), as one of the segments of non-urban contemporary Amazonian societies, cannot be allowed to pass without a critical analysis of the notion, which is precisely the task I set myself here.

I will analyze the constituent elements of the notion of 'traditional people' and their implications from two angles: firstly, I will reconstruct how the notion was configured within the sphere of international conservation and, secondly, I will sketch the formulations of contemporary Brazilian authors who have helped mould its content within the debate on biodiversity conservation in the tropics. In so doing I will draw from and dialogue with other authors who have approached this theme from their own specific perspectives (Vianna, 1996; Cunha & Almeida, 1999; Adams, 2000a, 2000b; Schwartzman, 2000).

The analytic perspective adopted here belongs to the set of issues raised in studies on environmental history and political ecology that understand 'ideas as ecological agents' (Worster, 1991, p. 211; cf. also Drummond, 1991). The

[1] Or at least they were, given that, as we shall see later, one of the provisions of Law no. 9.985, of 18/7/2000, which set up the National System of Nature Conservation Units (SNUC), was to embed a statute for traditional peoples among its articles.

notion of 'traditional peoples' expresses a set of collective cultural values concerning the environment – perceptions, values and structures of meaning – that guides and determines certain environmental policies. A term incorporated into the vocabulary of our current dialogue with the natural world and into the lexicon of the governmental organs responsible for the environment, the notion carries weight in the symbolic disputes that constitute the micro-politics of the peasant struggles for access to the means of production and the political processes that influence territorial rights – a theme dear to political ecology (Moore, 1995). Insofar as cultural meanings and symbolic struggles are not mere expressions of the material base of the social formations, but constitutive forces that also shape the history and affect material transformations, it must be recognized that the notion of 'traditional peoples' is a key part of the *historical situation*[2] of the contemporary caboclo societies of the Amazon. On one hand, it represents the conceptual mould used to cast and present these hitherto sociologically invisible social groups, and, on the other, it is the object of the active historical self-positioning of these groups, which often appropriate the notion, situating it, and themselves before it, in order to define what counts as 'traditional population'. Their concerns expressed here reflect the micro-region and historically specific segment of the Amazonian peasantry amongst whom I conducted the field-research for my doctorate.[3] If we seriously consider the agency structure typical of the biophysical environment, the fact that we are speaking about the Lower Rio Negro is especially significant, as almost the entire literature on caboclo societies focuses on rural groups from the floodplains and whitewater rivers of the Middle and Lower Solimões/Amazonas and their tributaries. If Nugent (1993) was right in suggesting that the caboclo societies of the Amazon suffer from a relative socio-political invisibility, and if the social peasant formations of the region are problematic from an anthropological perspective, then this is even truer for those located in the Lower Rio Negro. Originally from the whitewater river plains, today they occupy an interstitial environmental and epistemological space between the studies on the adaptive strategies and caboclo social systems of the whitewater

[2] In the sense in which Oliveira Filho (1988) coined the phrase and considering the way in which Viveiros de Castro (1999) applied it.

[3] The goal of the research was to conduct an anthropological analysis of the strict-use conservation units – national parks, biological reserves and ecological stations – in the Brazilian Amazon, which are defined as instruments of environmental policy. Through a comparative ethnography of two such units – Jaú National Park and Anavilhanas Ecological Station -, both located in the Lower Rio Negro micro-region, in the rural zones of the municipalities of Manaus, Iranduba, Novo Airão and Barcelos, I attempted, among other things, to estimate the effects the creation of these two protected areas had had on the local social landscape (Barretto Filho, 2001). The research was carried out with support from the following sources, for which I would like to extend further thanks: Grant 66AB, Ford/Anpocs Programa de Dotações para Pesquisa em Ciências Sociais/1997; Pre-doctoral Grant # 6289 from the Wenner-Gren Foundation for Anthropological Research; and Apoio CSR 103-98 Nature and Society Program WWF/Ford.

floodplains, on one hand, and, on the other, the analyses of the adaptive responses of the indigenous peoples of the Upper Rio Negro to the environmental pressures resulting from the general oligotrophy of the blackwater ecosystems.[4] In other words, they are neither originary, 'genuine' others like the natives from 'upriver', nor 'others' yet more 'incomplete' and 'invisible', albeit neglected by contemporary ethnographic research on the Amazonian caboclos. Fertile ground, then, for the generalization of the use of the notion of 'traditional peoples'.

The Invention of 'Traditional Peoples' within International Conservation

Recognition of the existence of 'traditional lifestyles' in the sphere of international conservation occurred within the dual context of the official incorporation of the principle of zoning within the processes for establishing protected areas and of the advent of concerns to relate the in situ conservation of biodiversity with local-scale socio-economic development in the management of these areas. These changes are more visible when it comes to formulations and orientations produced in the international forums for discussion on the statutes, goals and management methodologies of the protected areas.[5]

Long before the problem of human occupation of protected areas and consequent use of protected resources, whether temporarily or permanently, by local populations came to the attention of nature conservation planners, the general principle of national parks as announced at the First World Conservation Congress on National Parks (Seattle 1962) already admitted the possibility of there existing, in practice, exceptions to the general rule of ample legal protection against the exploration of a park's natural resources or any other damage caused by man. These 'exceptions', related to private rights to habitation, agriculture, cattle raising, mineral prospecting or hunting that may have already existed on the lands prior to the creation of the park, should be managed as such – that is, as exceptions, with aspirations towards their reduction and eventual elimination (Amend & Amend, 1992, p. 459). Given the

[4] Moran takes these adaptive responses as described by contemporary ethnographic and ethnobiological research as a general adaptive pattern typical of the 'ecology of human populations' of the blackwater rivers (Moran, 1990, p. 168 onwards).

[5] I am referring here to the general assemblies of the International Union for the Conservation of Nature (IUCN) and the World Congresses on National Parks and Protected Areas, promoted by the IUCN National Parks and Protected Areas Commission in conjunction with WWF International, the European Union and Inter-American Development and World Banks. The World Congresses have been held every decade since 1962 – when the first edition took place in Seattle – with a view to promoting more effective development and stewardship of the world's natural habitats so that they can make their maximum contribution to sustaining human society. The most recent Congress was held in Durban, South Africa, in 2003. Cf. Barzetti (1993) and IUCN (1992).

difficulties involved in earmarking an entire area for a single category, for the first time ever it was proposed that the parks be divided into different zones, each with its own set of permitted or prohibited activities.

Continuing with its attempt to establish a globally-accepted definition of national park, the International Union for Nature Conservation (henceforth referred to as IUCN), at its eleventh general assembly in Banff, Canada, in 1972, officially incorporated the principle of zoning into the definition of national park. This decision was ratified at the second World Congress on National Parks and Protected Areas, held at Yellowstone that same year. The annexation of zoning to the concept of a national park brought with it the recognition that human communities with specific cultural characteristics comprise 'anthropological zones' inside the ecosystems to be protected.[6] As such:

> the concept of national park is expanded with the addition of the *principle of zoning, which recognizes as part of the ecosystem all peoples inhabiting the area* (designated protected anthropological zones) that practice forms of agriculture specifically adapted to that ecosystem and whose cultural heritage deserves conservation and protection (Amend & Amend, 1992, p. 461 – my italics).[7]

Admitting that overly rigorous management runs counter to the principles of conservation, as it does not fit with the reality encountered on the ground in most protected areas throughout the world, the Amend suggest that it is essential to 'strengthen *the principle of zoning*, which in well-defined contexts *can allow for the long-term presence of human settlements*, blended to the maximum into the management of the park' (1992, p. 462).

In short, originary, autochthonous cultures practicing traditional forms of agriculture supposedly adapted to the ecosystem and whose cultural heritage is considered valuable were recognized so long as they accepted the zones in which they were to be managed. It is important to note, as did Nitsch, that 'zoning is negative, nothing more than a broad form of prohibition, applied simultaneously to every corner [of the] territory in question' (1994, p. 502). It is an aspect of administrative policing, representing a heavy-handed intervention in the ordinance and regulation of the appropriation of space. The definition of autochthonous and originary peoples as parts of the ecosystems to be protected subsumes cultural diversity under the heading of conservation in a clear expression of *enforced primitivism*.[8]

In similar fashion, by admitting that the establishment of protected areas could result in the expulsion or forced re-settlement of ethnic groups, the 12th IUCN general assembly, held in Zaire in 1975, urged public policy formulators and implementers creating such areas to avoid culturally and economically

---

[6] Those are: zones of natural environment with autochthonous human cultures; zones with traditional forms of cultivation; and zones of special interest.

[7] All italics in cited passages, including legislation, are my additions.

[8] 'The automatic assumption that indigenous peoples will accept or even welcome cultural stasis as a condition of their involvement in conservation management' (Clad, 1984, p. 69).

scattering groups whose presence does not affect the ecological integrity of the area: 'the establishment of protected areas should not bring as a consequence the displacement of Indians or the disruption of their traditional lifestyles *always and whenever these groups do not affect the ecological integrity of the area*' (apud. Amend & Amend, 1992, p.461).

The third World Congress of National Parks and Protected Areas, held in Bali, Indonesia in 1982 is considered by analysts and commentators (Brito, 2000; Diegues, 1996) to have been marked by a concern with the development in the definition of the concept of national park. Reflecting the ideas expressed by Kenton Miller, the key figure at the Congress, in his classic manual on park planning in Latin America (Miller, 1982), the formulations drafted in Bali postulated that parks should play a fundamental role in national development and in conservation. The third Congress also

> reaffirmed the rights of *traditional societies* to their own social, economic, cultural and spiritual determination; [and] *recommended that* those responsible for the planning and management of protected areas *study and make use of the traditional skills of the communities affected by the conservational measures* and that management decisions be taken in conjunction between the societies that traditionally managed the natural resources and the authorities in charge of the protected areas (Diegues, 1996, p. 100, my italics).

One of the products of the third Congress was a manual on protected area management in the tropics (Mackinnon, Mackinnon, Child, & Thorsell, 1990). This manual contains a chapter dedicated to the theme of integrating protected areas with regional development programs and another on the relationship between protected areas and the local populations. Symptomatically, these chapters came under the heading 'Obtaining support for protected areas' (Part B) and not under 'Bases for the establishment of protected areas' (Part A).

Chapter 5 suggests some principles for integrating

> the running and management of protected areas with other rural forms of soil use [..:] deals with some of the positive and negative relations that could exist between a protected area and its environs, suggests how these conflicts can be mitigated and speaks of the benefits to the local communities (Mackinnon et al., 1990, p. 81).

They sought to demonstrate how protected areas improve the perspectives of 'sustainable development' and proposed some ways they could be integrated with the process of development in various areas: silviculture, use of wildlife and other natural resources, tourism, agriculture, public works, job creation, etc. – besides the environmental service of regulating climate and water and soil cycles. In doing so, they try to seduce those responsible for formulating and implementing development policies toward conservation through protected areas.

In Chapter 6, the manual notes that governments have been creating protected areas without having adequate knowledge about the populations living within their limits and that 'the success of stewardship largely depends on *the acceptance and support of the neighboring communities*' (1990, p. 109, my italics). The argument is that the inhabitants could profit in many ways from the protected areas, despite the need to impose certain restrictions on their exploration in order to ensure the fulfillment of the prime goal – conservation. As such, 'as a form of compensation in cases where the establishment of a reserve implies loss of traditional harvesting privileges and rights on the part of the rural communities the proposal is that these be given preference in terms of job creation and the development of buffer zones.[9] The proposals of compensation and substitution, which were starting to be drawn up at this stage, came to constitute the bedrock of the so-called Integrated Conservation and Development Projects (ICDPs).

The ICDPs were created to replace the 'fences and fines' approach as a strict-use management technique given the conflicts sparked by the latter's routine as a way of minimizing 'human impacts' and discouraging 'illegal' activities inside the PAs: 'it has to be recognized that the radical elimination of human occupation explicitly mentioned in the documents does not fit with the realities found in the majority of the world's parks' (Amend & Amend, 1992, p. 462). Thus the park formulators and planners began to propose the inclusion of the support and cooperation of 'local populations' as a sine qua non to the long-term success of park management. Pictured by some as 'the vanguard of what will undoubtedly be a broad array of initiatives attempting to link conservation and development' (Brandon & Wells, 1992, p. 557),[10] the rubric ICDP subsumes a wide-ranging and diversified cast of potential measures for enhancing the socio-economic development of human groups on a local scale, generating alternative sources of income that do not threaten the flora or fauna of the protected lands. As Lima observes (albeit in another tone), 'integrated projects present the intent to promote better living conditions for the population as a 'return' for their share of sacrifice and as an incentive for them to accept the conservation proposal' (Lima, 1997, p. 288).

Going back to the original manual of the Bali Congress, the text stresses that:

> it is imperative that socio-economic studies be carried out on the communities that could be affected by the management decisions taken inside the parks. These studies ought to ascertain the ethnic diversity of the communities and their social structures, including the locations and proximity of kinship groups to the ritual grounds or centers for the exchange of goods and any other activity important to the community. This focus will afford a deep understanding of the autochthonous

[9] In these zones, the natural resources inside the protected area can be directly exploited on a seasonal, temporary or permanent basis in a controlled, managed and ecologically sound way – through shepherding, harvesting of non-timber forest products, fishing, etc. – so as to reduce dependency on the area's usable products.

[10] Assuming the interdependence of human and non-human systems and that the challenges of development and conservation are inextricable (IUCN, 1984), the core philosophy of the ICDPs reveals itself in the language used to describe them: 'community-based programs employing 'participatory' methods to simultaneously 'empower' rural residents and conserve threatened species [and biodiversity]' (Barrett & Arcese, 1995, p. 1073).

communities and make it possible to avoid misunderstandings and potential problems in establishing management policies in a park (Mackinnon et al., 1990, p. 111).

It is impossible not to see in the manual's recommendations a certain analogy to the stratagem of 'community development' as proposed by Bastide, who divides the process as follows: the first stage consists in carefully studying the population that will be affected, acquiring knowledge of all dimensions and sectors of its culture – 'which is precisely the task of the ethnologist'. According to Batside, it is with the task of the ethnologist that this 'painstakingly minute study of the social or cultural specificities of the population' ought to begin (Bastide, 1979, p. 1 and following pages). The second step of the strategy involves:

> Discover[ing] *factors* in the culture under study that could facilitate development, especially those that *help the stakeholders understand* first and foremost *their interests in the change* and then to *become enthusiastic about it*; in short, *participate in the work of the social experts or engineers* (Bastide, 1979, p. 109, my italics).

If we swap 'development' and 'change' for 'conservation' and 'protected area' we cut to the root of what the conservation planners are proposing.

However, the recognition that these native institutions do actually work is subordinate to the practical interests behind the studies and the clear concern to assuage conflict situations and harmonize the asymmetrical relations that prevail in the taking of management decisions. Within the scope of this conservation model, this is the task the social scientists have been called on to fulfill. What predominates here is the 'Cartesian model' of the relationship between theory and practice in the social sciences – which implies the control of social forces by reason and consequent dominion over man as part of a planned action (Bastide, 1979, p. 1 and following pages). Protected area planning that is sensitive to the cultural dimension justifies itself in function of the desire to implement these conservation units at a low social cost by reducing conflicts and asymmetries. As Gray observes, 'the local peoples are studied in order to make them change their forms of production so that [planners] can push ahead with the parallel goals of conservation and profit for the engineers' (1992, pp. 26–7).[11]

[11] At this stage, the debate on peoples in parks had already drawn a bigger audience than just the conservationists. The research director of the NGO Cultural Survival and editor of the homonymous magazine, in an editorial entitled 'Parks and People' (vol. 9, no. 1, Feb. 1985), criticized the expulsion of human populations from protected areas and laid out what he saw to be the functions of such parks: 'Protected areas could ensure the survival of habitats as well as the indigenous inhabitants. Reserves can either preserve traditional lifestyles or slow the rate of change to levels more acceptable to and controlled by local residents. Indigenous inhabitants can benefit from the protection of their rights to traditional areas, as well as [from] the sale of goods or [the] income generated from tourism' (Clay, 1985, p. 2). Diegues endorses this argument for leaving 'traditional populations' inside protected areas: 'so that they *can more adequately absorb the socio-cultural changes* taking place in broader technological and industrial society. This *buffering factor would give these populations more time and opportunity*

In 1986 the IUCN organized a conference in Ottawa, Canada, entitled 'Conservation and Development: Putting the World Conservation Strategy into Practice'.[12] According to Diegues, this was the first time 'the situation of traditional peoples living inside parks' was given clear and explicit treatment (1996, p. 103). Recognizing the particular relationship these peoples have with nature, Workshop 3, which discussed 'traditional peoples' and sustainable development, invited governments, NGOs and other institutions to guarantee: a) '*participation* in the control of the use of shared resources'; b) '*consultation and agreement* [. . .] in the establishment and maintenance of parks'; and c) 'that national governments pay due attention to the needs and aspirations of the traditional peoples whose territories will be affected by the creation of parks and reserves' (Diegues, 1996, p. 103, my italics). This same source also reveals that the conference made more incisive recommendation to the effect that the lifestyles of traditional peoples that chose to remain within the perimeter of a park should not be altered, and that they should not be re-settled elsewhere without their consent.

It would seem, then, that the IUCN, UNEP, WWF and other international organizations had suddenly woken up to the rights of 'traditional peoples', including the right to 'self-determination'.[13] But who are these 'traditional peoples'? How are they defined, characterized and represented by these agencies? A set of factors connected, on one hand, with the socio-political dynamic, properly speaking, and, on the other, with the development of scientific research could help us understand this process, given that they continue, in

---

to recreate [. . .] their relationship with nature on the basis of the changes coming in from outside' (1996, p.101, my italics).

[12] In 1980, in response to a request from the United Nations Environment Program (UNEP), the IUCN, in conjunction with UNEP and the WWF, drafted and published the *World Conservation Strategy* (IUCN 1984). This document explicitly laid out the actions required both to increase conservation efficiency and to integrate it with development. By 'explaining the contribution the conservation of living resources makes to human survival and sustained development', the *Strategy* stressed the importance of maintaining the essential ecological processes, of preserving genetic diversity and of making sustainable use of species and ecosystems as priority requisites for: human survival and development; the reproduction programs necessary to the protection and consolidation of species; the scientific progress, technical innovation and security of countless industries that use living resources. Gray sees the embryo of the neoliberal bent that he would later criticise in *Biodiversity Conservation Strategy*, i.e., the modelling of conservation programs in accordance with the logics of development and the market: 'Explaining the economic value of biodiversity conservation, the organizations [. . .] hope to make their strategy attractive to such sources of international funding as the World Bank, national governments and, in particular, the private companies [that use living resources]' (Gray, 1992, p. 22).

[13] I will return to this point below in the second part of this article and in its final considerations, drawing attention to the limits of this horizon of 'self-determination' given modern aspirations to well-being and the levels of consumption manifested by some of these groups. To what extent will the international conservationist organizations be willing to recognise the autonomy of these groups to decide upon their own futures, especially if they choose a path that diverges from the one these organizations have envisaged for them?

parallel and in tandem, to interweave the social definition of the 'problem' of the 'traditional peoples' with the conservation of living resources.

Firstly, it is important to highlight the native and/or autochthonous resistance to the implantation of protected areas, mainly in Africa and Southeast Asia, where the conservation systems are identified with colonial rule. Native resistance to the conservationist agenda forced concessions from part of the formulators and implementers in the sense of 'sniffing out resistances' in order to be able to carry through on their goals. For the native social groups that inhabit these areas, there is a lot more to the continuous management of the habitats, ecosystems and species protected in reserves than the generation of income or provision of foodstuffs. Hunting, fishing, gathering and itinerant farming, as practices, are intrinsically knit together with other dimensions of the social life of these groups, and infused with an importance and host of other values that go beyond subsistence alone (see Murrieta and Winkler Prins in this volume).

Next, we must consider the result of the deepening of knowledge on the cultural specificities of some societies and the ecosystems they occupy, especially the tropical forests, to which the ecological branch of anthropology, ethnobiology and historical ecology have made fundamental contributions. Analyzing regeneration patterns in tropical forests after various scales of natural disturbances and/or catastrophes – tree falls, wind damage, fires, floods etc. – and comparing them with man-made disturbances, botanists and agronomic and forest engineers came to the following conclusions: i) the regenerative system of the humid tropics seems to be well-adapted to the activities of 'primitive man', especially shifting agriculture, because its effects are more similar to occasional, small-scale natural damage to forest fragments (Gómez-Pompa, Vázquez-Yanes, & Guevara, 1972, p. 763), and ii) so long as disturbances caused by human exploitation of forest resources imitate and/or reproduce the kind of damage caused by small-scale natural disturbances in terms of size, duration and frequency the functional integrity of the ecosystem tends to be protected (Uhl et al., 1989, p. 237). For Uhl et al. (1989), the sustainable use of forest resources by man should take natural disturbances as its master, that is, it should model itself upon nature. Gómez-Pompa et al. (1972) have long been following the practice of shifting agriculture in various tropical areas in which a mosaic pattern can be observed – large swathes of primary forest interspersed with fragments of disturbed forest of varying ages, depending on when they were abandoned. They affirm that the studies available on these waves of forest succession largely tend to agree that 'shifting agriculture has been a *natural way* to use the regenerative properties of the rainforest for the benefit of man' (Gómez-Pompa et al., 1972, p. 763) – in conditions of technological austerity and low population density, we might add. In other words, in addition to modeling ourselves on nature, we might also model our activities on the 'natural way' in which the natives deal with the tropical forest.

This means recognizing that the biodiversity we find in these environments today is the result of complex historical interactions between physical, biological and social forces. The current composition of mature/adult vegetation may well be the legacy of past civilizations, the inheritance of tilled fields and managed forests abandoned centuries ago – especially in he Amazonian Basin, where, according to the archaeological, ethnohistorical and ethnobotanical evidence, there has always been continuously high population density and human occupation (Gómez-Pompa & Kaus, 1992; Denevan, 1992a, 1992b; Roosevelt, 1994; Cleary, 2001).[14] Hence the IUCN *Manual* of 1986 errs when it states that 'by maintaining their traditional cultures, the autochthonic populations protect extensive zones of what are essentially natural ecosystems' (MacKinnon et al., 1990, p. 109). It would also be incorrect to speak of forests and/or 'natural' areas with regard to many of the locations earmarked for protected areas, when it would be closer to the truth to speak of 'cultural forests' (Balée, 1989a, 1989b, 1992). Summarizing and generalizing the argument, McNeelly observes that virtually all of the world's forests and grasslands have been affected by cultural patterns of human usage and the resulting landscape is a permanently changing mosaic of habitat fragments, managed or otherwise, whose diversity is reflected in their size, form and arrangement (McNeelly, 1993, p. 252).

This realization has translated into normative propositions like those put forward by McNeelly himself, for whom it is important that, having identified a particular ecological attribute worthy of protection, one should first consider the needs and desires of those who contributed to shaping that landscape and who will need to adapt to the changes protection would bring (McNeelly, 1993, p. 252). Posey et al., on the other hand, argue that the sophisticated and wide-ranging indigenous systems of perceiving, using and managing natural resources could make a significant contribution to alternative, 'human, productive and ecologically prudent' development strategies, constituting the logical product of applied ethnological research (Posey et al., 1984, p. 96). These authors express an understanding of the generality and extensiveness of the 'ingenious systems' of indigenous resource and knowledge management, which largely legitimizes the generic notion of 'traditional peoples'. Recognizing the true caboclo as the intellectual heir to indigenous ecological knowledge in many areas, the authors assure that everything that is said about the ethnoecological knowledge of the Indians applies *mutatis mutandis* to the caboclos. The migrant settlers, as newcomers with scant ecological knowledge of the region, are excluded from this universe, which includes only those who have lived in the territory for generations (Posey et al., 1984, p. 105). Clay, in the editorial cited in footnote 11, claims that:

---

[14] Among the most expressive contemporary ethnographic evidence are the studies in historical ecology on the foraging strategies of bands of hunter-gatherers, conducted by Balée in pre-Amazonian Maranhão (1989a, 1989b, 1992), and those by Posey on the secondary management of forest, brush, fields and scrubland by the Kayapó, and the adaptive value of their strategies for living with the environment in which they find themselves and their knowledge of its natural cycles and ecological processes (Posey, 1987).

> People have developed a number of ways to live in fragile environments. We know very little about how these systems evolved, how they function or how they might be adapted to make them more productive and ecologically sound. We know however that the key to understanding sustained activities in fragile environments begins with local residents. Their knowledge is valuable to the future of the earth's environment and peoples (1989, p. 5).

These elements and this perspective are discernible in the more recent publications by international organizations I have been outlining here. In the text *From Strategy to Action*, 1998, in which the IUCN recommended measures for implementing the proposals of the 'Brundtland Report' – *Our Common Future* -, we read:

> For a long time now we have been concerned about the loss of human cultural diversity, aware that part of the richness of human life around the globe stems from the interrelations between peoples and their local habitats. The loss of cultures, or the traditional knowledge of cultures that suffer brusque social change is a problem at least as serious as species loss (apud. Diegues, 1996, p. 104).

Although the naturalization of human cultures is variegated with 'interrelations' – which presuppose a shared history rather than a passive adaptation of human cultures to their habitats — by equating the loss of cultures to that of species, the text surreptitiously re-inserts the view of these peoples as societies of nature. In the document, the term 'traditional peoples' refers to 'minorities culturally distinct from the population in general and which are almost entirely removed from the market economy'. It therefore recognizes the need for 'a broader interpretation [...] in order to incorporate minority societies that have the characteristics of traditional groups – including a stock of traditional knowledge of their environment and its resources and which are not yet intimately connected to the market economy' (apud. Diegues, 1996, p. 104).

In an original approach to the issue, Vianna suggests that there are differences between the way the term 'traditional peoples' is employed in Brazil, referring to the so-called 'rustic societies' (see below), and the way it is used in the international discourse, which, she says, avails of a plethora of fuzzily defined terms that nonetheless express an underlying conceptual agreement insofar as they all come down to 'ethnic populations' (Vianna, 1996, pp. 107–109). This does not strike me as being the case, if we consider the various exhortations for a broader interpretation of the notion of 'traditional peoples' – as in the abovementioned documents – and the various concepts applied in other publications cited in this text: autochthonous peoples and/or cultures; ethnic groups; indigenous groups; indigenous inhabitants; natives; traditional peoples; traditional cultures; traditional societies; traditional lifestyles; autochthonous communities, rural communities, local communities and/or populations; neighboring communities to protected areas – to make just a preliminary list. The diversity of situations addressed reflects in the variety of terms employed. If some indicate aboriginality while others go for ethnicity, others still stick to the spatial scale – proximity to ecologically critical, fragile or protected areas. In addition, as she perspicaciously observes, 'traditional' is employed indistinctly as an adjective to describe 'type of management, type of society, form of resource use, land use, lifestyle, specific groups or cultural type' (Vianna, 1996, pp. 107–108). Which is why, unlike Adams and Vianna, I can see no way to coin a 'more precise conceptualization' for 'traditional population' (Vianna, 1996, p. 89), or formulate 'a scientific response to the problem' of human presence in strict-use reserves (Adams, 2000a, pp. 24 & 262) that could be somehow helpful to conservation initiatives. This is an ideological construct whose power resides precisely in the general nature of its meaning and in fluctuations of its use.

Nevertheless, like Vianna (1996, p. 107 onwards), it is possible to identify some elements in the characterization of these 'traditional populations'. Among the common criteria and references that sustain this sprawling conceptual umbrella are: the population's unique relationship with nature, grounded in its total dependence on natural cycles and, consequently, on a thorough knowledge of their bio-ecological processes, generating a body of technical wisdom and systems of natural resource use and management specially adapted to the localized ecosystems in which they live; their peripheral position in relation to the market economy, which, though due to specific historical processes, is often taken as an intrinsic, permanent and substantive characteristic of these peoples; and, lastly, the fact that they now occupy the final frontiers of the national economies of their home countries, and therefore represent areas as yet relatively untransformed in comparison with other regions subjected to intensive agriculture and urbanization – a fact that, in itself, is taken as confirmation of the inextricability of biodiversity and sociodiversity.[15]

In order to comprehend adequately the content at play in the notion of 'traditional peoples', we must first compare it with that of autochthony, as understood on the battlefield for interpretive hegemony constituted by the notion of development. Within this scope, the variable geometry of the notion of autochthony is tightly bound to the variability within the idea of development itself. Pietilä (1990) alludes to a 'new' perspective opened up by understanding the importance of household economies, non-remunerated labor – especially by women and children – and non-monetary 'subsistence'. Trainer (1990) stresses the strategies to maximize self-sufficiency and economic independence at work in peasant villages with a view to their attaining an increasingly lower level of involvement with the monetary economy. Stavenhagen (1985) defines the peasant household economies as a 'new' goal of social and economic planning and would like to see effective recognition of ethnic communities as functional intermediary organizations between the individual and politics. The Commission on Amazon Development and Environment (Comisión, 1992) advocates the attribution of market value to these communities' 'ingenious systems' for the appropriation and stewardship of natural 'resources' and their stores of

[15] 'It is universally recognized that *indigenous*, that is, '*tribal, native, ethnic, aboriginal or remote-dwelling*' people occupy as much as 19 percent of the world's surface and are, as such, stewards of a significant portion of the earth's fragile ecosystems' (Kempf, 1993, p. 5, my italics). This assertion, at the same time as it corroborates this intertwining, totally denounces the conceptual fuzziness and confusion to which I refer.

'traditional' knowledge under the categories of 'provision of environmental services' and contributions to 'new styles of natural resource use'.

As such, the contributions one can expect from the 'minority voices' on how to solve the problem of living resource conservation are not much different from those identified by Stavenhagen as the central elements of 'alternative development':[16] valorization of *local* resources – natural, technical and human – directed towards autonomy and self-sufficiency; fruitful use of existing cultural traditions; an approach to development that hinges upon an endogenous perspective; respect for the environment; and concern with furnishing 'basic needs'. In the same manner as indigenous lifestyles and cultural specificities play an important role in the discussion on alternative forms of development, the representation of these so-called 'traditional' social groups move *pari passu* toward the recognition of the need for their 'participation', 'consultation', 'agreement' and 'consent' as a condition for the success of conservation programs based in protected areas. In other words, one process cannot be dissociated from the other, lest we fail to understand either correctly. This is not so much a 'discovery' of 'traditional peoples' – as 'traditionality' is not something one can discover – but rather its construction as a *subject* – in all the senses of the word – of protected area management, understood as the political and sociocultural process by which nature, these populations, and our grasp of what both mean, are transformed.

In this way, these peoples are, as far as the literature is concerned, the best possible human custodians of global biodiversity, for a number of reasons: a) their unique relationship with nature, which translates into a technical wisdom and store of knowledge of the natural cycles and local ecosystems upon which they live; b) the fact that these ecosystems often represent the last remaining and longest-standing samples of critical and fragile ecosystems; and c) because they position themselves on the margins of the price-based market, articulated into smallholder production systems driven by subsistence and a labor-intensive, technologically austere and supposedly low-impact model of natural resource use. The battlements of the planetary wildland; this is why these groups have been called to participate in, and make their contribution or sacrifice to the struggle for the consolidation of protected areas, even extended the right to refuse to collaborate in the 'shared effort' to conserve biodiversity through protected areas – as the abovementioned Ottawa recommendations would seem to suggest.

For many authors who share this view, the characteristics of these 'traditional' social groups also constitute a passport to humanity's future survival in the modern world. Almeida, for example, suggests that 'the politically strongest message [the peasantry, Indians and other marginal and disenfranchised social groups can give] is that in which they present themselves as the pivots of general

[16] Alternative insofar as it is founded upon organizational principles hitherto always ignored by the prevailing development paradigm, but towards whose recovery anthropological expertise has a contribution to make.

interests'. For Almeida, once needs to 'awaken universal vocations in stakeholder groups', 'legitimately propose policies of universal scope' and adopt 'the language of groups that possess special competencies that coincide with the collective interests' (Almeida, 1992, pp. 116–117). But are we really talking about competencies that coincide with collective interests here? Do we not run the risk of holding these groups hostage to an ahistorical and exogenous definition of these 'policies of universal scope'? Does valorizing them because they retain knowledges and techniques useful to living in critical and fragile environments – as one of the above-cited documents says – not cast an instrumental motivation across our relationship with these peoples? Recognition of the two-way street between biodiversity and socio-diversity produces an imperative to protect both, opening the possibility of appropriating the systems of use and management of the 'traditional peoples'.

Here too, in our attempt to reach an adequate understanding of the indispensability of the 'participation' of these peoples to the success of conservation initiatives in the form of protected areas, it might be interesting to follow Hoben (1982), Escobar (1991) and Huizer (1993) in looking at the issue in the light of the socio-genesis of the notion of participation and the bottom-up tack as 'alternatives' to the top-down or trickle-down approach to development.

Hoben and Escobar both situate the growing participation of anthropologists in issues related to development within the context of the reform of external aid interventions and policies from the nations at the centre to those on the outskirts of the global system. The optimism and enthusiasm of the development agencies and governments of 'developing countries' began to ebb as the results of the top-down, capital-intensive interventions came in far below expectations. They soon realized that the technical solutions were running aground upon disconcerting resistance coming from socio-cultural factors. It was, in large measure, the failure of the trickle-down approach that imposed this need to 'sniff-out' resistances, in the words of Bastide. 'They began to realize that the poor themselves had to participate actively in the programs if these were to have a reasonable margin of success. Projects had to be socially relevant, culturally appropriate, and to involve their direct beneficiaries in a significant fashion' (Escobar, 1991, p. 663).

It was therefore concern for the success and efficacy of the development programs – and those of conservation through protected areas – that was the key factor behind the emergence of this 'new sensibility' toward socio-cultural factors on these programs – which ended up creating an unprecedented demand for the skills of the social scientist. If 'culture' became something inherently problematic in development (Escobar, 1991), the same can be said for conservation policies based on protected areas.

One example of this concern was the most recent document issued by the IUCN, Pnuma, and WWF joint venture. Referring to indigenous peoples in a chapter entitled 'Letting the communities take care of their own environment', the text describes them as:

> culturally different communities, with land rights and other rights based on historical occupation. Their cultures, economies and identities are inextricably linked to their traditional lands and resources. The component of subsistence economy remains at least as important as the component of money. [. . .] In addition, the Indians represent themselves to their own communities as the undeniable continuation of their past and as an extension of the world of nature (IUCN, Pnuma, & WWF, 1991, p. 67).

The same document indicates the existence of a future option for these peoples, that of either 'returning to their former ways of life or abandoning subsistence and joining society at large'. The 'third way' for the indigenous peoples would be to 'modify their ways of obtaining subsistence, combining the old ways with the new so as to be able to maintain identities whilst allowing their societies and economies to develop' (IUCN, UNEP, & WWF, 1991, p. 67). In this regard, they suggest three key providences: i) 'recognize the aboriginal rights of the indigenous peoples to their lands and resources', including their administration, and 'their effective participation in taking decisions that affect their resources and lands'; ii) 'ensure that the pace, rhythm and form of development minimize damaging impact on the environment, society or culture of the indigenous peoples, and that they receive an equitable share of the benefits'; and iii) guarantee the Indian peoples the cooperation of legislators, planners, administrators and scientists 'in a concerted effort to manage the resources and economic development'. Put in these terms, the Indigenous peoples emerge as 'alternatives *in* conservation' (1991, p. 67, my italics).

For some commentators, however, the fourth World Congress on Parks, held in Caracas, Venezuela in 1992, was the apex of the discussion (Brito, 2000; Diegues, 1996). The issue of human populations and protected areas was one of the central themes, as attested by the fact that the workshop 'Populations and Protected Areas' drew most attention (Diegues, 1996, p. 107). According to this same source, the presence of national delegates at this workshop, particularly from 'Third World' countries, was by far the highest of the entire Congress. This meeting recommended more respect 'for traditional populations, very often the bearers of centuries-old knowledge of the ecosystems in which they live; the rejection of relocation as an option; and, whenever possible, their insertion inside the perimeters of the created park' (Diegues, 1996, p. 108). The Congress demonstrated that 'the biggest problem with parks is *convincing the populations, especially local populations, of the benefits of protected areas*' (Diegues, 1996, p. 108, my italics). Ergo: a problem of persuasion, of sniffing out resistances.[17]

[17] Concerning this Congress, it is important to underscore that the sometimes contradictory conclusions and recommendations that emerge from these large meetings stem from the fact that they bear on thematically organized symposiums, and therefore orient specialists from different areas of expertise toward sectorial discussions. As such, the social scientists will normally discuss the socio-cultural and economic context and the human dimension of conservation planning, while the natural scientists will deal with the more 'technical' and supposedly more 'scientific' aspects of protected area management.

What we see, then, is a certain ambivalence between, on the one hand, recognition of the rights of the local populations to social, economic, cultural and spiritual self-determination, the interest in consulting with them and obtaining their consent or agreement and thus securing their effective participation in the decision-making processes; and, on the other, the stratagem of getting to know them better so as to be able to convince, persuade, make them change their way of producing and living so that they can evolve, subordinate and equip their management systems to the practical interests of managing protected areas; in short, making them accept an exogenous agenda. This ambivalence reflects the ambiguity intrinsic to the characterization of these traditional populations, which are portrayed, on one hand, as inextricably bound to nature, almost an extension of the natural world itself, affecting the local ecosystems in a way that mimics the impacts of natural disturbances, but, at the same time, also as societies that stand outside nature, peoples that possess deep empirical, objective technical – and therefore distanced – knowledge of the ecological cycles and processes of their environment and of the natural resources they so masterfully manage.[18] The ambivalence characteristic of proposals envisaging the integration of development and conservation in the management of protected areas finds no response other than the proposition of compensatory and/or replacement mechanisms for the social groups affected by the limitations imposed upon their appropriation of the freshly valorized natural resources. And then only if they recognize the abovementioned characteristics to at least some extent. The idea of participation as a form of mediating conflict only holds so long as it does not threaten the strategy of in situ biodiversity conservation.

The Appropriation of the Notion of 'Traditional Population' in Brazil

Shaped within the field of international conservationism, particularly in the sphere of discussions on the relationship between certain social groups and in situ biodiversity conservation, the notion of 'traditional population' consolidated in Brazil within the same debate and largely at the hands of the sociologist Antônio Carlos Diegues – repeatedly cited throughout this text as an important commentator on definitions postulated on the international circuit and 'pioneer in the discussion on populations and protected natural areas [in Brazil], and indeed one of those responsible for introducing the theme to the wider environmental debate' (Vianna, 1996, p. 20).[19]

---

[18] Which corresponds to a conjecture, a *parti pris*, more so than an empirically based formulation, fruit of specific studies – as Adams (2000a) shows in relation to the caiçaras.

[19] Diegues has had a great influence on the ideas and discussions on this theme, in the construction of the concept of 'traditional population' and in the formulation of the

In this process, the incorporation of the notion within the distinct conservationist agendas of civil society and the public authorities in Brazil was directly influenced by two other vectors. On one hand, Brazilian formulators of the notion of 'traditional population' drew from certain wells of Brazilian thought servile to anthropogeographic studies and eager to characterize regional Brazilian cultural types via the concept of 'rustic' societies and/or cultures. On the other, the notion gained fresh impetus and meaning given the various social movements that incorporated the environmental variable as an important dimension of their activism – especially social agents directly affected by Amazonian deforestation and whose political mobilization focuses on the social effects of public policies for the region – and take their shape from opposition to the actions of the State.[20]

Vianna distinguishes two historical perspectives from which we can understand the incorporation of the discussion on the role certain populations play in nature conservation:[21] one being the conservationist milieu, whether of civil society or public authorities'; the other being the 'rural social movements that ally these discussions with wider social issues' – the former incorporating the population into the ecological discourse and the latter incorporating this discourse into the populations (Vianna, 1996, p. 94–95). Understanding that there is a difference in the applications of the term 'traditional population' in Brazil, where it refers to 'rustic populations', and abroad, where it is exclusively used with reference to 'ethnic populations', Vianna suggests that 'the adoption of the international [conservationist] discourse in Brazil [...] saw the term come to designate populations that were non-ethnic, but which, like the indigenous populations, possessed characteristics considered positive to conservation' – thus opening the possibility of some of these groups using the resources of, and remaining settled in, protected areas (Vianna, 1996, p. 94). From the perspective I adopt, the novelty resides not in the generalization of the application of the notion, a movement we can see in the sphere of international conservation, but in its reorganization around the hegemonic narrative of miscegenation, an at once biological and cultural constitutive process of the Brazilian people.

legislation. Systematically mentioned in academic work on the subject and in official documents, Vianna considers him to be the mentor of many non-governmental organizations, environmentalist or otherwise, that defend these populations (1996, pp. 20–21).

[20] Viola and Leis refer to this segment as the 'socio-environmental sector', one of the five to eight 'sectors' they saw emerge and consolidate in the mid to late eighties as a result of the dissemination, diversification and complexification of the environmental movement – a period they refer to as 'multi-sector environmentalism', according to their chronological typology (Viola, 1987, 1988, 1992; Viola & Leis, 1995).

[21] For all I have said, I do recognize the role these social groups play not so much in normative formulations concerning the conservation of nature in general as in strict-use protected areas as instruments for in situ biodiversity conservation in particular.

Traditional Populations, Ethno-Conservation and the Narrative of Miscegenation

Diegues characterizes the 'non-indigenous traditional populations and cultures' in Brazil as 'peasants' – meaning rural, non-urban groups – and '*fruit of intense miscegenation* between the white colonizer – the Portuguese –, the native indigenous population and the black slave' (Diegues, 1996, p. 14, my italics). Arruda, in turn, speaking of the socio-cultural model of environmental adaptation, occupation of space and appropriation of natural resources that cemented among the rural population during the colonial process, attributes most of this group's traits to the cyclical and irregular nature of the incursions of national society into the hinterlands – sociological and historical criteria –, and the influences of the indigenous populations – techno-ecological and cultural criteria (Arruda, 1997, p. 353). The indigenous 'influence' was expressed fundamentally in the adoption of adaptive and cultivation techniques, artifacts and implements of material culture, and organizational forms for work and sociability (1997, p. 354). Though he does not mention miscegenation, the notion underlies Arruda's entire argument, as expressed in terms of influence and assimilation of cultural traits.

It is important to note in this context that the narrative of miscegenation – even if not explicit – casts many definitions about Amazonian caboclos, highlighting their socio-racial type as a 'mixture' of the whites with the Indians and stressing their continuity from an indigenous formative template – despite the wealth of evidence to the contrary available today, which makes salient countless discontinuities on all levels, from the demographic to the cultural (Nugent, 1993). Posey et al. (1984), for example, express this conception when they refer to the true caboclo – as if there were a false caboclo (the settler?)– as the 'intellectual heir' to the ethnoecological knowledge of the Indians in various areas. This notion also prevails in the collection edited by Parker (1985) – the last and now remote attempt at synthesizing ideas on the Amazonian caboclo in the anthropological literature – in which he even coins the neologism 'caboclization' to refer to the process of the Amerindian's transformation, which he identifies as having occurred mainly during the colonial period – 1615–1800 – as a result of policies for integrating the Indians into colonial society. The *Cabanagem*, for Parker, had therefore begun in a caboclized Amazonia (see Harris this volume).

The miscegenation narrative interweaves the biological and cultural series – more specifically, the techno-ecological adaptive series – presupposing continuities between the formative templates of the Brazilian people and the contemporary historically specific social groups. The constitution of the 'rustic culture' model and the formation of regional sub-cultures can be largely explained in light of the idea of inheritance, descendence, succession and mixing, on the cultural and biological planes, of the three racial matrixes, plus two other important elements: a) the isolation from the dominant colonial undertakings – monoculture, cattle-raising and mining etc. – in which they developed, in interstitial, peripheral and sparsely populated areas, abundantly endowed with natural resources, pursuing subsistence economies whilst also supplying the colonial companies; and b) the historical and ecological peculiarities of these regions, which would have contributed to cementing the various ways of living and of being Brazilian, or even the local variants on the so-called 'rustic culture'. In all its formulations, this caboclization gave pride of place to indigenous adaptive invention, from which the regional subcultures – particularly the caboclo – inherited the lion's share of their cultural core. This was, therefore, the link in the chain between the caboclos and the indigenous peoples that has ensured them their place under the umbrella concept of 'traditional populations', bearers of positive characteristics for nature conservation.

The references upon which the native formulators of the notion of 'traditional populations' draw support are countless, but chief amongst them are: Manuel Diegues Jr., who proposes the division of the country into nine cultural regions, each with its own 'lifestyle', the result of active forms of human adaptation to the diversity of physiographic aspects in Brazil;[22] Antonio Candido, who draws upon R. Redfield's concepts of part society, part culture to define *caipira* as a Brazilian cultural regional type, that is, as at once a race, a way of being and a style of living, characterized by forms of sociability and subsistence that rely on minimal solutions nonetheless sufficient to their survival and to the cohesion of rural neighborhoods; and Darcy Ribeiro, who explicitly employs the narrative of genetic and cultural miscegenation in typifying five historic-cultural regions, all variations of the same rustic Brazilian culture: the creole, the caipira (backwoods) – or *caiçara* on the coast, the sertaneja (grasslands), the caboclo and the 'southern Brazils', the *matutos* (woodsmen), gauchos (from the treeless plains) and gringos (foreigners).

In the case of Darcy Ribeiro, the artifice of the most recent effort to synthesize the constitution of the 'Brazilian people', the opposition between the traditional or archaic and the modern translates and in some way replaces the rural/rustic vs. urban dichotomy. For Ribeiro, Brazilian diversity is eminently rural, with the urban representing the forces of homogenization. As such, as Arruti (1995) points out, Ribeiro inverts the conventional meaning of that opposition by identifying in the rural peoples, in their diversity, multiplicity and richness, a great curiosity and appetite for novelties and for social change, in which they are obstructed by the conservatism of the urban or urbanized elites. 'Culturally impoverished in comparison with their European, African and Indian ancestors, the common Brazilian [read: rural, rustic, traditional] emerges as a *tabula rasa*, more receptive to the innovations of progress than the traditional European peasant, communitarian Indian or tribal Negro' (Darcy Ribeiro apud Arruti, 1995, p. 241). The source of our originality and identity

[22] The nine regions are: the agrarian coastal northeast, the Mediterranean northeast, the Amazon, the mining fields of the central plateaux, the Mid-West, the far south, the area of foreign colonization, the coffee lands and the industrial belt.

as a people therefore lies in the human and cultural diversity of the rural environment, 'generated by the creativity of a people that had to constantly adapt to different and successive biological and social contexts' (Arruti, 1995, p. 241).

By way of example, let us compare two definitions of 'traditional' cultures and/or societies.[23]

> [They are p]opulations of smallholder producers that formed during the colonial period, frequently in the gaps between monocultures and other economic phases. With this relative isolation, these populations developed particular ways of life that involved reliance upon natural cycles, a deep knowledge of the biological cycles and natural, patrimonial technologies, symbologies, myths and even a specific language, with accents and countless words of Indian and Negro origin. [...] Traditional communities are related to a type of economic and social organization predicated on reduced capital accumulation and non-use of a paid workforce. In [these societies], independent producers engage in economic small-scale activities such as fishing, gathering and crafts. As such, economically speaking, these communities are based around the use of renewable natural resources. One important characteristic of this petty mode of production is the knowledge the producers have of natural resources [...] *The conservation of natural resources is part and parcel of their culture* (Diegues, 1996, pp. 14–5 & 87, my italics).

And:

> [T]hose who present a model of land occupation and natural resource use that is predominantly subsistence-oriented and weakly articulated with the market, based on the intensive use of family manpower, low impact technologies derived from inherited knowledge and, as a general rule, sustainable. [...] human groups that historically reproduce their own ways of life in varying degrees of isolation based on modes of social cooperation and specific forms of relating to nature that are *traditionally characterized by the sustainable management of the environment*. This notion refers equally to *indigenous peoples* and to other segments of the national population that *lead particular modes of life adapted to specific ecological niches* (Arruda, 1997, p. 352 & 356, my italics).

[23] I shall limit myself here to authors whose formulations were produced within the context of the debate on how to resolve the problem of human presence in areas of strict use protection. Theoretically speaking, there are some bolder and more sophisticated initiatives, like that of Lima and Pozzobon (2001), who take the concept of ecological sustainability as their platform from which to construct a new socio-environmental classification for human occupation and cultural diversity in the contemporary Brazilian Amazon: 'The criteria of ecological valuing confers new bases for a [positive] political valuing of [hitherto depreciated] social segments and engenders a new ordinatory framework for Amazonian social diversity'. Their classification distinguishes nine 'socio-environmental categories of rural producer': indigenous peoples engaged in sporadic trade; indigenous peoples engaged in regular trade; indigenous peoples dependent on mercantile production; 'traditional' smallholders; 'traditional' landowners; recent landowners; migrants/borderlanders; large projects; and shifting farmer/hunter/gatherers. While embracing supposedly non-traditional groups, the taxonomy never departs from either its rural reference – including both indigenous peoples and large undertakings under the heading 'rural producers' – or from the 'traditional' bearing in predicating two socio-environmental categories – small producers and large landowners – and the 'traditional ecological culture of the caboclos' – characteristic of these two categories and of the indigenous peoples who depend on mercantile production (Lima & Pozzobon, 2001).

Following the definitions comes the list of characteristics of 'traditional populations' and some empirical examples: the caiçaras (from the coastal scrub), the caipiras (from the backwoods), the *vargeiros* (from the floodplains), the cattle-breeders and river-dwellers, the traditional fishermen – like the *jangadeiros* (who fish from the traditional jangada raft) – the small ranchers and farmers, the *quilombolas* (Brazilian equivalents of the Maroon colonies), the rubber-tappers and other extractivist groups, the caboclos, and, perhaps, even the indigenous peoples themselves (Diegues, 1996, p. 14; Arruda, 1997, p. 356). Arruda went so far as to list empirical examples of non-traditional populations: ranchers, holiday-homers, traders, civil servants, businesspeople, workers, owners of processing plants for palmheart and other resources, loggers, etc. One is reminded of the 'school primer archetypal gallery of political geography', to use Arnt's wry expression (1994, p. 11) – a gallery that contains the likes of the cow herdsman, jangada fisherman, gaucho, and many others.

It is important to observe two things. First, the context of the production of definitions of 'traditional populations': the debate surrounding human presence in protected areas. As defenders of the rights of social groups that have suffered the effects of the implantation of strict-use protected areas – such as forced relocation, economic disorganization and restricted access to natural resources -, the argument presented by the native formulators accentuates the positive characteristics these groups bring to nature conservation. Rather than raising the debate on human rights and social justice within the political arena, they have opted for a supposedly scientific-technical argument built around the representation of these groups as 'islands of socio-environmental harmony surrounded on all sides by a market-driven society' – as Lima so aptly puts it (2001a)[24] – which would therefore justify their continued access to and appropriation of resources inside protected areas. However, these formulators went even further than/in the work of conceptual development by actively participating in the political organization of these groups. Diegues attended the first Conference of the Residents of Protected Areas of the State of São Paulo in 1994, which saw the formation of the São Paulo Protected Area Residents Commission, which prepared a second Conference for the following year, drawing 75 representatives from 23 rural communities from five conservation units in Vale do Ribeira (Vianna, 1996, p. 103). We therefore find ourselves before the constitutive process of an ecological/political identity – a topic I will develop in the following item.

Second, by defining the situation of these groups as posing no threat to in situ biodiversity conservation – quite the contrary, in fact, the claim is that they actually contribute to the biological diversity of their home regions – the

[24] Among the authors of various different areas of expertise who expose and problematize the pseudo-scientific and ideological presuppositions behind certain works attempting to apply such concepts as traditional population and sustainable development – some of whom are frequently cited in this article – one could mention Adams (2000a, 2000b), Escobar (1991), Lima (1997), Murrieta (1998), Nugent (1997), Ribeiro (1992), Santos (1991) and Vianna (1996).

native formulators draw upon a tradition in Brazilian social thought that mixes biological and cultural notions in its definition of 'regional cultural types' and 'historico-cultural regions'. In so doing they help nourish conservative expectations as to the productive models of these groups and oversimplify the diversity of social situations that develop in the various different regions where protected areas are in place. But there is more to it than that. Vianna and other critics are quite right when they identify in the definition of traditional populations an idealization, naturalization and deep-freezing of these groups, that is, the expectation that the context of their occupation and situation will remain somehow immutable. But even this is not all. The establishment of a typology of 'historico-cultural figures' leads to the definition of social groups as per a combination of substantive traits that surreptitiously reinstates a notion of race and, consequently, the idea of a natural code whereby each species or type – differentiated in both space and time – occupies its particular rung on the evolutionalary ladder. It is therefore a notion that conspires against the groups' autonomy to decide their own futures in relation to modern aspirations to levels of consumption and definitions of well-being (Lima, 1997) whilst also implying an instrumental relationship that holds them hostage to a definition that comes from outside their society (Nugent, 1997; Murrieta, 1998).

Traditional Populations and Extractive Reserves

Another source of the notion of 'traditional populations' in Brazil was the set of 'new social movements' led by segments of the peasantry and indigenous groups in the Amazon that rose to prominence toward the end of the 80s, a period to which Almeida refers as 'the first-meeting years' (1994, p. 524).[25] As the author observes, 'the 'ecological crisis' endured by segments

[25] The following meetings were all held during the first five months of 1989: 1st Meeting of the Indigenous Peoples of the Xingu, held in Altamira/PA, in February, which formalized protests against the construction of the Cararaô (or Belo Monte) hydroelectric plant, the plan to use the river's waters and to flood the indigenous lands; the 1st Meeting of the Forest Peoples, in tandem with the 2nd National Meeting of Rubber-tappers, which took place in Rio Branco/AC in March and defined a wide-reaching program to demand immediate agrarian reform, the demarcation of Indian lands and the implantation of extractive reserves; the 1st National Meeting of Workers Affected by Dams, held in Goiânia/GO, in April, demanding not only a new policy for the electricity sector, with representation from the workers, but also agrarian reform, the demarcation of Indian lands and recognition of the territories occupied by the remaining Negros from the former maroon colonies. The 'meetings' in Altamira and Rio Branco led to the formation of the Forest Peoples Alliance in May 1989, formed by the Union of Indian Nations and the National Council of Rubber-tappers. The 1st National Meeting of Rubber-tappers had taken place in Brasília some four years earlier, in October 1985, and signaled the nationwide launch of the proposal for the creation of extractive reserves (cf. next footnote) and the creation of the National Council of Rubber-tappers, with the core task of ensuring the viability of their implantation (Menezes, 1994, p. 52).

of the peasantry and by indigenous groups in the Amazon has a clearly political and ideological dimension to it' that sets it apart from the 'ecological issue' broached by other sectors of wider society. It was, first and foremost, a movement to maintain conditions of life that existed before the governmental projects and programs – or those induced by the government – and to demand assurances of 'effective control over domains represented as territories fundamental to the identity' of these social agents (Almeida, 1994, p. 522). As such, they express a crisis in traditional forms of political relations in the Amazon by articulating historical alterities around political identities.

However, in promoting the defense of Amazonian ecosystems through their specific and localized struggles to secure the natural resources essential to their social and cultural reproduction, the immediate victims of Amazonian deforestation quickly managed to form transnational coalitions with environmentalist and conservationist NGOs in Brazil and abroad (M. Almeida, 1992). According to Hurrell (1992), this was the pivot of the politicization of the Amazonian issue on the international stage and one of the key constitutive elements of the international policy toward Amazonian deforestation. Heavy international pressure and a slurry of joint management proposals for the Amazon biome from foreign governments and national and international NGOs with connections to these social movements made the Sarney government respond internally with an environmental policy of a nationalist character that narrowed the channels of interlocution in the face of those movements' demands and concentrated authority in a single power center in order to discipline and centralize decisions on the Amazon. This was done through various measures, including the creation of the Program for the Defense of the Complex of Ecosystems of the Legal Amazon – *Programa Nossa Natureza* – through Decree no. 96.944 of 12/10/1988, and of Ibama, via Law no. 7.735 of 22/2/1989.

Nevertheless, reflecting the mobilization of those social segments and the penetration of their demands within the governmental sphere, and while the work of the Constitutional Assembly was still in course, Incra issued Edict no. 627 of 30/7/1987 creating the project for extractivist settlements, thus meeting the first of the rubber-tappers' demands. As the lights went out on the Sarney government in January 1990, the first extractive reserve[26] was created by Decree

[26] An extractive reserve (Resex) is an instrument of territorial policy, a type of governmental management of the territory, which exists because of the processes of social management previously in place (Little, 1994, p. 13). The concept first emerged among rubber-tappers – autonomous settlers – harvesting rubber for their own benefit and selling it on to intermediaries on local markets whilst also farming and raising cattle for their own consumption – a social configuration which Allegretti suggests consolidated in the 1970s, chiefly in the Acre River valley. It resulted from a 'well-defined sequence of strategies' adopted by the rubber-tappers with a view to 'ensuring land rights over areas of forest occupied by successive generations' – strategies that began with the 'stalemates' of 1973 (Allegretti, 1994, p. 22). The extractive reserve established a territory and regulated access to it, basically expressing the demands and experiences of one historically specific segment of extractivists and resulting from its political rallying, the transnational coalitions it managed to form, the changes of

no. 98.863 of 23/1/1990 – the Alto Juruá Extractive Reserve (Resex) in Acre –, covering 506,186 hectares, and with an estimated population of 6,000 people. A week later came Decree no. 98.897 of 30/1/1990, specifically concerned with extractive reserves. More extractive reserves followed on March 12, 1990, already under the Collor administration: Chico Mendes (Acre), Rio Cajari (Amapá) and Rio Ouro Preto (Roraima). Also in 1989, Ibama hired the consulting services of Funatura (The Pro-Nature Fund) to formulate the first draft of the Bill of Law for the National System of Conservation Units, now Law 9.985/2000.

It was due to the impact of Amazonian issues on the wider foreign policy goals of the Collor government that the region and environmental and Indigenous issues were given an emphasis quite unlike that of the previous administration. For the Chancellery, it was necessary to milk the environmental concerns of the industrialized nations as a way of driving the country's foreign policy objectives, a necessity reinforced by the indications that Latin America was progressively marginalizing itself in world affairs, given the changes underway in Eastern Europe and the sharp decline in North-South relations. The global environmental interdependence was seen as an exception to this trend, and one that the nations on the semi-periphery of capitalism endowed with vast natural patrimony could not afford to ignore. This would explain the set of environmental measures taken by the Collor government, both those high on pyrotechnics and of dubious efficacy so widely covered in the media and others that genuinely contributed to altering the correlation of local political forces in the Amazon. The immediate goal of these measures was to court the industrialized nations and show that Brazil was capable of exporting environmental public goods to the rest of the world, trading forest conservation for the financial, technological and institutional aid of its international partners (Hurrell, 1992, 419 and onwards). At least two environmental programs emerged during the period that represented the concretization of the principle of transferring resources to the so-called developing countries: the National Environment Program (PNMA), which became operational in 1991; and the Pilot Plan for the Protection of the Tropical Forests of Brazil (PP/G-7), established by Decree no. 563, of 5/6/1992.

Recognition of 'traditional populations' by public authorities, expressed in the first ever references to these groups in law and in the creation of governmental organs to deal with them, also derives from this busy period. In 1992, Ibama launched the National Centre for the Sustainable Development of Traditional Populations as 'a governmental response to the demands expressed by populations that traditionally and culturally depend on extractivism and renewable natural resources for their subsistence'. The concept was therefore attributed 'a certain flexibility [...] in virtue of the enormous diversity of communities that could and should be attended by our program'. The

---

approach toward conservation – mentioned earlier – and a complex web of interactions and intermediations involving social movements, NGOs and the government.

affirmation is followed by the predictable list of 'human types' to which the notion applied: peoples of the forest that live off rubber tapping and Brazil nut harvesting, babassu coconut breakers, fishermen and mollusk pickers on the coast, the *calunga* herdsmen 'made up of fugitive slaves, who have kept their maroon quilombos in the hinterlands of Goiás for over two hundred years' (apud Vianna, 1996, pp. 105 & 115–116).

Article 3 of São Paulo State Decree no. 32.412/90, concerning the Juréia-Itatins Ecological Station, defines traditional populations as 'all those permanently settled on and living off a plot of no more than 10 hectares of useful land'. §2 of the same Article states that '[. . .] only those who conduct their subsistence activities in a manner compatible with the conservation goals established for the Ecological Station' 'shall be considered traditional populations'. Resolution no. 11 of the São Paulo State Secretary for the Environment, issued on 25/11/1993 with reference to the Alto Ribeira Tourist Park, in the interest of creating regulations for its occupation, defines 'traditional inhabitants' as 'those who fulfill all of the following criteria: I – have been living in the area for more than 10 consecutive years; II – make autonomous use of the land for the purposes of subsistence; and III – conduct their activities in a manner compatible with environmental conservation, posing no threat to the attributes that warranted the creation [. . .] of the Park'. State Law no. 293 of 20/4/1995, which concerns the permanence of 'native populations resident in the conservation units of Rio de Janeiro State for more than 50 (fifty) years', defines these groups as on equal footing with the Indigenous groups 'as they *practically live in isolation* and are *dependent upon the local ecosystems, have their own forms of social organization*, customs, beliefs and traditions, as well as their *own relationship with the environment, allowing them to live in harmony with those ecosystems*'. The law confers upon them the 'real right to use the occupied areas, so long as they directly and as a matter of priority *depend upon* the local ecosystems for their subsistence' (my italics).

Not much different, then, from the definitions put forward by the international conservationists and the translations made of them by the native formulators.[27] Note how the incorporation of the notion within legal documentation and the stipulations of governmental programs dealing with these 'traditional populations' in protected areas fulfill the recommendations of those same international fora and constitute Brazil's own particular contribution to the matter.

In this sense, the promulgation of Federal Law no. 9.985, of 18/7/2000, which established the National System of Conservation Units (SNUC), brings special consequences. The Bill of Law (BOL heretofore) that gave rise to it did the rounds for over ten years, suffering countless revisions. Part of the reason for this delay was the fierce debate on the possibility of reclassifying existing strict-use protected areas with human populations living inside them under less

[27] For a more complete and exhaustive look at the definitions of 'traditional populations' in legal documents and official governmental and non-governmental reports, see Vianna (1996: 115 onwards).

restrictive protection categories, such as extractive reserves, for example. This debate had a lot to do with the acerbic discussions on the notion of 'traditional populations' and the various definitions presented in the numerous different versions of the BOL. At one particular point in time, one of the versions of BOL 2.892, based on a draft by Antonio Carlos Diegues, defined 'traditional population' as follows: 'culturally differentiated human groups, resident for at least three generations in a given ecosystem, historically reproducing their style of life in strict dependence upon the environment for their subsistence and using its natural resources in a sustainable fashion' (Item XV, Art. 2, Chapter I of the Preliminary Definitions). Such was the polemical nature of the subject and so fiercely contested the definition that this item was vetoed and omitted from the preliminaries of the promulgated Law.

However, the lack of a definition among the items of Article 2 does not prevent us from gleaning a definition latent in rest of the document, principally Articles 18 and 20, which deal with extractive reserves (Resex) and sustainable development reserves (SDRs), respectively – symptomatically two categories of protected areas for sustainable use in which the presence of social groups collectively and associatively managing the units is a pre-condition of their creation. Article 18 defines an 'extractive reserve' as 'an area used by *traditional extractivist populations* – whose *subsistence* is based on extractivism and, complementarily, on subsistence agriculture and the raising of small animals – with the express purpose of protecting the livelihoods and cultures of these populations and of *ensuring the sustainable use of the natural resources of the unit*'. Article 20, for its part, defines an SDR as 'a natural area that houses *traditional populations, whose existence is based on sustainable systems for using natural resources*, developed down through the generations, *adapted to the local ecological conditions* and which *fulfill a role in protecting nature and maintaining its biological diversity*' (my italics). It goes without saying that, were the letter of the law to be used as a means of characterizing these populations, ascertaining the sustainability of each 'system of natural resource usage' would be an arduous task requiring time-consuming and painstaking study case-by-case.

Nevertheless, as the anthropology of law teaches us, the 'wording of a law should not be considered exclusively in terms of its degree of effective applicability', as 'it produces other effects when [...] viewed as a *mechanism of codification* – objectively situating hitherto disperse notions and ideas – *and* as an *instrument of formalization* (in the sense of bestowing a given form)' (Lima, 1995, p. 202, my italics). As this same author observes, 'the simple existence of the law, as a limit and horizon of the possible ways of officially seeing and intervening in certain problems, opens them to objective control' – regardless of whether it comes into effect and/or contains the necessary devices to be enforced (1995, p.). In this manner, one of the immediate effects of the abovementioned definitions is to establish a new social statute, as part of the codified system of positive and negative attributes contained in the articles and items of SNUC. What we have here is a 'statute for traditional populations' that sets the tone – in the sphere of in situ biodiversity conservation – for how inclusive political society will relate to that segment as defined in terms of a generic difference that reduces the diversity of situations while decharacterizing socio-cultural configurations and specific historical trajectories.

If we consider that the act of categorization wields power in its own right and that the efficacy of performative discourse is directly proportionate to the authority of the deliverer, 'traditional populations' are largely instituted as a reality by the power of revelation and construction exercised by their objectification in normative and administrative discourse. Especially when we consider the symbolic effect of the purportedly scientific discourse from which the notion of 'traditional population' supposedly originates: the most neutral verdicts of science help modify the very object of science, especially when those judgments are incorporated into the normative and administrative devices by which the groups are constructed. As a general rule, the use of purportedly objective criteria in symbolic struggles for the recognition and presentation of groups leads to the consecration of one state of the divisions, and the perception of those divisions, in the social world (cf. Bourdieu, 1989).

Final Considerations and Perspectives

Over the course of this chapter I have sought to show how definitions of 'traditional populations' generally tend to situate the social groups they take as their subject – including the Amazonian caboclo societies – as pertaining to ecosystems that need to be protected and as being in a certain state of harmony with nature, like 'animal populations [. . .] controlled within natural parameters, regardless of human symbolic praxis' (Viveiros de Castro, 1992, p. 25). Through this approach we encountered some important referents of the term 'population': on one hand, the simplifying ecologist reference, which naturalizes these groups as part of the landscape; and, on the other, the atomistic demographist prism, which is the same orientation of the register – understood as a means of social control and the generation of knowledge to meet the needs of governmental interventions.

Furthermore, the use of the term 'traditional' to define and/or characterize these groups or their 'lifestyles' and forms of appropriating natural or territorial resources masks the absence of a semantic but necessary criticism. 'Traditional', 'archaic', 'backward', 'primitive' and other inaccurate and mystifying terms – some of which contemporary anthropology has preserved out of intellectual complacency and laziness in designating this type of society – present a symmetrically inverted image of Western modernity (Copans, 1989). They are classificatory categories constructed from the outside-in, that is, measured against how we perceive ourselves and our concerns rather than the ways this differentiated set of groups that we subsume under the umbrella term 'traditional' actually see themselves.

Lima (1997), also referring to Amazonian caboclo societies, observes that the generalization of the notion of 'traditional peoples' tends to simplify the diversity of social situations and implies an expectation that family smallholder production – the preferred mode of the environmental movement, precisely because it is seen as more conducive to models of sustainable use than capitalist production – will remain frozen in time. And yet, without appropriate reflection, the conservative expectation for sustainable use may well conspire against the autonomy of these groups to determine their own futures in terms of modern aspirations toward levels of consumption and of defining their ideas of well-being (cf. Lima, 1997, pp. 287–288).[28] As we have seen, the desire to protect these peoples arose from the expectation of cultural stasis and balance hoisted upon their smallholder and theoretically subsistence-based production, but only so long as it does not affect the ecological integrity of the area under protection.

However, as Brito (2000, p. 20) tells us, this attitude does not hold the same for those groups less readily defined as 'traditional', including the caboclos, which have moved towards modernizing their equipment and lifestyles. West and Brechin (1991, p. 5) also observed that one of the most telling pitfalls in approaching the issue of residents in protected areas is the limited context in which this occurs: the emphasis is always placed on the Indigenous peoples and 'traditional' societies in a bid to distinguish those who live in ecological harmony with their surrounding environment from those who do not – i.e., the 'modern' societies. According to the authors, not only does this view not correspond entirely to the complexity of the real situation, but the erroneous assumptions on which it is based distract our attention from the genuine crux of the issue: is human residence – whether 'modern' or 'traditional' – in protected areas or critical and fragile ecosystems compatible with the goals of conservation? Is it politically avoidable, given the socio-environmental conjuncture in many countries and regions, especially those in the Neotropics? (West & Brechin, 1991, p. 6).

Considering the abovementioned points, I would like to suggest a conceptual alternative that would enable us to recognize the historical and sociological specificity of the social groups living within protected areas or ecosystems considered critical or fragile, including the caboclo societies. First of all, we must divest the term 'population' of its naturalizing reference and the demographist cleavage and census-like orientation of its employment, that simplifies, atomizes and subjectifies. On the other hand, we must also shift the emphasis away from the temporal dimension implied by the polysemy of the word 'traditional' toward the spatial. It is often forgotten that protected areas, as a

[28] The recent history of Indigenous peoples in Brazil, bound as they are to a wrongheaded public image of the ecologically noble 'savage' and victims of juridical hermeneutics built around the constitutional concept of 'lands *traditionally* occupied by Indians' – an ambivalent adverb caught somewhere between time and mode –, urges extreme caution in the use of the term 'traditional' and its derivatives in legal wording.

strategy for in situ biodiversity conservation in critical and fragile 'natural' landscapes, entail operating on different 'levels and scales of the biospatial hierarchy', depending on the object and focus of the conservationist action – be it landscapes, ecosystems, communities, populations, species or gene pools – and the assessment made of its diverse social contexts (Soulé, 1991).[29] Lastly, it is imperative that we replace the generic – and supposedly technical and scientific – cultural label (anchored as it is in the concepts offered by the tradition of anthropological studies on regional subcultures – such as that which applies 'traditional' to lifestyle) with other terms that are culturally less dense, but by no means politically neutral.

So in order to skip the conceptual traps implied in the notion of 'traditional population', the most advisable course of action would be to opt for a definition that sheds density whilst retaining its political charge, like that proposed by West and Brechin (1991, p. 6): 'resident peoples and/or social groups' – without the need for a formal and specific definition geared towards ends set forth in regulatory laws. These would be individuals, families, communities and groups[30] – whether 'traditional' or 'modern' – that occupy, dwell in or regularly or recurrently use a given territory inside or bordering upon an established or proposed protected area. The advantage of the notion of 'resident peoples and/or groups' lies in the fact that its connotations are less dense and that the idea is defined more in terms of space than of time – unlike what occurs with 'traditional' – and eschews cultural labels, such as that by which 'traditional' refers to a style of life that is at once different and generic.

In saying this, I am not ignoring the issue of minority rights, the problems of cultural dispersal or any of the other approaches and/or strategies for biodiversity conservation and the protection of the territorial rights of ethnic and/or other groups. It so happens we must defend all of the peoples and social groups that are fighting to survive and reproduce, especially the most underprivileged and exploited amongst them, but also those more readily identified with society at large and/or with the dominant culture. This particularly applies to the Amazonian caboclos, as Nugent (1993) well remembers, when they are approached as an incomplete and pathological 'other' unfit for the attentions of anthropology, as their very existence subverts the observer/other distinction, or when represented as 'inauthentic others', whether because they derive from European colonizers and are therefore not autochthonous, or because they bear the marks of the pernicious influence of 'civilization'.

The definition I am here proposing is also a step toward breaking away from the conceptual and administrative mechanisms for the control and subordination of the processes of cultural change via management plans; mechanisms that

---

[29] Discussing the main 'tactics' and/or 'systems' of conservation that can be employed by different countries, Soulé (1991) defends a tactical pluralism in conservation planning, not exclusively based on protected areas.

[30] Groups understood here in the wider and more descriptive sense of any collectivity whose members share some similarity.

are tied-in with the expected temporal stasis of these cultures. However, I do depart with Cunha and Almeida's suggestion (1999) of a potential neo-traditionalist pact by which 'traditional populations' would come to mean all those who accepted the designation and the legal and institutional implications that go with it in terms of 'the sustainable use of natural resources'. Presupposing that they do not have a track record of predatory resource use, what would be expected of these groups under the pact is that they lead, henceforth, a style of life compatible with the conservation of biological diversity (cf. Cunha and Almeida, 1999, p. 5 onwards). Contrary to the 'traditional populations' category and the efforts to construct a technical cultural concept with legal reach – and which could contribute still further to essentializing the relationship between the social groups covered by the notion (caboclos included) and nature –, I feel it is important to ensure a legal definition that guarantees openness to the sociological identification and characterization of any agents within the focus situations – where there already are protected areas and/or plans to establish PAs – and with which diverse partnerships – not pacts – could be created for long-term biodiversity conservation, rooting the protected areas more firmly in the local social milieu whilst ensuring the equitable distribution of the costs and benefits of conservationist endeavor.

References

Adams, C. (2000a). *Caiçaras na Mata Atlântica: pesquisa científica versus planejamento e gestão ambiental*. São Paulo: Annablume/Fapesp.

Adams, C. (2000b). As populações caiçaras e o mito do bom selvagem: a necessidade de uma nova abordagem interdisciplinar. *Revista de Antropologia, 43*, 145–82.

Allegretti, M. H. (1994). Reservas extrativistas: parâmetros para uma política de desenvolvimentos sustentável na Amazônia. In: R. Arnt (Ed.), *O destino da floresta: reservas extrativistas e desenvolvimento sustentável na Amazônia* (pp. 17–47). Rio de Janeiro/Curitiba: Relume Dumará/IEA/Fundação Conrad Adenauer.

Almeida, M. W. B. (1992). Desenvolvimento e responsibilidade dos antropólogos. In A. A. Arantes and G. G. Debert (Eds), Desenvolvimento e Direitos Humanos: a responsibilidade do antropólogo (pp. 111–122). Campinas: EdUnicamp.

Almeida, A. W. B. (1994). Universalização e localismo: movimentos sociais e crise dos padrões tradicionais de relação política na Amazônia. In: M. A. D'incao, & I. M. Silveira (Ed.). *A Amazônia e a crise da modernização* (pp. 521–537). Belém: Museu Paraense Emílio Goeldi.

Amend, S., & Amend, T. (1992). Habitantes en los Parques Nacionales: una contradicción insoluble? In: S. Amend, & T. Amend (Eds.), *Espacios sin habitantes? Parques Nacionales de América del Sur* (pp. 457–472). Gland/Caracas: IUCN/Editorial Nueva Sociedad.

Arnt, R. (1994). *O destino da floresta: reservas extrativistas e desenvolvimento sustentável na Amazônia*. Rio de Janeiro/Curitiba: Relume Dumará/IEA/Fundação Conrad Adenauer.

Arruda, R. V. (1997). 'Populações Tradicionais' e a Proteção dos Recursos Naturais em Unidades de Conservação. In: *Congresso Brasileiro de Unidades de Conservação, 1, Curitiba. Anais*, (pp. 351–367, Vol. 1). Curitiba: IAP/Unilivre/Rede Nacional Pró-Unidades de Conservação.

Arruti, J. M. A. (1995). A narrativa do fazimento, ou Por uma antropologia brasileira. *Novos Estudos Cebrap*, *43*, 235–243.

Balée, W. (1989a). The culture of Amazonian forests. In: D. E. Posey, & W. Balée, (Eds.), *Resource management in Amazonia: Indigenous and folk strategies*. New York: New York Botanical Garden.

Balée, W. (1989b) Cultura na vegetação da Amazônia Brasileira. In: W. A. Neves, (Ed.). *Biologia e ecologia humana na Amazônia: avaliação e perspectivas* (pp. 95–109). Belém: MPEG.

Balée, W. (1992). People of the fallow: a historical ecology of foraging in lowland South America. In: K. Redford, & C. Padoch, (Eds.). *Conservation of neotropic forests: working from traditional resource use* (pp. 35–57). New York: Columbia University Press.

Barrett, C. B., & Arcese, P. (1995). Are integrated conservation development projects (ICDPs) sustainable? On the conservation of large mammals in Sub-Saharan Africa. *World Development*, *23*,1073–1084.

Barretto Filho, H. T. (2001). Da nação ao planeta através da natureza: uma abordagem antropológica das unidades de conservação de proteção integral na Amazônia brasileira. 2 vols. Ph.D. thesis, University of São Paulo.

Barzetti, V. (1993). *Parques y progresso: áreas protegidas y desarollo economico en America Latina*. Washington: IUCN/BID.

Bastide R. (1979). *Antropologia aplicada*. São Paulo: Perspectiva.

Bourdieu, P. (1989). *O poder simbólico*. Lisboa/Rio de Janeiro: Difel/Bertrand Brasil.

Brandon, K., & Wells, M. (1992). Planning for people and parks: design dilemmas. *World Development*, *20*, 557–570.

Brito, M. C. W. (2000). *Unidades de conservação: intenções e resultados*. São Paulo: Annablume/Fapesp.

Clad, J. (1984). Conservation and Indigenous peoples: a study of convergent interests. *Cultural Survival Quarterly 8*, 68–73.

Clay, J. (1985). Parks and people. *Cultural Survival Quarterly*, *9*, 2–7.

Cleary, D. (2001). Towards an environmental history of the Amazon: fom prehistory to the nineteenth century. *Latin America Research Review*, *36*, 65–96.

Comisión sobre Desarollo y Medio Ambiente en América Latina e Caribe/ONU (1992). *Amazonia sin mitos*, (pp. 65–74). Washington: Banco Interamericano de Desarollo/PNUD.

Copans, J. (1989). *Antropologia: ciência das sociedades primitivas*? Lisboa: Edições 70.

Cunha, M. M. C., & Almeida, M. (1999). Populações tradicionais e conservação. In: *Seminário de Consulta 'Biodiversidade Amazônia*. Macapá. Subsídio ao GT Povos Indígenas e Populações Tradicionais.

Denevan, W. (1992a) *The native population of the Americas in 1492*. Madison: University of Wisconsin Press.

Denevan, W. (1992b). The pristine myth: the landscape of the Americas in 1492. *Annals of the Association of American Geographers*, *52*, 369–385.

Diegues, A. C. (1996). *O mito moderno da natureza intocada*. São Paulo: Hucitec.

Drummond, J. A. (1991). A história ambiental: temas, fontes e linhas de pesquisa. *Estudos Históricos*, *4*, 177–197.

Escobar, A. (1991). Anthropology and the development encounter. *American Ethnologist*, *18*, 658–682.

Esterci, N. (2001). Pescadores e ribeirinhos: questões acerca da organização e da representação. In: *Amazônia: Diversidade Sociocultural e Políticas Ambientais*. Rio de Janeiro: Comunicação pública.

Gómez-Pompa, A., & Kaus, A. (1992). Taming the wilderness myth. *Bioscience*, *42*, 271–279.

Gómez-Pompa, A., Vázquez-Yanes, C., & Guevara, S. (1972). The tropical rain forest: a nonrenewable resource. *Science*, *177*, 762–765.

Gray, A. (1992). Entre la integridad cultural y la asimilación: conservación de la biodiversidad y su impacto sobre los pueblos indígenas. *IWGIA Documento 14*, Copenhague.

Hoben, A. (1982). Anthropologists and development. *Annual Review of Anthropology, 11*, 349–375.
Huizer, G. (1993). Development anthropology in global perspective. *Zeitschrift der Arbeitsgemeinschaft Entwicklungsethnologie e. V.*, Heft 1, 2. Jg: 66–82.
Hurrell, A. (1992). Brazil and the international politics of Amazonian deforestation. In: A. Hurrell, & B. Kinsgbury (Eds.) *The international politics of the environment: actors, interests and institutions* (p. 398–429). Oxford: Clarendon Press.
IUCN. (1984). *Estratégia mundial para a conservação: a conservação dos recursos vivos para um desenvolvimento sustentado*. São Paulo: CESP.
IUCN. (1992). Parks, protected areas and the human future. The Caracas Declaration. Caracas, (Mimeo).
IUCN; Pnuma; & WWF (1991). Cuidar do Planeta Terra: uma estratégia para o futuro da vida. São Paulo: UICN/Pnuma/WWF.
Kempf, E. (1993). *The law of the mother: protecting indigenous peoples in protected areas*, (pp. 3–11). San Francisco: Sierra Club Books.
Lima, A. C. S. (1995). *Um grande cerco de paz: poder tutelar, indianidade e formação do Estado no Brasil*. Petrópolis: Vozes.
Lima, D. M. (1997). Equidade, desenvolvimento sustentável e preservação da biodiversidade: algumas questões sobre a parceria ecológica na Amazônia. In: E. Castro, & F. Pinton (Eds.), *Faces do trópico úmido: conceitos e questões sobre desenvolvimento e meio ambiente* (pp. 285–314). Belém: Cejup.
Lima, D. M. (2001a). *Políticas ambientais e populações humanas na Amazônia*. In: Amazônia: diversidade sociocultural e políticas ambientais. Rio de Janeiro: Comunicação pública.
Lima, D. M. (2001b) *Projeto de expansão e consolidação de unidades de conservação na região Amazônica (ARPA)*. Estratégia para o trato com as populações indígenas e populações tradicionais atingidas por sua execução (versão incompleta), (Mimeo).
Lima, D. M., & Pozzobon, J. (2001) Amazônia socioambiental: sustentabilidade ecológica e diversidade social. In: I. C. G. Vieira, J. M. C. Silva, D. Oren, & M. A. D'incao (Eds.), *Diversidade biológica e cultural da Amazônia* (pp. 195–251). Belém: Museu Paraense Emilio Goeldi.
Lima, L. & Rola, A. (2001). Comunicação pública. In: Amazônia: diversidade sociocultural e políticas ambientais. Rio de Janeiro.
Little, P. (1994). *Gestão social e sustentável do território*. Brasília: IEA. (Mimeo).
Mackinnon, J., Mackinnon, K., Child, G., & Thorsell, J. (1990). *Manejo de áreas protegidas en los trópicos*. Gland: IUCN.
McNeelly, J. (1993). Afterword. People and protected areas: partners in prosperity. In: K. Kempf (Ed.), *The Law of the Mother: protecting indigenous peoples in protected areas* (pp. 249–257). San Francisco: Sierra Club Books.
Menezes, M. A. (1994). As reservas extrativistas como alternativa ao desmatamento na Amazônia. In: R. Arnt (Ed.), *O destino da floresta: reservas extrativistas e desenvolvimento sustentável na Amazônia* (pp. 49–72). Rio de Janeiro/Curitiba: Relume Dumará/IEA/ Fundação Conrad Adenauer.
Miller, K. R. (1982). *Planning national parks for ecodevelopment: cases and methods from Latin America*. Ann Arbor: Center for Strategic Wildland Management Studies/The School of Natural Resources/University of Michigan.
Moore, D. (1995). Marxism, culture, and political ecology: environmental struggles in Zimbabwe's Eastern Highlands. In: R. Peet, & M. Watts (Eds.). *Liberation ecologies: environment, development, social movements* (pp. 125–147). London: Routledge.
Moran, E. F. (1990). *A ecologia humana das populações da Amazônia*. Petrópolis: Vozes.
Murrieta, R. (1998). O dilema do papa-chibé: consumo alimentar, nutrição e práticas de intervenção na Ilha de Ituqui, Baixo Amazonas, Pará. *Revista de Antropologia, 41*, 97–150.

Nitsch, M. (1994). Riscos do planejamento regional na Amazônia brasileira: observações relativas à lógica complexa do zoneamento. In: M. A. D'incao, & I. M. Silveira (Ed.), *A Amazônia e a crise da modernização* (pp. 501–512). Belém: Museu Paraense Emílio Goeldi.
Nugent, S. (1993). *Amazon caboclo society: An essay in invisibility and peasant economy*. Oxford: Berg.
Nugent, S. (1997). The co-ordinates of identity in Amazonia: at play in the fields of culture. *Critique of Anthropology, 17*, 33–51.
Oliveira Filho, J. P. (1988). *'O Nosso Governo': os Ticuna e o regime tutelar*. São Paulo/Brasília: Marco Zero/MCT/CNPq.
Parker, E. (1985). *The Amazon caboclo: historical and contemporary perspectives*. Williamsburg: College of William and Mary.
Pietilä, H. (1990). Environment and sustainable development. *IFDA Dossier, 77*, 61–70.
Posey, D. (1987). Manejo da floresta secundária, capoeiras, campos e cerrados (Kayapó). In: B. Ribeiro (Ed.), *Suma etnológica brasileira*, (Vol. 1, B, pp. 173–184). Petrópolis: Vozes.
Posey, D., Frechione, J., Eddins, J., Silva, L. F., Myers, D., Case, D., & Macbeath, P. (1984). Ethnoecology as applied anthropology in Amazonian development. *Human Organization, 43*, 95–107.
Ribeiro, G. L. (1992). Ambientalismo e desenvolvimento sustentável: nova ideologia/utopia do desenvolvimento. Brasília: Depto. de Antropologia/UnB.
Roosevelt, A. C. (1994). *Amazonian Indians: from prehistory to the present*. Tucson: The University of Arizona Press.
Santos, S. H. (1991). Reflexões críticas sobre o Relatório Brundtland: as questões do 'desenvolvimento sustentável' e da 'nova ordem econômica internacional'. Niterói: ICHF/UFF.
Schwartzman, S. (2000). Indians, environmentalists and tropical forests: the curious history of the 'ecologically noble savage'. Washington, DC, (ms.).
Soulé, M. (1991). Conservation: tactics for a constant crisis. *Science, 253*, 744–750.
Stavenhagen, R. (1985). Etnodesenvolvimento: uma dimensão ignorada no pensamento desenvolvimentista. *Anuário Antropológico, 84*, 11–44.
Trainer, T. (1990). A rejection of the Brundtland Report. *IFDA Dossier*, 77, 71–84.
Uhl, C., Nepstad, D., Buschbacher, R., Clark, K., Kauffman, B., & Subler, S. (1989). Disturbance and regeneration in Amazonia: lessons for sustainable land-use. *The Ecologist, 19*, 235–240.
Vianna, L. P. (1996). Considerações críticas sobre a construção da idéia de população tradicional no contexto das unidades de conservação. MA thesis, University of São Paulo.
Viola, E. J. (1987). O movimento ecológico no Brasil (1974–1986): do ambientalismo à ecopolítica. *RBCS, 1*, 5–26.
Viola, E. J. (1988). The ecologist movement in Brazil (1974–1986): from environmentalism to ecopolitics. *Journal of Urban and Regional Research, 12*, 211–228.
Viola, E. J. (1992). O movimento ambientalista no Brasil (1971–1991): da denúncia e conscientização pública para a institucionalização e o desenvolvimento sustentável. In: M. Goldenberg (Ed.). *Ecologia, ciência e política*. Rio de Janeiro: Revan.
Viola, E. J. (1995) O ambientalismo multisetorial no Brasil: para além da Rio-92: o desafio de uma estratégia globalista viável. In: E. Viola (Ed.). *Meio ambiente, desenvolvimento e cidadania: desafios para as ciências sociais* (pp. 134–160). São Paulo/Florianópolis: Cortez/Edufsc.
Viola, E. J., & Leis, H. R. (1995). A evolução das políticas ambientais no Brasil, 1971–1991: do bissetorialismo preservacionista para o multisetorialismo orientado para o desenvolvimento sustentável. In: D. J. Hogan, & P. F. Vieira (Eds.). *Dilemas socioambientais e desenvolvimento sustentável* (pp. 73–102). Campinas: EdUnicamp.
Viveiros de Castro, E. B. (1992). Sociedades indígenas e natureza na Amazônia. *Tempo e Presença, 14*, 25–26.

Viveiros de Castro, E. B. (1999). Etnologia brasileira. In: Miceli, S. (Ed.). *O que ler na ciência social brasileira (1970–1995)* (pp. 109–223). São Paulo/Brasília: Anpocs/Capes.
West, P. C., & Brechin, S. R. (1991). *Resident peoples and national parks: social dilemmas and strategies in international conservation*. Tucson: University of Arizona Press.
Worster, D. (1991). Para fazer história ambiental. *Estudos Históricos, 4*, 198–215.

Chapter 6
The Domestic Economy in Mamirauá, Tefé, Amazonas State[1]

Deborah de Magalhães Lima

Abstract This article examines the economic practices of the residents of a conservation unit in Amazonia, the Mamirauá Reserve, where the model of 'Sustainable Development Reserve' was created. The characteristics of the domestic economy are presented first in terms of the consumptive orientation of households and the horizontal model of kinship based communities. Consumption patterns and exchange conditions are central in defining production choices, also affected by the strong seasonal variations of the Amazonian floodplain. The consumption orientation of Mamirauá's domestic economy is shown to be instrumental to the environmentalists' proposition of involving residents in a sustainable model of resource management. The second part of the article develops a quantitative analysis of patterns of income and commodity consumption as a baseline for monitoring market participation and the effects of ecological management on the residents' life conditions.

Keywords Kinship · Economy · Production · Consumption · Commodity · Money · Land · Floodplain · Terra firme

Introduction

This chapter deals with the domestic economy of the residents of the Mamirauá Sustainable Development Reserve, located on a vast floodplain area of the Middle Solimões River in the Amazon Region. This population could be defined as either *caboclo* or as a 'traditional population'. Among other acceptations used in this book, I will use the term *caboclo* to refer to the predominantly riverside Amazonian peasantry of colonial origin, while the designation traditional population will

D. de M. Lima (✉)
Department of Anthropology, Federal University of Minas Gerais, Pampulha, Belo Horizonte, MG, Brazil
e-mail: deb.lima@terra.com.br

[1] I would like to thank Edna Alencar for her readings and comments.

C. Adams et al. (eds.), *Amazon Peasant Societies in a Changing Environment*, DOI 10.1007/978-1-4020-9283-1_7, © Springer Science+Business Media B.V. 2009 be reserved for either those living inside sustainable use protected areas, and therefore engaged in conservation practices, or to any other regional, Amazonian population with a long history of living within the natural environment. The chapters in this volume by Harris and Barreto Filho made valuable contributions to the analysis of these categories, drawing attention to the complexity of their application.[2] This chapter, however, has more in common with the work of Castro and Futemma. Despite their different focuses, these three chapters deal with the domestic economies of riverine producers.[3] That said, studies on domestic economies do unearth elements that are indirectly relevant to the discussion of identity insofar as they concern two aspects present in the representation of this social group: the generally ignored relationship between the producers and the market and the practices related to the use of natural resources.

This study is concerned with the social organization of production, consumption and circulation in Mamirauá and the following pages will present a quantitative analysis of the relationship between the producer and the market by describing patterns of income generation and the consumption of commodities.[4]

The Base of the Domestic Economy in Mamirauá: The Household, Kin and the Community

The household is the core reference in the economy of Mamirauá. It is the locus of production, the reference of circulation and the nucleus of consumption. While not every domestic economy has the domestic group exercising such a coincidence between the functions of production, consumption and circulation

---

[2] On the social category *caboclo* and on the particular identity of the residents of the Mamirauá region, see Lima Ayres, 1992; Lima, 1999; Lima and Alencar, 2001; and Alencar, 2002.

[3] In addition to their different thematic focuses, there are also important regional differences to consider. For example, in the Middle Solimões region, the annual floods reach up to 12 metres, the 'restinga' (high levee) forest tracts are narrow and the floodplain 'lake' system is formed by hundreds of relatively small bodies of water, if compared with the large lacustrine formations of the Lower Amazon, where the variations in water levels are lower and the flooded forest far more extensive. In the Middle Solimões, a community's fishing territory consists of a mosaic of differently managed lakes, while in the Lower Amazon a number of communities may share the same lake (such as Lago Grande de Monte Alegre, or the Curuai) on negotiated terms. The Lower Amazon floodplain features large stretches of natural pastureland for cattle-raising as a source of income and important domestic savings (as shown by Castro and Futemma in this volume), while in the Middle Solimões, where there is no natural pasture, floodplain cattle-raising is sparse.

[4] As this study was conducted at the beginning of the implantation of the reserve, it can be considered representative of the domestic economy of the floodplain residents of the Middle-Solimões region. At the time, the few changes effected by the creation of the reserve were limited to an increase in fishing yields due to the closure of the lakes to 'outside' fishermen; a fact that was already palpable during the research. For more on the Mamirauá SDR and changes in the living conditions of its residents, access the following URL: www.mamiraua.org.br/.

(cf. Sabean, 1990, p. 98), in Mamirauá no division can be made between the 'economic unit' and the 'family unit'.

The main characteristic of the domestic economy in Mamirauá is that it is geared toward the domestic group's consumption, that is, the members of each household work to meet their own consumption needs. Production for sale and for domestic use feed, clothe, medicate and ensure the comfort of all those living under the same roof.

Financial incomes in Mamirauá are earned from the sale of three main product types: agricultural products (chiefly manioc flour, known locally as *farinha*, and banana), the sale of fish (dried, salted or fresh), and the sale of wood (firewood and light and heavy woods, referred to respectively as 'white woods' and 'wood of the law'). Other sources of income are salaries (for teachers, health professionals and paid services in the Mamirauá Reserve) and rural pensions.

In Mamirauá, those sharing the same household are invariably members of the same family unit and the production is organized on the basis of family relations. The domestic authority responsible for organizing the economic activities is usually a couple, which delegates the tasks amongst themselves and their children. The division of labour is determined in accordance with the sex and age of the members of the domestic group. This peasant form of organizing domestic production is centred on the ideal of the economic autonomy of the household, whose good reputation will depend on its capacity to fend for itself. Under normal conditions, there is no strict economic dependence among domestic groups.

The formation of a new domestic unit, which normally occurs after a marital union, automatically implies the constitution of a new economic 'cell'. When a couple marry, they generally establish their first residence alongside the parents of one of the spouses so that they can prepare to make their own start in life. The autonomy of a couple essentially consists in their working for themselves; where before they were part of the workforce of the parental household, they will now have their own independent agricultural production – their first plot. With this autonomy of labour and consumption, after roughly a year of living with one of the spouses' parents, the couple will set up their own household, an event that usually coincides with the birth of their first child (Lima & Moura, 1995).

Neither land inheritance, nor the transmission of material goods from one generation to another counts for very much in the reproduction of domestic groups. The most important legacy a couple can receive is the bonds of kinship. Horizontal relationships between living relatives form a network of mutual assistance that confers the right to the collective use of natural resources. This right is directly linked to the household residence, which is in turn facilitated by the presence of relatives at the locale where the couple wants to set up home. Rights of use can also be exercised at former points of residence, but this tends to depend on the presence of relatives at the site, both in terms of hospitality and in conferring legitimacy to continued rights of usage.

Kinship is the condition for inclusion and belonging to a local group (composed by members of the 'community', as the rural settlements of the region are known) and this belonging confers the right to appropriate the natural resources of the occupied territory. However, kinship is not the only means of establishing oneself in a community. A person unrelated to anyone in a group can request permission to join, in which case residence is the predicate for the right to use the natural resources. This definition of consuetudinary rights to appropriate the natural resources can be empirically observed in Mamirauá, where, in principle, each household enjoys the same rights of access to and use of the natural resources.

The absence of privileges or some legitimate social form of grading access to fishing lakes, arable lands or forests constitutes the basis for an ideal model of horizontal social relations among the residents of the same locale. Even in places where a given family is recognised as the 'founder', and may or may not have some legal title to prove 'ownership' (such as receipts for the payment of rural tithes or licenses), when it comes to the appropriation of natural resources, the members of this family have no entitlement over and above the other residents. Length of residence, however, does confer a certain distinction upon the families of the pioneers, though this may be expressed in such simple terms as special deference and respect, or perhaps the status of host to the patron saint of the site.[5] By extension, the families of incomers will usually hold lesser status, especially those that have no relatives at the locale or who do not immediately establish affiliations with one of the kinship groups. The hierarchy between households and individuals also corresponds to the order of generations within a kinship group.

With time, given the high rate of marriage between people from the same locality, the different groups end up becoming related through affinity and blood line, thus gradually dissolving the discrepancies in status resulting from length of residence. It is in this context that one can observe the result of the process of reproduction of domestic groups that characterizes rural society: the constitution of 'kinship communities' that hold the right to common use of the territory on which they farm, fish or gather.

Staking out a 'community's territory' is more precise in relation to the conjunct of individual areas of residence and plantation than it is to fishing zones of collective access to the group, which are defined by common agreement with neighbouring localities and are thus subject to conflict.[6] The boundaries of logging areas (in forest and *restinga*) tend to become fuzzier the further away they are from the settlement. Their extension is equivalent to the range of the residents' capacity

[5] In the case of internal disputes, the status of the older family may nominally lead to their word being taken by the group, though this will largely depend on the weight of the kinship network, which in such cases tends to be more numerous.

[6] These conflicts escalated with the advent of the lake preservation movement, through which the riverine communities defended the conservation of the lakes and the exclusive right to fish them. The riverine peoples stood against the entry of commercial and traditional fishermen from other localities. In fact, the lake preservation movement represents more a demand for territorial ordinance than environmental preservation. The equitable sharing of the fishing lakes amongst the localities is overseen by the Movement for Grassroots Education (MEB) and the Catholic Church. The Pastoral Land Commission (CPT) refers to the water zoning that resulted from the suits brought by community leaders at the regional assemblies as 'aquatic reform'.

to fell and extract native woods, though these areas are not thought of as properties, whether individual or collective, as they were not cultivated by anyone.

Exchanges between Households and Communities

Among the domestic groups, exchanges are generally non-monetary in nature. Donations of fish or game, help with work or an rally-rounds, the services of midwives and the remedies of healers are exchanges usually oriented by the principle of reciprocity and express an ideal model of horizontal social relations. They are characterized by an interval between receipt and repayment, which is determined by the necessity of the group that receives and by the capacity of the donor to do without. The exchanges are not recorded and the bestowals do not necessarily need to be equivalent. The volumes are not defined by market criteria, but by the rules of reciprocity of the group, and their observance is part of the code of conduct that defines what it is to be a 'good neighbour'.[7]

While monetary exchanges between households are rare in the settlements, there are usually one or two that offer a small stock of basic merchandise for local sale. Foodstuffs are normally given and constitute the main object of the verb 'vizinhar', a Portuguese verb that holds the regional sense of regularly exchanging foods, but when the fish or other item of exchange has a fixed monetary value, such as the pirarucu, its sale is considered legitimate. In this case, one neighbour can sell a piece of pirarucu to another without souring their relationship. In terms of exchange of hours worked, there is no record of cash payments for services rendered by one domestic group to another. Work done for another is usually considered help or a favour and will eventually be repaid in kind. However, on the terra firma uplands, it is not unusual to encounter a small, irregular labour market paying daily rates for hired extra manpower during manioc flour production. In these cases, the hired hands are residents from the same locality who had small yields and could therefore offer their services to larger producers on a temporary basis. This difference can be explained by the fact that manioc flour is the main agricultural produce on the uplands and equivalent in value to pirarucu or timber on the floodplains.

In terms of exchanges between localities in Mamirauá, there is practically no trade between them and even non-monetary exchanges are rare. In the flooded forest, the main item in regular circulation is 'seed', or bitter manioc stalks, as they become scarce after the heavier floods. However, despite the strong demand among the farmers of the floodplains, manioc is not part of the range of items of commercial exchange, perhaps because the manioc flour market in the *várzea* is still relatively recent in comparison with the extractive activities of fishing and lumbering. On the other hand, it could also be that the powerful symbolism of manioc makes it resistant to commercialization. Like the women,

[7] For more description of economic exchanges in riverine communities, see Lima Ayres (1992) and Harris (2000).

with whom manioc is associated in planting rituals, manioc circulation is responsible for the reproduction of their society.[8]

There is also little or no direct circulation of products between the floodplains and the uplands, though they could easily exchange manioc flour (farinha) for fish, as fish is scarcer in the uplands than it is in the flooded forest, and farinha more plentiful on the terra firma than on the floodplains, and these are, after all, their staple foods. Practically all exchange outside the localities is monetary and therefore part of the commercial network of the urban traders.

As a result of this configuration of economic exchanges, the small rural communities position themselves as satellites around the urban markets, each connected to the closest regional hub (in the case of Mamirauá, the towns of Tefé, Alvarães and Uarini), but without any real commercial interaction between them. The intra-community exchange networks are largely social in nature, based on relationships between people. In the case of manioc stalks, the fact that the item has no monetary equivalent, despite its ample circulation among households, indicates that the shift from the sphere of donations to that of monetary exchange is reserved for the traders and responds to the demands of the market they represent. On the local level, non-monetary exchanges prevail, and these present a largely open-ended character, as - befitting what are intended to be permanent relationships - the point is not to close these transactions, but to keep them open.

The ideal of domestic autonomy is also reflected in the collective identity of the settlements, especially those that, appropriately, call themselves communities. As the residents share equal usage rights to the territory, in principle they all have a part in the collective representation of the group. The identity of the local group is very strong, as can be seen from the power of the movement for the preservation of the communitarian lakes that rallied floodplain riverine communities from all along the Amazon and Solimões rivers. Collective tenure over a fishing area by a group of households, and legitimized by their common use of that resource, involves a great political effort and even the armed defence of the lakes. However, community identity is powerfully expressed in other ways too, such as in the religious rituals, whether Catholic or Protestant, and in the usual forms of identity through kinship, as seen in their recurrent representation as ‘kinship communities’. However, the strongest manifestations of local identity are those connected with economic praxis, as it is in the economic context, in which the domestic groups organize themselves to reproduce their numbers and the means of sustaining them, that the foundation of their particular and horizontal notion of collectivity is constructed.

Two Types of Consumption

In theory, the domestic consumption that determines production in Mamirauá can be divided into two types, according to origin: the consumption of items produced by the group, such as staple foods (fish and manioc flour), and the

---

[8] See Murrieta (1998, 2001) and Murreita and WinklerPrins in this volume.

consumption of articles obtained with proceeds from the sale of domestic produce, such as industrialized foods, tools, clothing, etc. In practice, however, separating direct consumption (home-grown consumption) from indirect consumption (purchased goods) is a merely formal operation, as the two are interdependent. The use of industrialized tools to produce foodstuffs is, for example, clear demonstration of dependency on commercial exchange to secure the means of production for direct consumption. Nevertheless, this formal distinction is of interest to us as a platform from which to analyze market involvement in domestic production.

There has been a gradual increase in the consumption of items acquired from the market among modern peasant populations. In the Amazon, as in Brazil as a whole, there never was a peasantry entirely independent of the market; in fact, it owes its very existence to colonial implantation of a market economy (Santos, 1989). This process of change in patterns of consumption stems from specialization and the need for higher yields for sale, a trend that is in-line with the development of the regional market. Also involved in changing definitions of consumption needs is the practical and symbolic allure of these goods. Clay pots, thatch roofs, wooden stoves are all examples of basic articles of local production that have been replaced (by aluminium pots and pans, gas stoves, etc.) and indicate increasingly deeper involvement of the domestic economy with the market. Though this trend is dominant, the shift away from locally produced articles and toward sale-bound surplus production is subject to fluctuations and periodical reversions.[9]

Economic Orientation and Environmentalism: The Economic Base of the Ecological Partnership

For the Mamirauá producers, nature is their main wellspring of consumption, both direct and indirect. This dependency results in the affinity of these producers with environmentalist proposals, which they translate in their own terms. For these communities, nature conservation means ensuring continued access to their means of survival in the area of the community's use. On the other hand, for the fishermen and lumberers who work the flooded forest without any fixed territory, moving from place to place after shifting resources, ecology, concerned as it is with the future consequences of present actions, runs counter to their immediate interests in the pursuit of the profits that sustain their activities.

In this sense, Weber's differentiation (1968, p. 14) between 'consumptive' economies (like that of the domestic producers of Mamirauá) and 'lucrative' economies (those of extractivist fishing and logging), distinguishes two social groups that, depending on how they reach their economic goals, react differently in the face of environmental degradation. For the former, what matters is ensuring

[9] A statistical characterization of domestic consumption, presented below, qualifies the degree of involvement of the domestic economy with the market in the 1990s, relevant to the monitoring of changes in patterns of local consumption.

consumption, for the latter, the profits.[10] On occasions, the two groups clash in violent confrontation or in political demonstrations for the preservation of natural resources; on others, they are simply forced to recognise, from the results of their practise, the different ways in which they are affected by environmental degradation. If continued exploration is in the interests of both, the capacity to shift about and the expansion of the area they take advantage of exonerates the corporate exploiters of the responsibility to preserve the integrity of the environment. On the other hand, the domestic producers are aware of the negative consequences of intensive exploitation, because they are directly affected by resource depletion.

The distinction between the consumptive and lucrative orientations is by no means insurmountable, and a household can switch from one to the other or work them both. In Mamirauá, when a producer adopts the lucrative model, he either specializes in local commerce, becoming a 'small local patron', or begins to sell on a small-scale in the town, where there is a higher chance of commercial success.

'Production' and 'Commodity' – Terms that Identify Commercial Exchange

The vocabulary the Mamirauá producers use to talk about their domestic economies makes an interesting distinction between the two classes of object involved in their commercial exchanges: items purchased and items sold. In the common economic acceptation, the term *mercadoria* (commodity) should refer to any commercialized item, but in the lexicon of the Mamirauá producers (and those of the Amazon in general), mercadoria is reserved for consumer items that are purchased, while those put on sale are referred to as production, or produce. Hence, when they discuss exchange conditions, their analyses are clearer because the identities of the objects in question are not lost somewhere along the chain of commercial transactions: 'products' are the fruits of their labour and 'commodities' are the objects they buy in from outside.

As the domestic economy in Mamirauá consists in organizing productive activities that result in yields partly destined for home consumption and partly for sale, so as to finance the purchase of commodities for the household, the strategies adopted for meeting domestic consumption needs are informed by the context in which the exchange between produce and commodity takes place. The equivalence between the commercialized volumes of produce and

[10] This distinction between the consumptive and lucrative orientations has the same analytical sense as Gudeman and Rivera's (1990) distinction between the 'household' and 'corporation'. However, in terms of any comparison between the riverine domestic economy of the Brazilian Amazon and the Colombian peasants studied by these authors, it is important to recall Almeida's observation (1992) that the orientation of consumption among the rubber-tappers of Acre, like that of the riverine communities of Mamirauá, does not involve the same concern with thriftiness presented by the Colombian peasants. In Acre, on the contrary, the capacity to exceed basic consumption needs is socially valorized and the incurrence of high debts with one's patron is viewed as a demonstration of courage and will to work.

commodity, defined by market prices, requires the attention of the Mamirauá producers in virtue of the central role it plays in the result of domestic production, the household consumption of merchandise. The planning and organization of production activities, the decisions taken concerning product consumption or sale, and even the choice of productive activity itself, are examples of economic behaviours that are informed by knowledge of trends in market prices.

While the volume of production destined for domestic consumption can be directly estimated by the producer, production earmarked for exchange will depend on the consumption of merchandise whose volumes are determined outside the domestic sphere. In the Amazon, while barter has been the main means of exchange since colonial times, the exchange of products for commodities is based on monetary values even when cash is not directly involved in the transaction. Money only arrived in the Amazon in the mid eighteenth century, though its circulation was further delayed by the golden age of debt-bondage during the second half of the nineteenth and first half of the twentieth centuries (Santos, 1980).

Perhaps the physical absence of money contributed to the domestic producers having to pay such attention to the relationship between the volumes of produce delivered to the patron and the merchandise received in exchange. For example, in the 1930s a crate of Brazil nuts was equivalent to six kilos of sugar. This type of tabling still exists today, but as the range of merchandise consumed has grown, memory will largely dwell on the most recent volumes of produce (manioc flour, fish or timber) and not to the quantity of each commodity acquired through its sale. Even so, in 1998 one could hear various complaints in the Middle Solimões that a crate of banana was only 'worth' a single pack of biscuits.

The Two Circuits for Exchanging Products for Merchandise in Mamirauá

In Mamirauá there are two circuits through which produce can be exchanged for merchandise. The oldest is the traditional system of debt-bondage. This begins with the merchandise received as credit on later yields. The other circuit, and the most common form of commerce among the domestic economies, is the sale of produce in return for money which is then used to purchase commodities.

The debt-bondage system, which centres upon the figure of the patron, still tinges the local perception of the economy today, even when selling directly to urban markets. In the debt-bondage system, the patron (as the middlemen supplying merchandise as credit are still known) is the personification of the market itself, as he retains the right to stipulate prices and therefore to determine the volumes required for exchange with a margin of profit. The client's dependence on this figure derived from the need to consume merchandise and the fact that there were none of the options of supplier that exist today. In this exchange the patron generally conducts two operations, profiting from each: he purchases domestic yields for resale, and resells merchandise to the producers. This double-edged profiteering also occurs with the modern figure of the small patron. For example, Mr. Zozó, a resident of Barroso, earns 30% on transactions involving merchandise and from 10 to 15% on produce.

Even if the patron (like Mr. Zozó) profits less from the resale of produce than with merchandise, this dual profiteering is a practice widely condemned by the producers. In the local perception, a 'fair' exchange would be the payment of merchandise with produce, which would actually be the case if the patron did not profit on the resale. This is the criterion that determines a 'good patron'. As Mr. Sabá, from Punhã, recalls of the 'days of Alceu Gama', whom he considered a good patron:

> Now Alceu... bought Brazil nuts, bought at the same price as he sold for. He earned nothing on the product, he profited on the merchandise. Today, the traders profit on the produce and the merchandise, while the product [price] is way down low...[11]

With the freedom of trading without the patron as middleman, the produce is exchanged for money and the money used to purchase merchandise. This is the normal circuit of exchange for domestic economies, which Marx described as: Commodity 1 → Money → Commodity 2. According to the local lexicon, the equation would be Produce → Money → Commodity. Under the debt-bondage system, the order of exchange is reversed, as the circuit begins with the supply of merchandise as credit to be paid down later in produce: Merchandise → Debt/ Patron → Produce. The exchanges are parts of the same transaction, with the patron as pivot. With debt-bondage removed from the equation, the exchange reverts to the normal order, with money taking the place of the middleman. Thanks to the existence of a more competitive market and the separation of the stages of the transaction, the producer can choose the best price going.

In Mamirauá, debt-bondage is still the predominant form of commerce for the two extractive products, fish and timber. Given transportation costs, trading with middlemen is also the main option for those living in more remote areas, farther away from the urban centres. For those living closer to towns, on the other hand, agricultural produce, mainly manioc flour, can be sold directly to urban markets. These groups recognise the personal and dominating commercial relationship with the patron, expressed in the observation that 'the patron is now our production, our money'. This affirmation can be understood as their way of establishing functional equivalence between the patron and cash. Both constitute means of access to merchandise, with the difference that debt-bondage imprisons the producer and imposes a personal dependence upon the middleman.

Household *Gastos* (Expenses) and Consumption that Exceeds 'Despesas' (Expenditure for provisions)

Besides the distinction between production and merchandise, there is another division in the lexicon of the domestic economy based on the destination given to the produce and, consequently, to the consumption it enables. Baseline

---

[11] Interview by Alencar (1994).

(minimum) production, which meets the basic needs of the household, is described as production for 'expenditures', for 'costs', for 'rations', or for 'the house'.[12] These terms refer to the acquisition and restocking of merchandise such as foodstuffs, cleaning materials and other articles of immediate consumption. Both items of consumption derived from local production and those exchanged for merchandise can fall within this consumer goal and be included under the category of gastos (expenses for provisions). As such, when manioc flour production is entirely consumed it can serve as provisions (*despesa*), or, when some manioc is sold and the proceeds are used exclusively for the acquisition of basic foodstuffs – household staples - , the purchase is called a '*rancho*' (ration). This set of terms can be translated into economic jargon as ways of referring to 'staple provisions', the group of non-durable consumer items that need to be frequently restocked.

In addition to baseline production, there is also production earmarked for the acquisition of some durable consumer item, such as a motor, a stove, an aluminium roof, etc., that will result in an increase in household patrimony. This expenditure beyond provisions is not given any specific terminology. It is quite common for the householder to programme production to surpass the baseline with a view to acquiring a consumer item of this nature.

Among those who deal with middlemen, both the volume and type of consumption, that is, the designated end for the merchandise, whether for baseline provisions or beyond, are programmed in advance of production. The circuit of exchange based on debt-bondage imposes a specific destination for all production – namely to liquidate the debt with the patron for merchandise received in advance.

The 'commitment' the producer must honour with the patron is agreed at the moment the merchandise is ordered by the smallholder, as this will determine the volume of production to be delivered in exchange. It is also at this moment that the producer decides whether the merchandise he is about to order will be used for provisions or if he is going to acquire an asset of value, such as an electric saw, a motor or some other such item, in which case he will have to work harder to pay off the debt. When working with fish, for example, a patron like Mr. Zozó from Barroso, mentioned earlier, will set a delivery deadline of 10 to 20 days between the delivery of the merchandise, when he goes up-river, and his return, when he comes down-river to collect his payment in fish. In the case of timber production, the deadline is longer. The merchandise is delivered in March or April and the account is settled after the timber *fábrico*,[13] which normally occurs between July and August.

---

[12] Garcia Jr. (1983) presents the notion of *gasto* (costs) as used by the rural smallholders of Pernambuco. The notion signifies domestic consumption and is distinguished from *mais que o gasto* (more than costs), which refers to the part of production yields that goes toward buying other consumer items. For these, *gasto* corresponds to direct consumption and *mais quo o gasto* to indirect consumption. The Pernambucan smallholders also have a concept of profit that refers to both the liquid produce, i.e. total production minus production costs, and profit considered as a gift, exemplified by a tree's production of fruit, that is, a gain that does not come from human effort (Garcia Jr., 1983: 109; 143; 203). The lexicon of the riverine communities of the floodplains contains no such notion of profit.

[13] The term used to refer to the time of year when a given harvest occurs.

Extractive activities with fish and timber are usually the means of obtaining household patrimony, while agricultural produce, such as manioc, is usually exclusively used for direct consumption or for the purchase of other provisions.

In the flooded forest, manioc flour production is seasonal, with none of the continuous plantations found in the uplands. The manioc is planted when the waters are low or lowering, in August or September, and allowed to mature over a period of six months, when the first stalks are cut (which the locals call 'shaking' the plantation). When the waters rise and threaten to flood the plantations, the residents of the *várzea* try to harvest all the manioc they can and make as much manioc flour as possible so as not to lose any yield.

Between February and March, when they start to clear their plots, the residents will already know their annual farinha production. At this point they will decide how much to keep for domestic consumption and how much to sell. They try to estimate the reserves they will need to at least get through the winter, when the waters are high, but families often need to buy manioc flour to tide them over until the ripening of the following year's crop. The seasonality of production is illustrated below through quantitative data.

The economic context experienced by the residents of Mamirauá can be illustrated by the situation in which one particular resident of Vila Alencar found himself in the month of November, during the dry season. The crop planted in September of the previous year had been 'shaken' in March, after seven months' growth, and the harvest was completed nearly two months later, at the end of April. Due to rising water levels, the manioc had to be harvested in a hurry, working flat out over a period of days until it was all in. The yield was 16 sacks of manioc flour, eight of which were sold. In November, this resident only had half a sack of manioc flour left out of the eight he had kept in reserve. He would therefore have to buy manioc flour to keep him going until the ripening of the next crop, planted only three months earlier, in August. He could only start to plan the harvest once he had some signs of how the waters were going to behave. So long as there was no manioc to eat, he decided to fish to buy the basic household provisions. In the winter of that same year he had bought a motor with the proceeds from the sale of 50 logs. In this example we see a resident who sold manioc he would later lack in his own reserves, whereupon he had to buy in manioc and other provisions with money earned from selling fish yields, but nonetheless sold timber felled specifically for the purchase of an 'extra' asset over and above his baseline provisions.

Limits of Consumption

What limits household consumption, whether baseline or above, is not necessarily satiety of demand, but the productive capacity of the domestic group and the prevailing market conditions. The simplicity and paucity of material assets and the residents' self-description as 'poor' testify to the limits that impinge upon consumption. The adjective 'poor' and its opposite 'rich' are categories of identity that can basically be whittled down to the producer and the patron. The relationship of debt-bondage unifies them and assigns different positions in terms of access to merchandise. In this regional context, being poor means experiencing limited consumption, restricted access to the market and to the benefits of the modern world, such as health and education - but not indigence or hunger.

How much the domestic group can consume depends as much on external factors as upon its own efforts, including the personal choice of economic strategy. The availability of manpower and work tools, access to natural resources and the environmental conditions are the key local variables that determine the household's production capacity, while the most important external variable affecting consumption is the ratio of exchange, the relationship between the prices of products and those of commodities.

Local restrictions upon production derive from the fact that manpower is basically family-based, the tools are predominantly simple, and that the environment imposes seasonal limitations on production. Not only does the activity of middlemen (the patrons) reduce domestic capacity to consume, but both these households and those that sell on the urban markets depend on the external scenario in order to be able to determine the ratio of exchange between their produce and the commodities they consume.

It is still quite common to hear rural extentionists working in the Amazon declare that in the peasant economy, like that at Mamirauá, only surplus production is commercialized, that is, only what is left over after all the household needs have been met. This interpretation that peasant trade is merely an afterthought rests upon the romantic notion that these producers are somehow independent of the market.[14] In this context, the only real surplus is not the slice of production earmarked for sale, but the sporadic consumption of items that exceed the baseline requirements of the domestic group. Besides, this commerce often involves the production of items specifically destined for the market and not consumed by the population. On other occasions, rather than selling surplus, the commercialized produce is actually taken from the domestic stores in order to finance the purchase of essential merchandise. This is especially true of manioc flour, as seen earlier, as other products, such as timber and even some species of fish (like the red-bellied pacu and the pirarucu), are specifically produced for sale. Proceeds from articles produced for the market are used both to subsidise baseline requirements and to purchase more durable assets. According to the consumptive logic of this economic orientation, only the latter can truly be considered extras.

---

[14] Nugent (1993) rejects the pertinence of the concept of 'subsistence' to the characterization of what could be described as an autonomous area of the Amazonian peasant economy. For the author, the opposite is true. The conditions needed for subsistence production only exist where there is a relationship between the domestic economy and the capitalist market.

Quantitative Analysis[15]

The following data on economic production in Mamirauá is based on the results of two socio-economic studies: one isolated sample of 71 households from 24 different localities, interviewed in October 1991, and a monitoring survey that tracked the monthly budgets of 59 households in six localities in Mamirauá (see appendix) over the course of 18 months between 1994 and 1995. The economic indicators directly assess the participation of domestic groups in the market economy and, indirectly, the pressure to use natural resources for commercial production.

Average incomes and expenditure were taken as indicators for the participation of the domestic economies in the market. The study also analysed levels of economic disparity amongst households and locations, and documents the effects of seasonality on domestic budgets.[16] All of the budgetary variables analysed are estimates made for the households, the atomic economic units in Mamirauá.

Household incomes come from two main sources, the sale of produce and from pensions and salaries. The sale of produce is the most important of these sources, representing some 62% of annual monetary income, while monies from pensions and salaries (civil service payments to teachers and health workers, as well as services rendered to the Mamirauá reserve) had grown to 38% of annual family income by the period 1994/1995.

Incomes Generated by Domestic Production

The two studies (1991 and 1994/1995) produced different estimates for the gross annual incomes derived from the sale of domestic production. In 1991, produce sales generated an average gross annual income equivalent to US$ 566 per household. The results for 1994/1995 were higher, with gross average incomes estimated at US$ 888. These estimates correspond to an average monthly income of US$ 47 (1991) and US$ 74 (1994/95).

---

[15] A team from Project Mamirauá helped generate data between 1991 and 1996, a period in which I coordinated the socio-economic research programme. I would like to thank all who participated, especially Marise Reis, Margareth Diógenes and Inês Sousa, for their fieldwork, Aline Azevedo and Edila Moura, for their technical collaboration, and the interns Reginaldo Dias (in memoriam), Fred Cahete, Marília Souza, Márcia Macedo and the Project Mamirauá Database team, for their contributions.

[16] It is important to remember that monetary income does not completely reflect either the total production of the domestic group or the total consumption of the household, as there are various spheres of domestic consumption, the main being that of foods, which are provided directly by family labour. Due to these research limitations, data was collected only for those economic indicators related to market production, which is relatively easier to measure than consumption not monetarily quantified. Data on the consumption of natural resources can be found in the collections by Padoch et al. (1999) and Queiroz and Crampton (1999).

Despite the methodological differences, we believe the disparity in income estimates between the two periods is genuine and reflects real improvement in the economic context. During the period in which the second study was conducted, the Real Plan (decimalization of the Brazilian currency) brought down inflation and, from 1995 onwards, the residents of rural areas began to reap the benefits of better commercial conditions. In addition, the data also reflects the positive effects of the creation of the reserve, as it closed the lakes to outside fishermen and boosted the contribution of fishing as a source of income (see below).

Comparison with monetary incomes for smallholders in other regions of the Amazon floodplains enables us to observe the diversity of income patterns. In Mamirauá, the average income is much lower than that which Anderson and Ioris (1992) estimated for the extractive producers of Ilha de Combu, a floodplain area located only 1.5 km from Belém. In this region, the sale of extractive produce (mainly palmheart, followed by rubber and cocoa) results in gross annual incomes of US$ 3,000 per family.[17] The study conducted by Chibnik (1994) on the economy of the Upper Amazon floodplains near Iquitos in Peru revealed conditions much more similar to those at Mamirauá. The author cautiously estimates gross average income at US$ 628, considering oscillations within household incomes over time and variations from location to location. In the region of Santarém in the Lower Amazon, McGrath, Castro, Câmara, and Futemma (1999) put the gross average incomes of the domestic groups of Ilha de Ituqui at US$ 950.

Methodological differences aside, both studies demonstrate great diversity in income patterns among floodplain populations. Considering the broader picture provided by these studies, which covers a territory from the Upper Solimões to the Amazon Estuary, average gross incomes from the sale of domestic produce range from US$ 500 to US$ 3,000.

As shown by the analyses below, even considering the Mamirauá region alone, there is still a wide spectrum of income values among households from the same locality, between different localities and even within the same household over the course of a single year. The study tracking domestic budgets over the period 1994/1995 demonstrated that average gross annual incomes per locality varied from US$100 to US$ 2,000. The variation is even more expressive if we consider the budgets of individual households; where one could bring in a mere US$ 56 in a single year, another could make US$ 5,800 in product sales. This would indicate enormous differences in economic strategies between households, resulting in considerable differences in their market participation and greater or lesser emphasis on in-house consumption.

[17] However, Combu must be considered an exception. In addition to its proximity to Belém, which enables the residents to sell their palmheart production directly, the economic context of domestic production is very different to that at Mamirauá. The producers tracked on the study did not have free access to natural resources. They were residents on a property and had to pay 50% of their production to the landowner. The real value going toward their domestic budgets was therefore US$1,500.

Sources of Income

In Mamirauá, the products that generate monetary incomes come from agriculture (mostly the cultivation of manioc to make farinha, followed by banana plantations), fishing and timber extraction. The weight of each on income generation varies from household to household and community to community and from year to year. This difference influences the abovementioned variations in income generation, as concentration on fishing or logging produced higher incomes than predominantly agricultural production. The diversity of economic strategies is another characteristic of the domestic economy on the floodplains, as also identified in other studies (Anderson & Ioris, 1992; Chibnik, 1994; Harris, 2000; Hiraoka, 1992; Lima Ayres, 1992; Nugent, 1993; Padoch & De Jong, 1992).

Comparison between the data sets from the two studies reveals that there was a change in the order of importance of the products destined for sale. In 1991, the domestic production that contributed most to domestic income generation was agriculture (representing 44% of incomes obtained from the sale of produce), followed by timber extraction (37%) and fishing (12%). However, by 1994/95, fishing had overtaken agriculture as the key source of income, representing 67% of the monies obtained from domestic produce, followed by agriculture (17%) and timber extraction (11%). Box 1 (annexed) presents the species contemplated in the categories 'fishing', 'timber', 'agriculture' and 'other products' sold by the 59 households accompanied over the period 1994/95.

The greater domestic income share represented by fishing during the period 1994/95 is due to the favourable environmental conditions that prevailed at the time, as the floods of the previous years had been relatively high, allowing fish to migrate to the lakes. It also reflects the positive results the closing of the lakes to commercial fishermen brought to the Mamirauá users.

Fishery production basically revolves around the pirarucu (*Arapaiama gigas*), red-bellied pacu (*Colossoma macropomum*) and some siluriformes (Pimelodidae), which together account for more than 80% of the species sold. The wood species most commonly commercialized during the period were Pimientillo (*Xylopia frutescens*) and Kapoc (*Ceiba pentranda*). The main agricultural product for the same period was manioc flour, while alligator meat was the most commonly sold item.

No comparison between the two studies is possible for the participation of pensions and salaries, as they focus essentially on domestic production, i.e., on the use of resources and the incomes generated by their sale. In our studies, data on salaries and pensions received by householders was only collated in 1994/95.

Of the 48 households in five localities whose domestic budgets were tracked for over 12 months, 19 (39.58%) did not receive any salaries or pensions, while 9 (18.57%) received irregular salaries and 20 (41.67%) could count on regular salaries or pensions. In relation to the 20 houses with composite incomes, the salaries and pensions accounted for 72% of total income, with the remaining 28% coming from the sale of domestic produce. With the 19 households

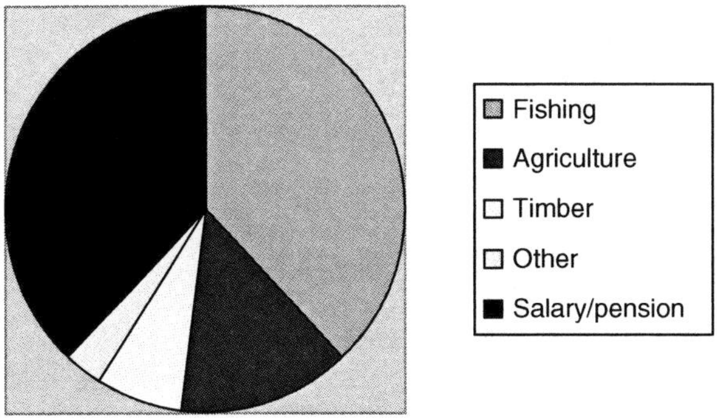

Fig. 6.1 Composition of domestic incomes in 59 households

without salaries or pensions, incomes generated exclusively from the sale of domestic production yielded average monthly incomes of US$ 47, compared with the average monthly incomes of the salary/pension earners of US$ 199, with the sale of domestic produce accounting for US$ 76 of that. In other words, the domestic yields of households with composite incomes was almost four times those based on production sales alone, with the latter accounting for 1.6 times more than for non-salary/pension earners.

If we calculate all of the budgets from the survey to obtain a general estimate of average annual salary for the period 1994/95, we will see that the contribution of salaries and pensions to annual domestic income was US$ 552. Add to that the sums brought in by the sale of produce and the average annual domestic income rises to US$1,440.

The importance of salaries and pensions in the composition of domestic income is therefore highly expressive (Fig. 6.1). In some localities (particularly Vila Alencar and Jarauá), this result was influenced by the activities of Project Mamirauá, as some of the researchers had hired locals as field assistants on a temporary or full-time basis. However, the most prevalent sources of salary were teaching and health posts, with salaries paid by the municipal government, and the rural pensions claimed from the State by residents over 60 years of age and those with disabilities.

These data shows the importance of rural pensions to the domestic groups of Mamirauá. The sale of domestic produce continues to be the main source of income, accounting for roughly 62% of domestic income. However, this is a low rate if we consider that it represents by far the greatest work effort of the adults and children of the household.

Patterns of Consumption

The main category of consumption that comes directly from domestic production is food. The staple daily meal (fish and manioc flour) is entirely home produced. Other foods consumed are fruits gathered from the surrounding forest and those cultivated on the family plot.[18] Besides foodstuffs, various artefacts and household utensils are made from forest materials.[19]

Commodity consumption includes both articles not produced locally and items that are, with even fish and manioc flour being purchased occasionally. Though the range of merchandise is much broader than that of domestic produce for home consumption, the volumes bought are much lower. Pasta or rice, or even a packet of biscuits are rarities. Box 2 (appendix) illustrates the diversity of articles purchased by the households.

The expenditures of the residents of Mamirauá were studied based on a sample of 754 monthly domestic budgets. The analysis of this data estimated average monthly purchases at US$70. This expenditure goes almost entirely toward buying foodstuffs.

As with the home-produced items, commercial articles purchased during the scope of the study were classified according to their finalities.[20] The category 'rancho' (rations) includes the staple items of consumption, the basic household provisions, while 'gastos' (expenditures) includes food items, cleaning materials, beverages and tobacco products, items of domestic and personal use, personal hygiene and medication. The sums spent on these items account for 59% of total purchases.

Items outside the sphere of expenditures generally involve clothing and household goods, which includes everything from basic utensils to articles of higher value, classified here as domestic assets. Spending on these items amounts to 20% of total outlay.

Production costs, which includes spending on tools and maintenance of equipment, constitutes a mere 7% of total outgoings. Fuel expenses, for example, represent 8% of the total. There are few purchases of building and school materials, representing a mere 2% and 1% of expenditure, respectively. Leisure, community contributions and unspecified or 'other' expenses each account for 1% of total spending. Figure 6.2 illustrates this breakdown in expenditure.

---

[18] The diet in Mamirauá contains such fruits as abiu, avocado, bacaba, bacury, palmheart and various types of banana (baré, long, spike, guariba, blue guariba, apple banana, miranha, inaja, pacova, silver, reboio, São tomé, frog, tree-bunch, and urucury), cherimoya, Brazil nuts, cocoa, coconut, cupuaçu, guava, ingá fruit-bean, jackfruit, lemon, mango, melon, orange, patauá palm fruit, pupunha, sapota, sugar-cane, tucuma and watermelon. Also grown for home consumption are corn, gerimum pumpkin, manioc, necklace pod bean, peach tomato, squash, sweet pepper, tobacco, tomato and water yam, among other products.

[19] Such as assorted baskets, axe handles, bellows, brooms, canoes, clay basins, clay pots, clay stoves, dough boxes, earthenware, egg whisks, fishing rods, gourd cups, hats, huts, jacitara cylinders, lashes, machetes, oars, oil presses, mats, sieves and straw-thatched walls and roofs. In order to make these objects, the following are gathered from the forest: arumã, assorted vines, bamboo, bast fibres, cauaçu, cauixi to make ash, croton, guava, straw from the babaçu, cabbage palm, coco-palm, jacitara, jaurai palm, rasp palm, ubim, fava beans, lath, silk cotton and various types of white and heavy woods.

[20] Almeida (1992, p. 122) analyses the system used by rubber-tappers in Acre to classify merchandise. The foundation for classification is, primarily, the distinction between durable and non-durable merchandise, and, secondarily, the distinction between basic (essential) and non-basic ('luxury') items.

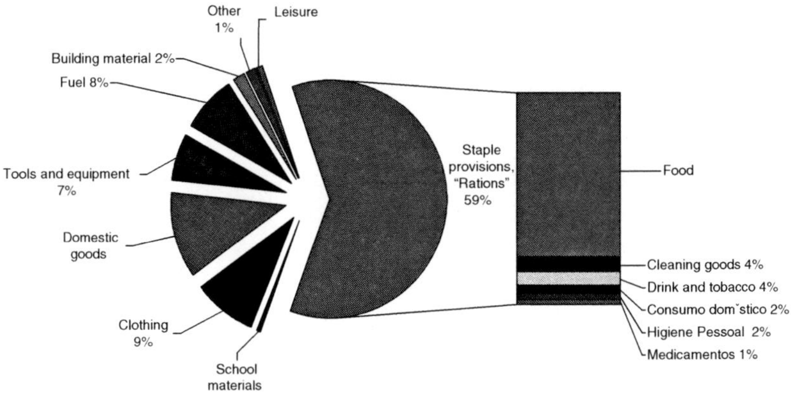

1. Leisure, 1%
2. Other expenses, 1%
3. Building materials, 2%
4. Fuels, 8%
5. Tools and equipment maintenance, 7%
6. Domestic assets, 11%
7. Clothing, 9%
8. School materials, 1%
9. "Rations", 59%

a) Foodstuffs, 47%
b) Cleaning materials, 4%
c) Beverages and tobacco products, 4%
d) Domestic consumption, 2%
e) Personal hygiene, 2%
f) Medication, 1%

Fig. 6.2 Composition of monthly domestic expenditure for 754 households in five communities

Though 'rations' are responsible for the lion's share of domestic outlay, the merchandise purchased is very simple and the patterns of consumption extremely regular, with very little variation between households. The five items most commonly bought are, in order of importance: sugar, coffee, bars of soap, cooking oil and powdered milk.

Table 6.1 presents the 15 items that feature most frequently among household purchases, along with the quantity and average monthly values spent on these articles during the period 1994/95.

Total production costs represented a small portion of overall expenditure throughout the period of the study (a mere 7%). Of the productive activities, fishing accounted for the highest expenses through tackle and fish storage.[21] The second highest costs were incurred on spare parts and motor

[21] Hooks, harpoons, harpoon heads and chord, arrow heads, polystyrene boxes, ice, fishing rods and line, nets, nylon and spears.

Table 6.1 List, in descending order, of the 15 items most frequently bought and their average quantities and purchase prices (59 households, 1994/95)

Product bought	Occurrence in purchases (%)	Monthly quantity	Price in US$
Sugar	61	10 kg	8.13
Coffee	56	414 g	3.65
Bars of soap	53	1.6 kg	2.46
Oil	40	1 can	2.29
Milk	40	285 g	2.74
Matches	36	8 boxes	1.24
Tobacco	33	2 oz	1.63
Salt	32	7 kg	3.44
Wheat	32	1.5 kg	1.91
Flour	30	10 kg	10.72
Powdered soap	24	0.4 boxes	0.69
Biscuits	24	528 g	1.31
Rice	24	0.7 kg	0.78
Gasoline	22	4 litre	2.92
Batteries	22	1 unit	1.30

maintenance.[22] These are transport-related costs that feature in all productive activities and subsequent commercialization. All told, these two categories represent 82% of direct production costs.

Agricultural[23] production and hunting[24] costs account for 14% of total expenditures. No expenses from logging were recorded, either because these are often covered by the patron, or because they were listed under 'motor maintenance'.

Economic Variability on the Reserve

The main cause of economic variability among the different localities is the location of the village and the abundance of natural resources in the vicinity. As fishing and logging are generally more profitable than agricultural activities, the communities with readier access to these resources tend to obtain higher incomes.[25] Other factors that can influence local income patterns include

[22] Burnt oil, caulk, drums, gaskets, grease, hosing, lubricating oil, motor bottoms, oars, pebble pumps, spanners, spark plugs and turbine vanes.

[23] Emery wheels, grating benches, grindstones, hoes, jacitara cylinders (for manioc juice), machetes, ploughs and sieves.

[24] Cartridges, gun caps, gun powder and pellets.

[25] Though the financial return on fishing is higher than for agriculture, the latter is more socially valued in Mamirauá. Agriculture is considered 'work' because of the human involvement in production and because it requires more time and energy than fishing. Full-time dedication to fishing is frowned upon for precisely the opposite reasons. Harris (2000)

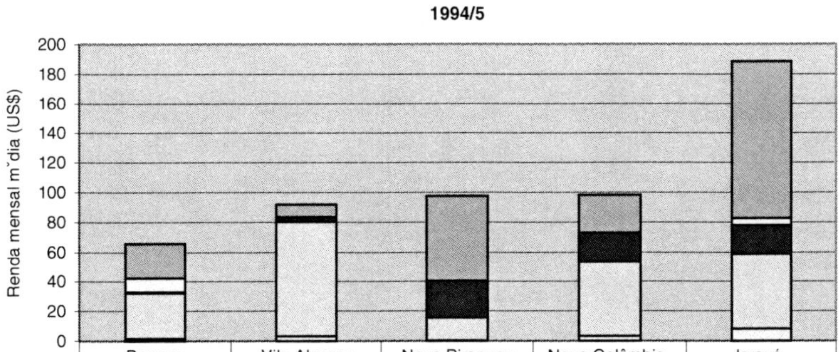

	Barroso	Vila Alencar	Novo Pirapucu	Novo Colômbia	Jarauá
Fishing	23	9	57	26	106
Wood	9	0	0	0	5
Agriculture	1	3	25	19	19
Salaries and pensions	31	77	15	50	50
Other products	2	3	0	3	8

Fig. 6.3 Composition of average monthly incomes for households in the five communities

individual aptitudes, family traditions, access to work tools, manpower, community organization, cooperation between households and trade conditions.

Figure 6.3 shows the variations in patterns of income generation in five communities; Barroso, Vila Alencar, Nova Colômbia, Novo Pirapucu and Jarauá during the period 1994/95. The estimates of average monthly incomes are broken down into sale of produce (per product type) and salaries and pensions.

The highest monthly income estimate was for Jarauá: US$ 194, nearly three times the lowest estimate – US$ 67, for Barroso. While the remaining three locations presented similar averages – in the region of US$ 90 – they varied in the ways they generated this income. In addition to differences in the weights of fishing, logging and agriculture among these communities, the contributions of salaries and pensions also varied, from 15% in Novo Pirapucu to a staggering 85% in Vila Alencar.

The income generated in Jarauá from the sale of production alone (US$ 143) was higher than total incomes for the other four localities in the sample. This disparity concerning Jarauá largely stems from its location near a lake system abundant in fish and the enterprising behaviour of the residents. Barroso, on the other hand, is situated far from the nearest urban centre, so its residents depend almost exclusively on middlemen in order to sell their extractive produce.[26] In

discusses these conceptions of value among the residents of a floodplain community in the Lower Amazon.

[26] Extractive production (fishing and logging) limits agricultural production and causes the residents to include the purchase of flour among their domestic expenses. Hence the vicious circle: agricultural production is hampered by the extractive activities needed to pay the patron for merchandise that includes the flour they cannot afford to produce themselves because they have to devote their time to extractivism.

Vila Alencar, where salaries and pensions are the main source of income, on the other extreme to Jarauá, the sale of production yielded a paltry US$ 14.

While the study sites presented significant differences in the configuration of their sale of produce, there was little variation in the order of the relative contributions of each core product.

As one would expect, variations in consumption patterns between the areas surveyed largely came down to their different levels of income. The lowest monthly outlay was in Barroso (US$ 45), followed by Vila Alencar (US$ 47), Nova Colômbia (US$ 60), Novo Pirapucu (US$ 60) and Jarauá, with much higher outgoings (US$ 123). Foodstuffs were the main expense at all of these locations, but while this accounted for over 60% of total costs in the first four localities, the share was as low as 37% in Jarauá.

The percentage of monthly income destined for foodstuffs is a clear indicator of living conditions - the higher this percentage, the closer the community is to the survival baseline. A lower percentage of outlay going toward food supplies indicates that survival needs are being met whilst leaving room for other expenses. If we take consumption patterns in Jarauá as an example, where spending on food is 37% of an average monthly expenditure of US$ 123, we can deduce that a monthly average income of US$ 200 would afford the local population comfortable living conditions by the consumption standards of Mamirauá.

In relation to spending on production-related costs, those localities more deeply engaged in agricultural activities (Vila Alencar, Nova Colômbia and Novo Pirapucu) used 4% of their budgets to purchase work tools and pay for equipment maintenance. In Barroso and Jarauá, where extractive activities predominate, expenses on this category were higher, at 9 and 12% of domestic budgets, respectively.

Income Variations among the Households

Variations in household incomes were analyzed for a selection of 48 homes that kept the most complete monthly records. These households were distributed across six income brackets based on the lowest incomes for the period (R$ 100 was on parity with US$ 100 at the time). The distribution of household incomes is presented in Fig. 6.4 below.

As we can see from the graph above, 58% of these households had average monthly incomes below the minimum salary rate for the period (31% below half the minimum salary and 27% over the half but below the full minimum). Nineteen percent of these households obtained average monthly incomes of between one and two minimum salaries, while 23% earned more than two minimum rates. In reality, however, the income of any given household varied greatly over the short and long-terms. Over the long-term, a family's income pattern will vary depending on the stage of life of its members, influenced by such demographic factors as the number of people old or young enough to

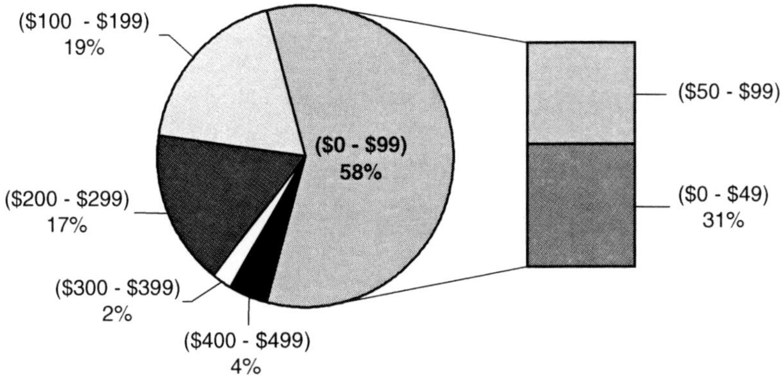

Fig. 6.4 Distribution of average monthly income (N = 48 households in 5 localities)

participate in the productive tasks, the number of dependents and the age of the couple at the head of the household. Over the short-term, domestic incomes suffer the impact of the seasonal changes inherent to the floodplain environment, as we shall now see.

Effects of Seasonality on Domestic Incomes in Mamirauá

In the Middle Solimões, the difference in water levels between the lowest ebb and the highest flood is somewhere in the region of 11 to 12 metres. As in other regions of the *várzea*, in Mamirauá, the movements of the waters determine the productive calendar. The alternateness of the wet and dry seasons affects production conditions and this is reflected in the household's patterns of income and consumption.

The year is divided into winter and summer, as they call the rainy and dry seasons, but the dates of the beginning and end of each season and the height the floods reach are not entirely predictable. Summer generally occurs between August and February, with winter ranging from March to July. The plantations are '*desmanchadas*' or 'cleared' (harvested) when the waters rise, which is when the sale of manioc flour is most frequent. Timber is felled toward the end of summer and the logs are transported and payment made in the winter. Fishing is more intense during summer, which is also when the plantations are sowed.

The effects of seasonality on the domestic economy in Mamirauá are striking, with stark variations in incomes from the sale of manioc flour, fish or timber from month to month. Though there is a certain complementarity between the three *fábricos*, income and, consequently, expenditure are still affected by environmental changes. Figure 6.5 shows how income falls in winter and peaks at high summer. This oscillation is largely owing to variations in the

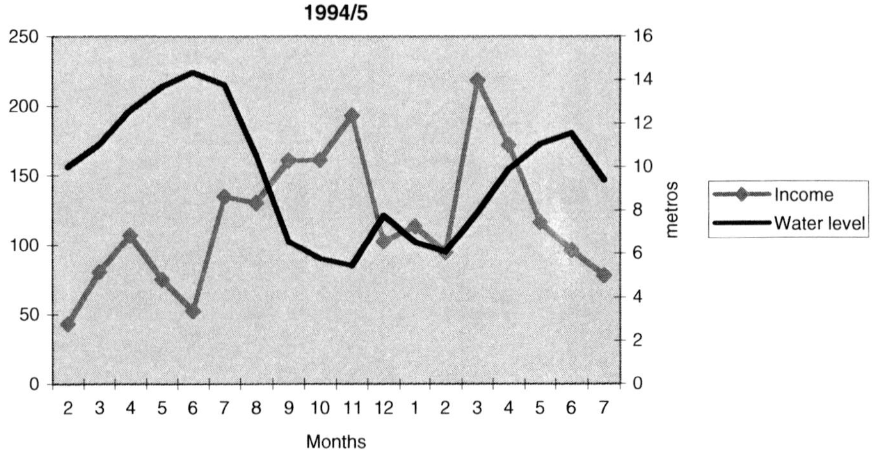

Fig. 6.5 Variations in water levels and average monthly incomes in five villages

proceeds of fishing, the most influential activity for these households during the period of study.

The unpredictability of the rise and fall of water levels makes conditions of life on the floodplains unstable. Agriculture, fishing and logging cannot be planned with precision, as the environmental conditions can hinder as much as they can help each activity in different ways. In agriculture, the size of the plantation is limited by its elevation and the availability of manpower to bring in the harvest before the floods. Each year the plantations have to be remade because of the flooding, and if the inundations are particularly severe, fruit trees and livestock can be lost. Timber, which is felled in summertime, depends on whether the waters reach the landing patio where the logs have been prepared for transport. If the waters do not come, the wood is lost. In relation to fishing, the lakes need to be fed by high waters in order for fish stocks to replenish. If the floods are weak, fish yields may fall. The result of this scenario is the constant burden of imposed risk and an unending search for better economic conditions.

Final Considerations

The domestic economy in Mamirauá is an economy of exchange in the sense prescribed by Weber (1968, p. 12) of being 'oriented, in general terms, by the fact that some goods are offered in return for certain others'. It is not, therefore, a natural economy, where money is unknown, or even a natural barter economy, in which exchanges of non-monetary goods are calculated in monetary terms. The unit of this domestic economy is, as the name suggests, the household, which consists in a nucleus of family relatives working toward the production and reproduction of their kin and livelihoods. In its relationship with the market, the household sees the prices of what it produces and consumes set for it as per a ratio monitored by the identification of the relative proportions between produce and commodity. The monetary income from the sale of products goes toward 'expenses' or 'provisions' for the household, the main category of this domestic economy in terms of consumption, though this will also be supplemented by home-grown produce. Dependence on an unfavourable pricing system, which keeps household incomes low, impairs the accumulation of domestic patrimony and shifts the emphasis onto non-accumulative consumption. Expenditure is characterized by a very simple composition based around staple items, the 'rancho' (rations), which is basically made up of foodstuffs, despite the fact that the staple meal (fish and manioc flour) is produced locally. Economic planning beyond this essentially replenishment-based pattern of consumption revolves around the prior identification of a specific item of value to be acquired (such as a motor or gas stove) and on the balance of prices, which will inform the choice of productive activity to be used toward that end. Besides such target-specific economic planning, the varying performance of the household economy also depends on the following variables: availability of, and access to, natural resources; proximity to urban markets and the level of dependence upon middlemen; and the composition or life stage of the domestic group (seen as rural pensions tend to reformulate the situation of households with elderly family members). As this is a floodplain area that experiences a great variation in water levels, the household economy is also subject to the annual cycles of production. The seasons are responsible for stark differences in income, with highs during the dry season and lows during the rains.

References

Alencar, E. (1994). *Memórias de Mamirauá*. Manuscript.

Alencar, E. F. (2002). *Terra caída: encante, lugares e identidades*. Tese Ph.D. Dissertation, University of Brasília, Department of Anthropology, Brasília.

Almeida, M. W. B. (1992). *Rubber tappers of the Upper Juruá River, Brazil: the making of a forest peasant economy*. Ph.D. Dissertation, University of Cambridge, Cambridge, Inglaterra.

Anderson, A. B., & Ioris, E. M. (1992). The logic of extraction: management and income generation by extractive producers in the Amazon. In: K. H. Redford, & C. Padoch (Eds.), *Conservation of neotropical forests: working from traditional resource use*, (pp. 175–199). New York/Oxford: Columbia University Press.

Chibnik, M. (1994). *Risky rivers: the economics and politics of floodplain farming in Amazonia*. Tucson: The University of Arizona Press.

Garcia JR., (1983). A. *Terra de trabalho*. Rio de Janeiro: Paz e Terra.

Gudeman, S., & Rivera, A. (1990). *Conversations in Colombia: the domestic economy in life and text*. Cambridge: Cambridge University Press.

Harris, M. (2000). *Life on the Amazon*. Oxford: Oxford University Press.

Hiraoka, M. (1992). Caboclo and Ribereño resource management in Amazônia: a review. In: R. H. Redford & C. Padoch (Eds.), *Conservation of neotropical forests: working from traditional resource use*, (pp. 134–157). New York/Oxford: Columbia University Press.

Lima Ayres, D. (1992). *The social category 'caboclo': history, social organization, identity and outsiders identification of the rural population of an Amazonian region*. Ph.D. Thesis, University of Cambridge, Department of Social Anthropology, Cambridge, England. 342.

Lima, D.; Moura, E. (1995). *A reprodução social de grupos domésticos em comunidades do Mamirauá*. Projeto Mamirauá – Mamirauá Project – Report n°6, manuscript.

Lima, D. (1999). A construção histórica do termo caboclo: sobre estruturas e representações sociais no meio rural Amazônico. *Novos Cadernos do NAEA*, *2*, 5–32.

Lima, D., & Alencar, E. (2001). A lembrança da história: memória social, ambiente e identidade na várzea do Médio Solimões. *Lusotopie 2001*, 27–48.

McGrath, D., Castro, F., Câmara, E., & Futemma, C. (1999). Community management of floodplain lakes and the sustainable development of Amazonian fisheries. In: C. Padoch, J. M. Ayres, M. Pinedo-Vasquez, & A. Henderson, A. (Eds.). *Várzea: diversity, development, and conservation of Amazonia's whitewater floodplains*. New York: NYBG Press.

Murrieta, R. S. S. (1998). O dilema do Papa-Chibé: consumo alimentar, nutrição e práticas de intervenção. *Revista de Antropologia da Universidade de São Paulo (USP)*, *41*, 97–150.

Murrieta, R. S. S. (2001). Dialética do sabor: alimentação, ecologia e vida cotidiana em comunidades ribeirinhas da Ilha de Ituqui, Baixo Amazonas, Pará. *Revista de Antropologia da Universidade de São Paulo (USP)*, *44*, 39–88.

Nugent, S. (1993). *Amazonian caboclo society: an essay on invisibility and peasant economy*. Providence/Oxford: Berg.

Padoch, C. & De Jong, W. (1992). Diversity, variation and change in ribereño agriculture. In: K. H. Redford & C. Padoch, (Eds.), *Conservation of neotropical forests: working from traditional resource use*, (pp. 158–174). New York/Oxford: Columbia University Press.

Padoch, C., Ayres, J. M., Pinedo-Vasquez, M., & Henderson, A. (Ed.). (1999). *Ecology, conservation and development of Amazonian várzea*. New York: The New York Botanical Garden Press.

Queiroz, H., & Crampton, W. (1999). *Estratégias para manejo de recursos pesqueiros em Mamirauá*. Brasília: Sociedade Civil Mamirauá.

Sabean, D. W. (1990). *Property, production, and family in Neckarhausen, 1700–1870*. Cambridge: Cambridge University Press.

Santos, R. (1980). *História econômica da Amazônia (1800–1920)*. São Paulo: T. A. Queiroz,.

Santos, R. O (1989). *Genius* de uma economia: reflexões e propostas sobre o desenvolvimento da Amazônia. In: *Populações Humanas e Desenvolvimento Amazônico, Série Cooperação Amazônica 3*, (pp. 13–92). Belém: OEA, UFPa.

Weber, M. (1968). *História geral da economia*. São Paulo: Editora Mestre Jou.

Chapter 7
Patterns of Resource Use by Caboclo Communities in the Middle-Lower Amazon

Fábio de Castro

Abstract In his chapter, Fábio de Castro provides a detailed analysis of the diversity in the economic strategies deployed by riverine households in the Lower Amazon. The empirical data used in his analysis was collected by means of the Community Statistical Census (CEC), a participative methodology based on community meetings, having encompassed 172 communities and 8,570 households in the region under focus. The economic strategies of the Lower Amazon floodplain populations studied by the author combine four main economic activities: fishing, agriculture, cattle-raising and waged work/pension. A large part of Castro's article is dedicated to the analysis of the combination of these four main activities in distinct communities and in households of a single community, seeking to verify which factors influenced in the different combinations of these four economic pillars. In both cases, the greatest determinants of this combination variation are the activity's 'economic purpose', 'each family unit's structure' and the 'degree of access to resources'. The socio-economic picture that emerges from this exercise is a picture of great heterogeneity and not the homogeneity suggested by several authors.

Keywords Household · Community · Floodplain · Agriculture · Fishing · Cattle ranching

Introduction

Studies on resource use in the Amazon are usually focused on three main categories of social groups (Amerindians, *caboclos*, and settlers) and two ecological systems (floodplain and upland) (Adams, 2002; Moran, 1993). The

F. de Castro (✉)
Centre for Latin American Research and Documentation (CEDLA), Amsterdam, The Netherlands
e-mail: f.decastro@cedla.nl

C. Adams et al. (eds.), *Amazon Peasant Societies in a Changing Environment*,
DOI 10.1007/978-1-4020-9283-1_8,

European colonization in the Amazon, the establishment of the Directorate (state led development during the colonial era), and the implementation of the National Integration Program (PIN) are important historical milestones used to explain the distinctions among the three social groups, respectively. The Amerindians, who inhabited the floodplains at the time of the Europeans colonization, were nearly annihilated during the sixteenth and seventeenth centuries due to epidemics, massacres, and slavery, forcing the surviving these groups out to remote upland areas (Roosevelt, 1989). The establishment of the Directorate, in 1757, led to the intensification of the intermarriage between Amerindians, Europeans, and later with Afro-Brazilians, and in turn, to the establishment of the caboclo populations who gradually re-settled along the rivers (Moran, 1974; Parker, 1985). Later on, migrants from Northeast Brazil, drawn to the Amazon during the rubber boom in the late nineteenth century, contributed to the caboclo culture (Weistein, 1983). Finally, the government-sponsored development projects initiated in the 1960s, triggered another immigration wave to the Amazon who settled in farm lots along the roads (Bunker, 1985; Moran, 1981).

The brief description above of the human occupation history in the Amazon holds a few assumptions regarding the relation between broad social and ecological categories. For example, caboclo populations are often defined as inhabitants of riparian communities who rely their livelihood on floodplain resources whereas Amerindians and settlers are described as upland residents who base their production on upland resources. These generalization hides local social variability across social groups and use of ecological systems. The broad landscape categories of floodplain and upland also limit the understanding of the role of socioenvironmental factors on the economic strategies performed by the different social groups. The assumption of high productive floodplain soils in comparison to low productive upland soils masks significant intra-regional environmental heterogeneity (Moran, 1995; Murrieta et al., 1999), as well as variability in economic responses by local households (Padoch & de Jong, 1992). Moreover, assumption of floodplain populations as aquatic-base resource users vis-à-vis their upland counterpart as land-based resource users overlooks the remarkable importance of the ecological and economic connections between both ecological systems (Futemma et al., 2002; Winkler Prins, 2002).

The focus on aggregrated social and ecological systems is a reflex of theoretical development of the human ecology of the Amazonian populations. Based on the assumption of ecosystems relatively isolated from external social influences, several authors have applied the cultural ecological approach to test how local factors influenced the adaptive process in the natural resources use by the Amerindian populations (Hames & Vickers, 1983; Posey & Balee, 1989). In contrast, the close connection with market demands, public policy, and technological change, led the application of the political economy approach to suggest connections between external factors and the mal-adaptive behavior regarding forest resource use by the migrant settlers (Moran, 1981; Schmink & Wood, 1992). Both theoretical approaches helped to unveil a range of factors influencing different patterns of natural resource use in the Amazon. However, the use of different theoretical approaches to study resource use patterns by Amerindians and settlers has masked the variability of response within each group according to the dialetics between local and non-local factors. Amerindians are treated as hunter-gatherers and subsistence horticulturalist societies, adapted to the local environment and isolated from a broader society. Settlers are often described as migrant farmers-ranchers, mal-adapted, and influenced by external pressures only.

Unlike the studies focusing Amerindians and settlers, the theoretical approach used to explain the behavior of caboclo populations towards resource use relied heavily on historical and ecological contextualization. The clear connection between the local environment and external pressures, as well as the historical continuity linking the caboclo populations with Amerindians and settlers, are features limiting the placement of this group in a stereotypical social category. These particular features are also responsible for the invisibility of the caboclos, until recently, in the development of explanatory models of human-environment relationship in the Amazon (Parker, 1985; Nugent, 1993). Only in the 1990s, the interest on the caboclo populations became more evident to develop analytical experiments in order to fulfill the theoretical demand for a diachronic, integrated, and regionalized approach to understand the complexity of the resources use patterns in the Amazon (e.g., Brondízio & Siqueira, 1997; Castro, 2000; Futemma, 2000; Lima, 1992; McGrath et al., 1993; Murrieta, 2000, 2001; Nuggent, 1993).

The caboclo population as a specific social category has been fundamental to provide a better understanding of the historical process of human occupation in the Amazon. The acknowledgement of this social group, however, has occasionally let to the idealization of a 'traditional' group sharing history, social organization, and economic strategies (Barreto Filho, this volume). The distinction between this social group and Amerindians and settlers should not conceal the diversity in its history (Guzman, this volume), occupation patterns, and economic strategies ranging from Brazil nuts harvesting and rubber tapping in the Western Amazon (Allegretti, 1990), to agriculture in the Upper Solimões (Lima, 1992), fishing in the Lower Amazon (McGrath et al., 1993) and açaí harvesting in the estuary (Anderson & Ioris, 1992; Brondizio & Siqueira, 1997).

In sum, the generalization between broad social groups (e.g., caboclos), and ecological systems (e.g., floodplain) must be balanced with social and ecological subcategories which influence in the variability of local responses (Murrieta et al., 1999). Moran (1995) calls the attention for the disaggregation of ecological categories in order to avoid incorrect generalizations regarding productivity of the Amazonian ecosystems. Likewise, the disaggregation of social groups is necessary to restrain from assumptions related to resource use strategies.

This article aims at providing an analysis of the diversity in economic strategies across caboclo household in the Lower Amazon who rely on floodplain resource for their livelihood.

Methodology

This article is based on the economic activities of the households who relied at least partially upon the floodplain system for their livelihood. The study area encompasses 172 communities located in an area of approximately 4,500 km^2 in the Lower Amazon, distributed in four municipalities (Santarém, Monte Alegre, Alenquer, and Óbidos) (Fig. 7.1).

The data was obtained between 1993 and 1997, by applying the Community Census Statistics (CEC). The CEC is participatory methodology developed by Isaac et al. (1999) and is based on community meetings of approximately 4 hours long in collective spaces such as a school, community shelter or simply outdoors.

The CEC was carried out by an interdisciplinary team of researchers and practitioners and encompassed activities such as participatory community mapping, and questions on infrastructure, history, and social activities in the community. Further, the community map was divided in different sectors and smaller groups of residents living in each sector provided information on household structure and socioeconomy in that sector.

The number of participants in the CEC's ranged from a few individuals to almost the whole community. Group discussions among the participants, combined with the approachable questions ensured the reliability of the information. The intensity of each economic activity, however, could not be quantified due to categorical responses (yes/no). This chapter is based on the questions related to the household economic activities (see Castro, 2000 for a more detailed discussion including community infra-structure and history).

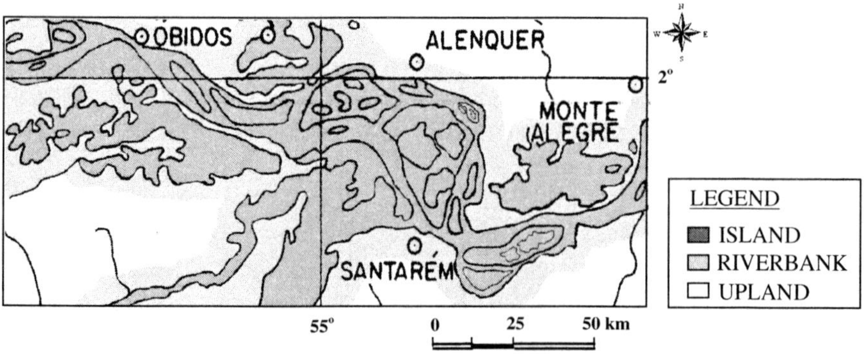

Fig. 7.1 Study area

Results

Economic Activities

The economic strategies of floodplain populations combine four major activities: fishing, agriculture, ranching and wage-based activities (Fig. 7.2). Each activity has a distinct pattern throughout the year, involves different members of the household, and fulfills specific purpose in the household economy. Fishing represents the main commercial activity, whereas agriculture is subsistence-oriented. Cattle ranching provides the households with a capitalization system, and wage-based income offers additional financial support for small expenses and investments on cattle, boats, and house building.

Fishing is currently the main activity in the floodplain, in which 73% of the households are engaged for either subsistence or commercial purposes. Subsistence fishing provides the main protein source for the floodplain population with estimated average fish consumption of 379 g/capita/day observed in 17 of the studied communities (Cerdeira et al., 1997).

The relatively low percentage of households performing commercial fisheries (29%) reflects the vague definition of commercial and subsistence fishing. Although fishing strictly for consumption and eventual distribution of surplus to relatives and neighbors do occur in some cases (Murrieta, 2000), fish production is commercialized whenever possible. Furthermore, the significance of fish commercialization can vary throughout the year and between years in the same household. Therefore, the percentages of commercial fishing in the

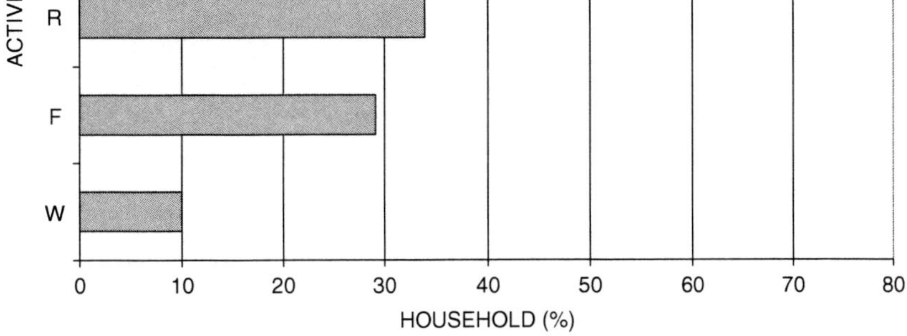

Fig. 7.2 Relative frequency of the major economic activities performed by households in the studied area (S = subsistence fishing; A = agriculture; R = ranching; F = commercial fishing; W = wage labor/pension) (number of households = 8570)

region could be interpreted as a trend of households engaged in a more intensified commercial fishing.

Agriculture is the second most common activity in the studied area, performed by approximately 60% of the households. The intensity of farming activity could not be determined due to categorical responses. However, the analysis revealed a few important farming activity patterns. For example, annual crops predominate in the farming strategy while perennial crops are mainly cultivated in communities with direct access to uplands (see next section) (Fig. 7.3). Bitter manioc, maize, and beans are the three most frequent crops in the floodplain, each one cultivated for a different purpose. Bitter manioc, processed into flour and other by-products, has a key role in the local diet (Adams et al., this volume). The local population cultivates a special variety of manioc whose growth cycle of approximately six months allows the use of seasonally flooded terrain. Maize is used primarily to feed small livestock and can conveniently be stored for several months and used throughout the year. Beans became one of the most important cash crop in the floodplain since the mid-1980s. Fast cultivation cycle (about three months), suitability to well drained lowland soils, uncomplicated storage, and relatively high market price are some of the advantages of this product in the region (Futemma, 1995; WinklerPrins, 1999).

Both small and large livestock are found in the floodplain. The former consists mainly of chicken and duck, raised by 35% and 15% of the households, respectively. Small livestock play a significant role in the local diet during fish scarcity periods, and cash raising in emergency situations

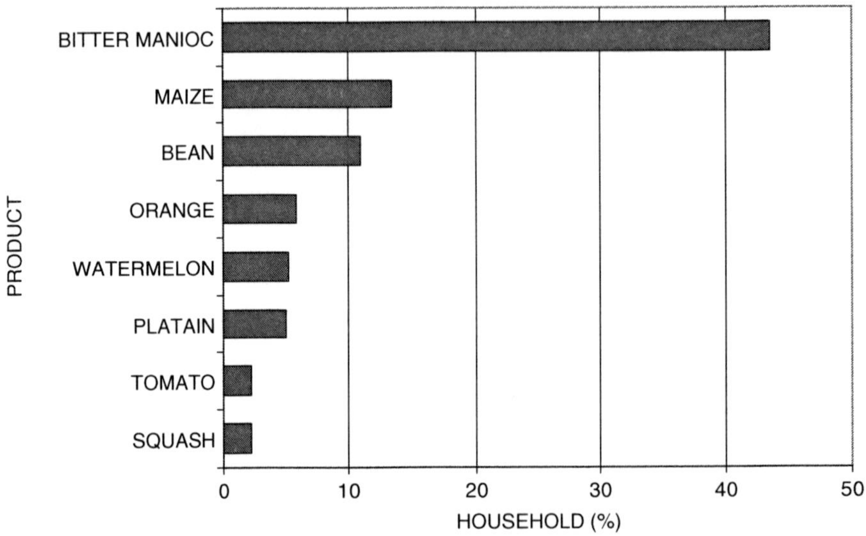

Fig. 7.3 Agricultural products cultivated by more than 2% of the floodplain households in the studied area (number of households = 7429)

(Adams et al., this volume, Murrieta & Dufour, 2004). The low labor demand associated with small livestock enables children, elderly people, and women to perform this activity in combination with other household tasks. In contrast, large livestock is labor intensive and limited to adult males. Cattle are the main large livestock in the floodplain, raised by 27% of the households, followed by water buffaloes at smaller scale (2%). Although cattle raising has been a traditional activity in the floodplain since the nineteenth century, the community residents engaged in this activity only after 1950, and has become more widespread after 1980 (Castro, 2002). Cattle are particularly important for the residents as a way to save money. The price of meat, which follows (or surpasses) the inflation rate, relatively high liquidity, and the abundant, accessible grassland are some of the factors driving floodplain residents to invest their surplus income in this activity. In addition, the insertion of the local residents in this activity has been facilitated by a partnership system developed in the region. The so-called cattle ranching partnership is an informal agreement in which outside cattle owners send their herd to the floodplain and the local residents who take care of the herd receive half of the born calves as part of their payment.

The increasing cattle activity in the floodplain is clearly observed in the distribution of cattle herds among the households who own cattle in the region (27% of the sample). The cattle ranchers can be divided in three groups according to the size of their herd: small-scale ranchers (16%) with up to 10 heads, medium-scale ranchers (8%), between 11 and 50 heads, and large-scale ranchers (3%), with more than 50 heads. Figure 7.4 reveals a large percentage of small-scale ranchers (60%) who own only 10% of the total cattle herd, in contrast to large-scale ranchers with opposite values (10% of the households who own 62% of the total cattle herd). Middle-scale ranchers is of particular interest as they represent an emergent group among the local residents who has been successful in accumulating capital during the jute boom and currently own small cattle ranches. This group usually occupies a social position of local elite in the communities. Often, the middle-scale ranchers are seen as local patrons who hold influential economic, social and political positions in the community such as owners of small shops, middleman, motor boat owners, and community leaders.

Water buffaloes raising is also an important activity which has been recently introduced in the region. This activity has been promoted by governmental agencies due to suitability of this animal to the floodplain environment (Lourenço Jr., 1983). In contrast to cattle, buffalo raising is performed exclusively by large-scale ranchers whereas community residents have strongly reacted against this activity. Destruction of fences and crop fields and soil compactation - which affects grass regeneration - are some of the impacts of buffaloes in the area mentioned by local residents.

Wage-based activities represent a major income source among floodplain residents and include mostly teachers, carpenters, cattle caretakers, and retirees. Similar to the Mamirauá case (Lima, this volume), wage-based income represents

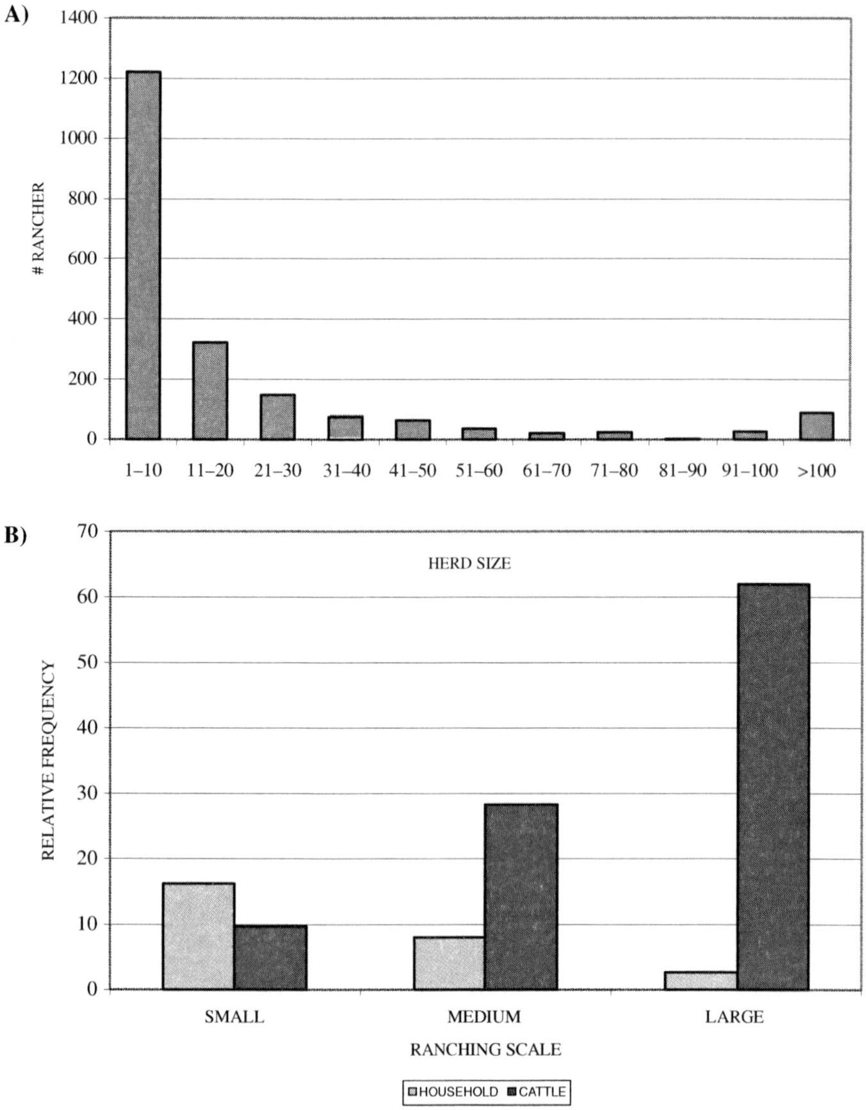

Fig. 7.4 Cattle ranching in the study area. (**A**) Distribution of the herd size by floodplain ranchers; (**B**) Rancher categories according to herd size (number of ranchers = 2029)

an opportunity for the households to overcome environmental risks and to increase their ability to capitalize. Pension is particularly important among households with elderly members. Women above 60 years old and men above 65 years who are registered in the rural workers Unions have the right to a monthly payment of approximately US$100. Among the 7,066 households distributed in the 172 communities analyzed in this study, 29% included at least one retiree (a total of 2,342 individuals). Only 30% of this total live in houses with one or two residents, who are probably elderly couples or widows/widowers, respectively, whereas the remaining live in larger households. Pension has re-defined the socioeconomic role of elderly people in the household, from a burdensome role to a significant complementary income in the household, impinging on the degree of economic investment in floodplain communities (Futemma et al., 2002).

Economic Strategies of Households

The four main economic activities in the region discussed in the previous section can be combined in different strategies by the floodplain households. The economic strategies may vary according to the number of activities performed, from specialized in one single activity (40%) to diversified in two (32%), three (26%), and four activities (2%) (Fig. 7.5). The importance of each activity in the economic repertoire can also vary, leading to heterogeneous economic strategies within and across communities. Three factors strongly influence in shaping such a diverse economic strategy, namely: (1) the economic purpose of the activity, (2) the household demographic structure, and (3) the degree of access to the natural resources. Each factor is discussed in more detail below.

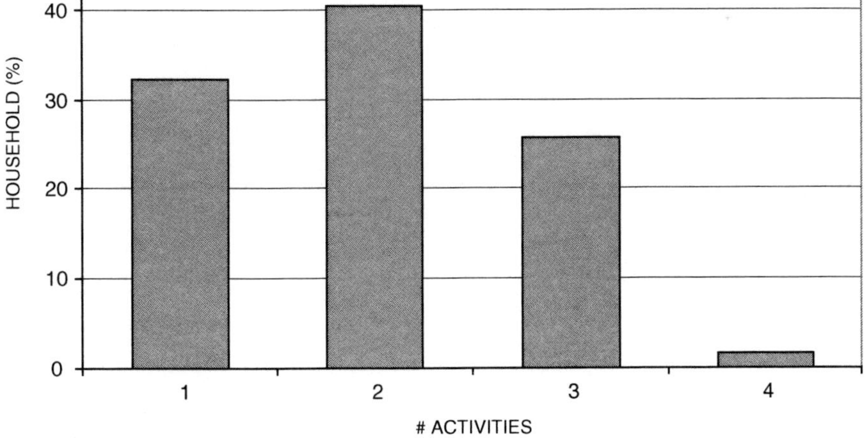

Fig. 7.5 Average number of economic activities performed by the floodplain households in the study area (number of households = 7770)

Economic Purpose

From the studied households, 61% holds a subsistence-oriented economy while the remaining developed cash-oriented strategies relied mostly upon commercial fishing (31% of the households) and wage-based sources (8% of the households) (Fig. 7.6).

Specialization is more common in fishing (23%), followed by agriculture (5%) and ranching (2%). The dual purpose of consumption and income source of fishing helps to explain the predominance of this activity among the floodplain population. As discussed before, the category 'commercial fishing' reflects only part of this activity since a fraction of the catch of the so-called 'subsistence fishing' is often commercialized.

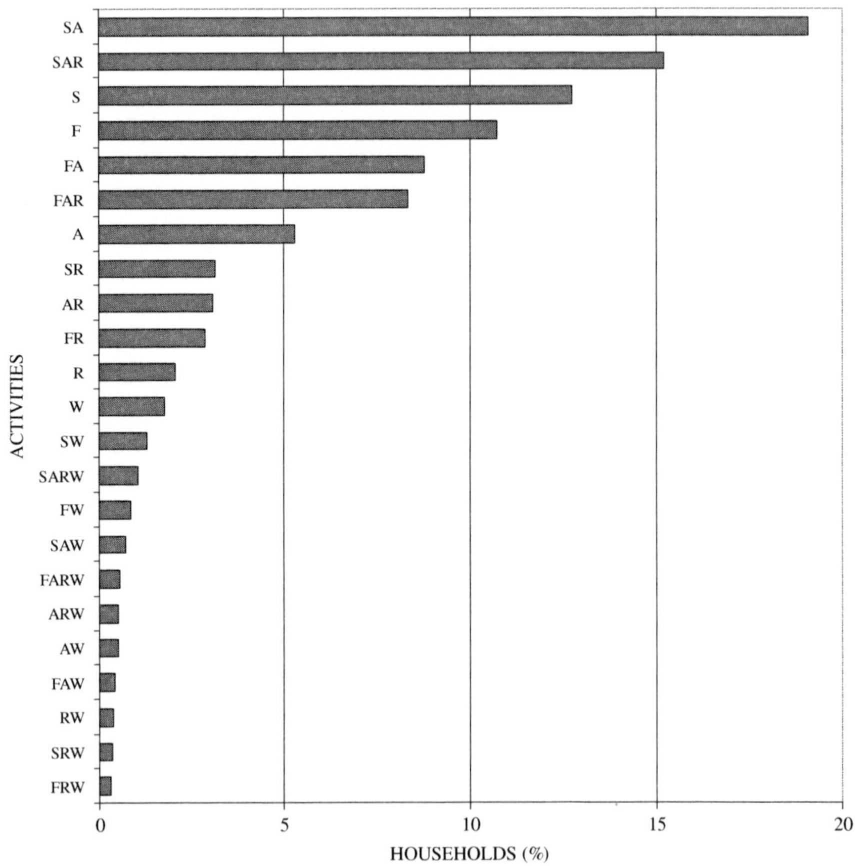

Fig. 7.6 Economic strategies performed by the floodplain households in the study area (S = subsistence fishing; A = agriculture; F = commercial fishing; R = ranching; W = wage labor/pension) (number of households = 7614)

The degree of diversification of economic activities seems to depend on the intensity of commercial fishing in the household. The more the fishing activity is oriented toward commercialization, the higher is the tendency toward economic specialization in the household. This pattern is clearly observed in Fig. 7.6, which shows that commercial fishing (F) occurs more frequently as a specialized activity (11% of the households) than as combined with other economic activities such as agriculture (9%) or agriculture and livestock (8%). Oppositely, subsistence fishing (S) is more common in combination with other economic activities such as agriculture (19%) or agriculture and livestock (15%), than as a specialized activity (13%).

Fishing (commercial or subsistence) and agriculture are the most common combination of two activities, performed by 28% of the households. This pattern is expected as both activities serves for both subsistence and commercial purposes. The combination of fishing (commercial or subsistence), agriculture and livestock is carried out by only 24% of the households (Fig. 7.6) who have been able to accumulate capital in order to invest in ranching activities.

Household Composition

In addition to the economic purpose, the number of activities performed seems to be related to the household composition. For example, households with larger number of adult members tend to add more activities to their economic repertoire. Likewise, the intensity of cattle raising activity tends to increase according to the number of adult males and retirees in the household. Futemma et al. (2002) observed a similar relation between cattle raising activity and the number of adult males in the household. The authors suggest that this pattern may be related to the fact that cattle ranching implies tasks typically performed by male labor force in the region. The seasonal transportation of the herd between the floodplain and the upland, construction of stilt-raised corrals during the flooding season, and pasture cultivation are some of the activities which demand mobility and physical strength. In addition to adult males, the presence of elderly members influences the ability of a household to engage in cattle ranching through the provision of money from pension to invest in this activity.

Degree of Access

The third factor influencing the definition of activities carried out by a household is the degree of access to the floodplain resources. The degree of access to ecological systems and natural resources vary enormously across communities according to their geographic locations. The floodplain communities can be divided into three geographic categories – island, riverbank and upland communities. Island communities are located on the floodplain islands with no direct access to upland soils. Riverbank communities are located between the floodplain and upland soils and thus enjoy direct access to both systems. Upland communities are located on higher terrains, close to the floodplain system. Among 114 communities analyzed, island, riverbank and upland communities are represented by 28%, 60%, and 12%, respectively.

The differences in the pattern of economic activities between communities of each category are remarkable. Commercial fishing is far more important in island communities – where approximately 60% of the households are involved – when compared to riverbank communities (20%) and upland communities (10%) (Fig. 7.7).

Figure 7.8 reveals the importance of different economic strategies in the three categories of communities. The island and riverbank communities present similar degree of engagement in agriculture and ranching. However, the variety of products cultivated in riverbank communities is much larger (36 items) when compared to the island and upland communities (16 and 18 items, respectively).

The degree of specialization and diversification of activities is another important difference between the three groups of communities. In general, island communities are more specialized in commercial fishing (22%), in contrast with communities from other two categories (between 8% and 10%) (Fig. 7.8). While commercial fishing appears in most of the economic strategies in the island communities, subsistence fishing predominates in the diversified strategies in the riverbank and upland communities (Fig. 7.8). In contrast to fishing, specialized agriculture is more noticeable in upland communities where

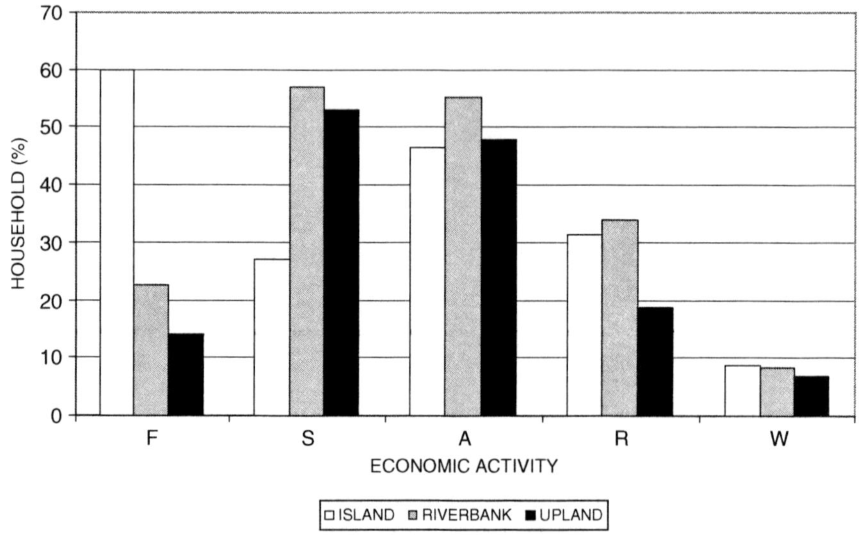

Fig. 7.7 Relative frequency of household involvement in the major economic activities according to geographic location (F = commercial fishing; S = subsistence fishing; A = agriculture; R = ranching; W = wage labor/pension) (number of households = 5883)

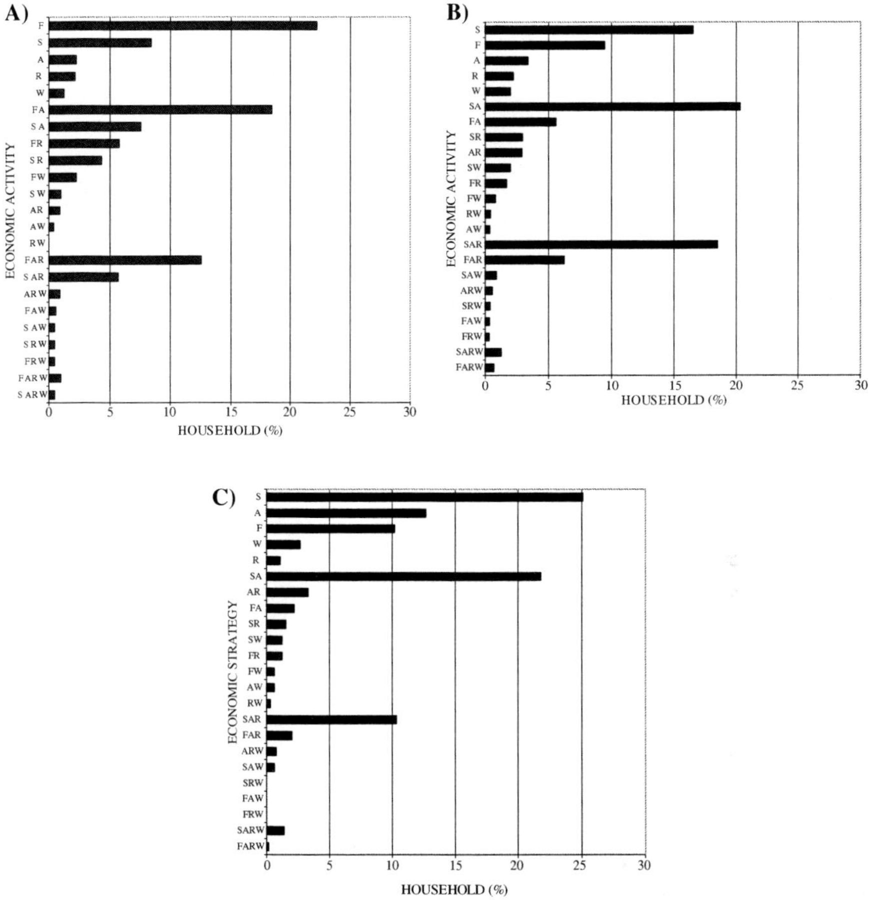

Fig. 7.8 Economic strategies performed by the floodplain households in the studied area, according to geographic location: Island (A), Riverbank (B), and Upland (C). (S = subsistence fishing; F = commercial fishing; A = agriculture; R = ranching; W = wage/pension)

10% of the households are engaged in this strategy, compared to only 3% in the other communities. Beans cultivation, however, prevails in the island communities where lowland is more abundant. Finally, specialized ranching shows no difference across all communities where only 1% of the households is engaged in this strategy at the time of this survey. These households are probably represented by large-scale ranchers living in the area.

Discussion

The caboclo's economy has long been based on a combination of resource use activities and wage-based income. Such a strategy has enabled these populations to promptly adjust their socioeconomic repertoire to new realities and assimilate new economic opportunities. The remarkable economic flexibility is one of the key features shared by the caboclo communities, which has enhanced its social resilience during several economic cycles in the region (Schmink, 1985). Despite this similarity, the caboclo society varies significantly in terms of activities included in their economic repertoire, and in the patterns of change in their economic strategy, according to their socioecological context.

The regional landscape, the occupation history of the group, and some exogenous social factors (e.g., market demand, technological innovation, and policy measures) are some of the key factors to understand the regional patterns of resource use strategies, such as the rubber tappers in the Upper Solimões, the *varjeiros* in the Lower Amazon, and the palm fruit harvesters in the estuary. Variability defined at finer scale, however, strongly influences different local responses across households located in the same region such as spatial distribution of ecosystems, household structure and community organization.

The spatial and temporal variability of the floodplain ecosystem offers highly diverse economic opportunities and constraints (Denevan, 1984; Hiraoka, 1992; Padoch & Jong, 1992; Lima this volume). The ability of a household to benefit from the opportunities and to overcome the constraints is related to its level of access to valuable resources and to its ability to explore them. The caboclo communities encompass residents enjoying different local political power which influences considerably in their access to particular landscape units or resources (Castro, 2000). By the same token, the household structure has direct influence in their ability of adopt a diversified economic strategy or to incorporate a specific activity in their economic repertoire (Futemma et al., 2002). Family relations is also an important factor defining the rules of access to natural resources, as pointed out by Futemma (this volume) and Lima (this volume). Therefore, while generalizations of economic activities are valuable to describe regional socio-environmental patterns, a refined analysis on the diversity of economic strategies is fundamental to investigate local processes leading to different responses by local households.

This study revealed that the economic strategy of 8,570 caboclo households distributed in 172 communities along the Lower Amazon is based on four main activities – fishing, agriculture, cattle ranching, and wage-based activities. At the regional level, the caboclo populations of the Lower Amazon share similar history of land occupation and resource use (Gentil, 1988; McGrath et al., 1993), enjoy access to a similar ecological landscape (Goulding, Smith, & Mahar, 1996), and have been influenced by similar exogenous social pressures (Castro, 2002). The apparent socioenvironmental similarity of this population has inspired Furtado (1993) to adopt the term varjeiros (from *várzea* = floodplain) to refer to caboclos living on the floodplain communities along the Lower Amazon. According to the author, varjeiro is a social category locally used and it describes the close cultural association with the dynamic landscape pattern, and their mixed, subsistence-oriented production system. The varjeiro category is, therefore, strongly related to the multiple subsistence economy (polyvalent) as opposed to another caboclo population in the region represented by the urban commercial fishers who rely upon one single economic activity (monovalent). This present study challenges the varjeiro model, revealing that this group sometimes may not adopt a polyvalent strategy, and fishing may not be for subsistence only. Approximately one third of the sampled households carry out only one activity and 11% are specialized in commercial fishing (Fig. 7.6). Lima (this volume) argues that the tendency to idealize the caboclo economy based on the commercialization of the production surplus is not supported by empirical data. In the Lower Amazon, varjeiros specialized in commercial fishing are usually associated with boat owners, i.e., are part of a larger production system. These individuals are possibly part of an incipient group of specialized commercial fishers who adopt the univalent commercial fishing strategy at times which can develop into a permanent strategy upon displacement to the urban center. In other words, although varjeiros may have access to other resources, each household adopts a dynamic resource use strategy that swings between specialization and diversification, according to the socio-environmental opportunities and constraints. Furtado (1993) suggests that the specialization of commercial fishing by urban fishermen is a consequence of the increased migration of varjeiros to the urban center where the access to other natural resources is more limited. Although this model has explanatory value, it is possible that the specialization of the commercial fishing start in the community and the migration to the urban center be the consequence (and not the cause) of this process. Often, the shift from polyvalent to monovalent economic strategy is related to local processes based on the household structure, access to resources, and local political structure. Therefore, the assumption of polyvalent subsistence-oriented economy for the floodplain household may hide complex, dynamic local processes.

Another problem related to the urban/varjeiro fisher dichotomy emerges in cases when the distinction between both groups is unclear. Floodplain residents frequently migrate between the community and the towns nearby according to the economic opportunities, optimizing the access to economic resources in both systems. Therefore, specialization in commercial fishing is only part of a dynamic economic strategy subject to change upon new economic incentives in the community or urban centers. In this regard, the level of access to urban center (in terms of distance or transportation availability) plays a key role in shaping variability in intra-regional economic strategies (Lima this volume).

The assumption of polyvalent subsistence-oriented economy by the floodplain residents not only overlooks the social connection between community-urban center, and the economic flexibility of the households between mono and polyvalent strategies, but also fails to consider the role of the variability in the spatial distribution of landscape units influencing the opportunities and constraints for resource access and use.

The disaggregation of the studied communities in the Lower Amazon in three categories according to their geographic location – upland, riverbank, and island - reveals that floodplain and upland populations use resources from both systems whenever possible. Residents of upland communities located close to floodplain lake systems also enjoy access to aquatic resources. By the same token, residents of floodplain communities located on the riverbank are benefited by the direct access to upland resources. Therefore, the 'upland' and 'floodplain' dichotomy seems inappropriate to define level of access to natural resources, and its relation with resource use strategies. In fact, economic strategies performed in riverbank communities are often more comparable to upland communities than to those observed in the island communities.

According to the household economy analysis, the island community residents face limited access to the upland-based resources, and they rely their activities mainly on the floodplain system. On the other hand, the riverbank community residents are particularly privileged by the physical access to both upland and floodplain resources. As a result, these communities present the most diverse economic strategies close to the idealized varjeiro household, including fishing, agriculture, and ranching.

In addition to the theoretical relevance of the diversity of economic strategies among the caboclo communities and its relation to the local socio-ecological opportunities and constraints, the remarkable differences between riverbank and island communities sheds some light on the dynamics of response to new economic opportunities in the region (WinklerPrins, 2002). Recently, the residents of riverbank communities have been engaged in the three new economic activities: commercial fishing, logging and cattle ranching. Commercial fishing is an extractive activity that provides prompt cash returns. Logging represents a new income source where forest resources are available (Futemma et al., 2002). Cattle ranching is particularly favorable in riverbank communities, due to access to the cost-free and high quality natural grassland on the floodplain during the dry season and the access to ever-dry terrain during the flood season. In short, the mixed caboclo economy is observed mostly in the riverbank communities where the privileged geographic location benefit their residents to combine upland-based activities (e.g., perennial and semi-perennial crops, forest extractivism), floodplain-based activities (e.g., fishing, fast-growth crops), and activities which are better conducted under integrated use of upland and floodplain systems (e.g., cattle ranching).

In contrast, despite of its rich nutrients, the use of the floodplain soils is limited to part of the year and the crops are highly vulnerable to flooding-related risks (Chibnik, 1994; Futemma, 1995; WinklerPrins, 1999). Moreover, the seasonal variation in the water level creates troublesome residence conditions in the island communities during the flood season (Lima this volume). In other words, the ecological advantages of the floodplain system related to the soils and transportation are offset by the environmental risks associated to the flooding pattern. Since the jute bust in the 1960s, the island community residents were left with a few economic alternatives, limited to annual crops, fishing, and small livestock. In contrast to riverbank populations, island populations have been able to engage only in two of the three emergent economic activities in the Lower Amazon: commercial fishing and cattle raising. In comparison with commercial fishing, cattle raising in the island communities implies extra costs during the flooding season in building and maintenance of stilt-raised corrals, or alternatively to transfer the herd to the upland. Consequently, commercial fishing has become the most viable economic alternative for this population (Figs. 7.7 and 7.8). For this reason, specialization in commercial fishing may also be adopted by residents of island communities and not limited to urban fishers as usually assumed.

The differences between island and riverbank communities discussed are not only relevant theoretically. They shed some light on the debate about the viability of community-based management of floodplain lakes in the Amazon. The claim for local management systems in the region has unveiled an important alternative strategy for a sustainable use of local resources. However, the tendency to treat the varjeiro as a homogeneous socioeconomic group has contributed to the invisibility of subgroups coping with uneven access to landscape units, to valuable resources, and ability to perform economic alternatives. As discussed elsewhere (Castro, 2000), the reliance on commercial fishing by the island households is directly related to the limited access to other economic alternatives. Island populations are pressured by the expansion of floodplain cattle ranching, on one side, and by the expansion of commercial fishing in the community lakes by better equipped fishers, on the other side.

Consequently, more than any other group, the island community inhabitants face the dilemma between unrestricted exploitation of fishing resource to supply short-term needs, and the establishment of local management institutions to ensure the long-term maintenance of the fishing resources (Castro & McGrath, 2003). The lack of economic opportunities increases the incentive to violate fishing rules (McGrath et al., 1994; Castro, 2000). Therefore, the contextualization of local management systems including the local processes influencing the economic responses by the households is fundamental to assess the potential and limitations of this management model.

Conclusion

While studies focusing indigenous populations and settlers have usually based on local and regional analysis, respectively, the analytical focus of the studies of caboclo populations have relied mainly on the integration of both scales. The household economic strategy and the degree of access to natural resources proved to be powerful analytical methods to carry out a multi-scale analysis of resource use by the caboclo population. Similar to other studies in this volume which reveals variability in food consumption (Adams et al.), income (Lima) and institutional and resource access (Futemma), the socioeconomic analysis of the caboclo households in the Lower Amazon uncovers the heterogeneous patterns of economic activities performed by the local population.

Pre-defined aggregated social and ecological categories carry along assumptions regarding natural resource use which masks refined categories under distinct local socioenvironmental influences. Subgroups of varjeiros, for example, become visible when degree of access to floodplain and upland resources is assessed. The analytical focus on the households' economic activities and their degree of ecological access to natural resources helped to unveil how local factors influence different responses by groups under similar pressure of regional factors, and to explore the different processes influencing local decisions. The aggregate analysis revealed a mixed economy based on four major resources, influenced by regional socioenvironmental processes, by the ecological landscape pattern, by the occupation history, and external pressures related to market demand and technological innovations. The disaggregated analysis based on three categories of communities uncovered a striking distinction in their pattern of economic activity according to the local landscape variability. Further disaggregation such as between island communities (Castro, 2000), between households in the same community (Castro & McGrath, 2003; Murrieta & Dufour, 2004), or even between seasons in a single family (Lima this volume) is a powerful methodological tool to explore other factors influencing the variability of responses toward resource use patterns.

Integrated analysis including floodplain-upland systems is not only of theoretical relevance. Conservation policies must take into account the degree of access to and value of resources in the household economy. Reliance on fish resource as main income source in the floodplain is, in some cases, related to socioenvironmental factors limiting the adoption of other economic activities by the households. Thus, the support to local management systems may not be enough to attain the expected goals of conservation and local development if the access to other floodplain resources shows major discrepancies among the local participants of this endeavor.

In sum, the acknowledgement of the socioenvironmental diversity in the region is of key importance to avoid generalized and incorrect assumptions associated to pre-defined broad social and ecological categories such as caboclo and floodplain. A systemic analysis of resource use and decision-making processes at the household level, and the flexibility in the definition of social and ecological units according to the research question are important methodological strategies to avoid conceptual problems. The visibility of the caboclo society should not happen at the expense of the invisibility of more disaggregated social categories of major theoretical and practical relevance.

References

Adams, C. (2002). Estrategias adaptativas de duas populaçőes caboclas (Pará) aos ecossistemas de várzea estuarine e estacional: Uma análise comparative. PhD Dissertation – Bioscience Institute, University of Sao Paulo, Sao Paulo.

Allegretti, M. H. (1990). Extractive reserves: An alternative for reconciling development and environmental conservation in Amazonia. In A. B. Anderson (Ed.) *Alternatives to deforestation: Steps toward sustainable use of the Amazon Rain Forest*, (pp. 252–264). New York: Columbia University Press.

Anderson, A. B., & Ioris, E. M. (1992). The logic of extraction: Management and income generation by extractive producers in the Amazon. In K. H. Redford, & C. Padoch, (Eds.). *Conservation of neotropical forest: Working from traditional resource use*, (pp. 175–199). New York: Columbia University Press.

Brondízio, E. S., & Siqueira, A. D. (1997). From extractivists to forest farmers: Changing concepts of caboclo agroforestry in the Amazon estuary. *Research in Economic Anthropology*, *18*, 233–279.

Bunker, S. (1985). *Underdeveloping the Amazon*. Urbana: University of Illinois Press.

Castro, F. (2000). Fishing accords: The political ecology of fishing intensification in the Amazon. PhD Dissertation. Indiana University, Bloomington, USA.

Castro, F. (2002). From myths to rules: The evolution of the local management in the Lower Amazonian floodplain. *Environment and History*, *8*(2), 197–216.

Castro, F., & McGrath, D. (2003). Moving towards sustainability in the local management of floodplain lake fisheries in the Brazilian Amazon. *Human Organization*, *62*, 123–133.

Cerdeira, R. G. P., Ruffino, M. L., & Isaac, V.J. (1997). Consumo de pescado e outros alimentos pela população ribeirinha do Lago Grande de Monte Alegre, PA – Brasil. *Acta Amazonica*, *27*, 213–227.

Chibnik, M. (1994). *Risky rivers: The economics and politics of floodplain farming in Amazonia*. Tucson: The University of Arizona Press.

Denevan, W. (1984). Ecological heterogeneity and horizontal zonation of agriculture in the Amazon floodplain. In M. Schmink & C. H. Wood (Eds.), *Fronteir expansion in Amazonia*, (pp. 311–317). Gainesville: University of Florida Press.

Furtado, L. G. (1993). *Pescadores do Rio Amazonas: Um Estudo Antropológico da Pesca Ribeirinha numa Área Amazônica*. Coleção Eduardo Galvão. Belém, Brazil: MCT/CNPq, MPEG.

Futemma, C. R. T. (1995). *Agriculture and Caboclo Household Organization in the Lower Amazon Basin: Case Studies*. Master's thesis, Tulane University, New Orleans, La.

Futemma, C. R. T. (2000). Collective action and assurance of property rights to natural resource: A case study from the Lower Amazon Region, Santarém, Brazil. PhD. Dissertation, Indiana University, Bloomington, USA.

Futemma, C., Castro, F., Silva-Forsberg, M.C., & Ostrom, E. 2002. The emergence and outcomes of collective actions: An institutional and ecosystem approach. *Society and Natural Resource*, *15*(6), 503–522.

Gentil, J. M. L. (1988). A juta na agricultural de várzea na área de Santarém – Médio Amazonas *Boletim do Museu Paraense Emilio Goeldi, Série Antropologia*, *4*, 118–199.

Goulding, M., Smith, N., & Mahar. D. J. (1996). *Floods of Fortune: Ecology and Economy along the Amazon*. New York: Columbia University Press.

Hames, R. B., & Vickers, W. T. (Eds.) (1983). *Adaptative responses of native Amazonians*. New York: Academic Press.

Hiraoka, M. (1992). Caboclo and *ribereño* resource management in Amazonia: A review. In K. H. Redford & C. Padoch (Eds.), *Conservation of neotropical forests: Working from traditional resource use*, (pp. 134–157). New York: Columbia University Press.

Isaac, V. J., Mitlewski, B., & Oliveira, P. R. S. (1999). Metodologia para primeiros contatos, levantamentos participativos e multidisciplinares e incentivos à cooperação juntos aos grupos alvo de projetos de desenvolvimento. In *Recursos Pesqueiros do Médio Amazonas: Abordagem Socioeconômica*, (pp. 13–29). Coleção Meio Ambiente. Série Estudos de Pesca 21. Brasília: Brazil: IBAMA

Lima, D. (1992). *The social category caboclo: History, social organization, identity and outsider's social classification of the rural population of an Amazonian region (Middle Solimoes)*. Ph.D. Thesis, University of Cambridge.

Lourenço Jr., J. B. (1983). A pecuária na Amazônia. In J. Hebette (Ed.), *Ciência e Tecnologia para Amazônia*, (pp. 53–78). Belém, Brasil: Núcleo de Altos Estudos Amazônicos (*hereinafter* NAEA), Universidade Federal do Pará.

McGrath, D., Castro, F., & Futemma, C. (1994). Reservas de lago e o manejo comunitário da pesca no Baixo Amazonas: Uma avaliação preliminar. In M. A. D'Incao & I. M. Silveira (Eds.), *A Amazônia e a Crise da Modernização*, (pp. 389–402). Coleção Eduardo Galvão. Belém, Brazil: MPEG.

McGrath, D., Castro, F., Futemma, C., Amaral, B. D., & Calabria, J. (1993). Fisheries and the evolution of resource management on the Lower Amazon floodplain. *Human Ecology*, *21*, 167–195.

Moran, E. F. (1974). The adaptive system of the Amazonian caboclo. In C. Wagley (Ed.), *Man in the Amazon*, (pp. 136–159). Gainesville: University of Florida Press.

Moran, E. (1981). *Developing the Amazon*. Bloomington: Indiana University Press.

Moran, E. (1993). *Through Amazonian eyes: The human ecology of amazonian populations*. Iowa City: University of Iowa Press.

Moran, E. (1995). Disaggregating Amazonia: A strategy for understanding biological and cultural diversity. In L. E. Sponsel (Ed.), *Indigenous peoples and the future of Amazonia: An ecological anthropology of an endangered world*, (pp. 72–95). Tucson: The University of Arizona Press.

Murrieta, R. S. S. (2000). The dilemma of the 'Chibé'-eaters: Food choices, ecology and everyday life among peasant communities in the Lower Amazon, Brazil. Ph.D. Dissertation. University of Colorado, Boulder.

Murrieta, R. S. S. (2001). Dialética do sabor: Alimentaçao, ecologia e vida cotidiana em comunidades ribeirinhas da Ilha de Ituqui, Baixo Amazonas, Pará. *Revista de Antropologia*, *44*, 39–88.

Murrieta, R. S. S., Dufour, D. L., & Siqueira, A. D. (1999). Food consumption and subsistence in three caboclo populations on Marajo Island, Amazonia, Brazil. *Human Ecology*, *27*, 455–475.

Murrieta, R. S. S. & Dufour, D. L. (2004). Fish and farinha: protein and energy consumption in Amazonian rural communities on Ituqui Island, Brazil. Ecology of Food and Nutrition. *43*, 3, 231–55.

Nuggent, S. (1993). Amazonian caboclo society: An essay on invisibility and peasant economy. Oxford: Berg.

Padoch, C., & Jong, W. (1992). Diversity, variation, and change in ribereño agriculture. In K. H. Redford & C. Padoch (Eds.), *Conservation of neotropical forests: Working from traditional resource use*, (pp. 158–174). New York: Columbia University Press.

Parker, E. (Ed.) (1985). *The Amazon Caboclo: Historical and contemporary perspectives*. Studies in Third World Societies No. 32. Williamsburg, Va.: Department of Anthropology, College of William and Mary.

Posey, D. A., & Balee, W. (Eds.) (1989). *Resource management in Amazonia: Indigenous and folk strategies*. Advances in Economic Botany, vol. 7. Bronx, N.Y.: The New York Botanical Garden Press.

Roosevelt, A. (1989). Natural resource management in Amazonia before the conquest: Beyond ethnographic projection. In D. A. Posey, & W. Balee (Eds.), *Resource Management in Amazonia: Indigenous and Folk Strategies*, (pp. 30–62). Advances in Economic Botany, vol. 7. Bronx, N.Y.: The New York Botanical Garden Press.

Schmink, M. (1985). São Felix do Xingu: A caboclo community in transition. In E. Parker (Ed.), *The Amazon caboclo: Historical and contemporary perspectives*, (pp. 143–166). Williamsburg, V.I.: Studies in Third World Societies no.32.

Schmink, M., & Wood. C. H. (1992). *Contested frontier in Amazon*. New York: Columbia University Press.

Siqueira, A. D. (1997). The ecology of food and nutrition: Patterns of land use and nutritional status among caboclo populations on Marajó Island, Pará, Brazil. Ph.D. Dissertation. Indiana University, Bloomington, USA.

Weistein, B. (1983). *The Amazon rubber boom, 1850–1920*. Stanford University Press.

WinklerPrins, A. M. G. (1999). Local soil knowledge: A tool for sustainable land management. *Society and Natural Resources*, *12*, 151–161.

WinklerPrins, A. M. G. (2002). Recent seasonal floodplain-upland migration along the Lower Amazon River. *Geographical Review*, *92*, 415–431.

Part III
Resource Management

Chapter 8
Agriculture Intensification, Economic Identity, and Shared Invisibility in Amazonian Peasantry: Caboclos and Colonists in Comparative Perspective

Eduardo Brondizio

Abstract The main focus of this contribution is to demonstrate the erroneous interpretation that has been carried out around the agricultural production systems of small-scale Amazonian rural producers, especially when analysed under the light of intensification. This mistaken judgement supports a depreciative vision of these social groups and their contributions to the regional economy. Both land use patterns, *caboclo* and *colono*, are often based on the co-existence of intensive and extensive activities that, simultaneously, minimise risk and guarantee the consolidation of rural property, as well as the expansion (or retraction) of activities geared towards the market. The author suggests the adoption of the term 'small-scale producers' to refer to these populations, which would contribute, in Brondízio's opinion, to the creation of a more positive socio-economic identity for these populations within the region's agricultural and resource economy. The author closes his analysis as he concludes that the redefinition of these populations' identity as small-scale rural producers would provide a great step ahead in the sense of overcoming prejudices incorporated by regional and national societies.

Keywords Economic identity · Rural development · Açaí palm · Land use · Agriculture intensification

E. Brondizio (✉)
Department of Anthropology, Indiana University, Bloomington, IN, USA
e-mail: ebrondiz@indiana.edu

C. Adams et al. (eds.), *Amazon Peasant Societies in a Changing Environment*,
DOI 10.1007/978-1-4020-9283-1_9, © Springer Science+Business Media B.V. 2009

Introduction: Views of Agriculture Intensification and Shared Invisibility in Amazonian Peasantry

Obvious differences between Amazonian caboclos[1] and recent colonists[2] render unnecessary any elaboration of their socio-cultural and historical particularities.[3] At the same time, these diverse groups share striking similarities as they are lumped together under the rubric of 'Amazonian peasantry.' Regionally, both terms (but particularly caboclo) carry several distinct, usually derogatory meanings. Whereas academic definitions take into account variation in historical context, ethnic background, geographic, and contemporary socioeconomic identity, the colloquial usage of the terms caboclo and colonist[4] share similar socio-cultural and economic prejudices. In essence, as small-scale rural producers, whether caboclo or colonist, they share a lack of economic, political, and infrastructure support. This chapter aims to discuss the existence of commonalities underlying their condition of 'invisibility.' While trying to value their historical and socio-cultural particularities, I attempt to discuss, particularly, the implications of misinterpreting their agricultural systems for the construction of an economic and social identity of these rural Amazonians. I attempt to show that the so-called 'invisibility' of Amazonian peasantry (whether economic, political, technological, or social) is in part a result of the dominant views of what is considered an agricultural system as it relates to its agronomic, aesthetic, economic, technological, and social efficiency and characteristics. A core element in this equation is how the 'process of intensification' of agricultural production is defined, particularly given its comparative nature and its implications for understanding social and socioeconomic changes throughout the contemporary history of the region. Although for different reasons – and that I will attempt to illustrate by means of field data and examples – both caboclo and colonist production systems tend to be disregarded in terms of

[1] Particularly, in this paper, the term Caboclo refers mostly to the riverine and inter-fluvial rural populations of the Amazon estuary represented by the study case discussed here.

[2] The term colonist is used in this paper to refer to the families arriving in the region since the late 1960s through government-sponsored and spontaneous migration to areas previously occupied by Indigenous groups or Caboclo settlements. In particular, in this paper I concentrate on colonists settled as small farmers (that is, in lots varying from 50 to 150 hectares) during the last 3 decades as a result of government incentives for colonization. Whereas clear differences exist between these more recent migrants and caboclos, one may find it difficult to distinguish between communities occupying areas of century old colonization, like the Bragantina region, and "caboclo" communities interweaved within and around them.

[3] As noted by Pace "In all these definitions [n.a.: pointing to the same citations used in this paper to refer to Caboclo studies] it is acknowledge that caboclos are nontribal – not Native American – and non-settlers – not migrants who have come to Amazônia since the 1950s…"(1997:82).

[4] In the colonist case, examples of terms with derogatory connotations include "Arigó" (particularly for those of Northeastern origin) and "Quiçassa," (a term also used for abandoned areas in Western Amazônia), among others.

their socioeconomic relevance and effectiveness when compared to exogenous, large scale, high input and capital-based agriculture. To some extent, this argument reinforces Nugent's idea of 'manufactured invisibility' where these farmers are placed in a condition of 'social pathology,' a stumbling block that impedes regional development (1993). This view tends to emphasize the substitution of local land use strategies for external technology based on energy and capital intensive systems primarily focused on export-oriented agriculture. This tendency is liable to neglect investment to improve existing socio-economic and physical infrastructure that would, in turn, enhance local production systems without displacing rural families or threatening local resource basis (and, consequently, the local economy and food security).

Explanations of land use intensification are usually based on conceptual models using parameters such as fallow cycle, or variables based on factors of production, for instance, labor, energy, technology, and/or capital – the so-called 'input factors.' Alternatively, 'output factors,' such as the maintenance of productivity over time, are often used as a complementary measure of agro-pastoral intensification (for review see Brondizio & Siqueira, 1997). However, models of fallow cycle offer limited explanation to agricultural systems in frontier areas (by colonists) where land occupation is primarily based on cycles of progressive expansion of the used area, as well as in caboclo's swidden agroforestry where a clear distinction between the 'productive' and the fallow period is not obvious. Both caboclo and colonist patterns of land use are often based on the co-existence of intensive and extensive activities that simultaneously minimize risk while guaranteeing farm consolidation and expansion of market activities. By the same token, another element underlying our views of agricultural systems involves a subtle link between 'agronomic' and 'aesthetic' arrangements. Dominant views of productive agricultural systems include elements of field homogeneity and shape, types and composition of plant species and crop varieties. It also includes particular patterns of land allocation representing the domesticated, technologically driven production and a farmer's ability to keep it 'clean.' These characteristics, usually borrowed from temperate areas, generally defy even the most productive farm lots in riverine or frontier Amazônia. The rigid boundaries drawn between different food production systems usually place forested areas (as in the case of agroforestry systems) in the 'fallow,' 'unproductive' or, at best, the 'agro-extractive' category.

This chapter builds upon two previous works that look at land use trajectories and intensification in caboclo (Brondizio & Siqueira, 1997) and colonist areas (Brondizio et al., 2002). Both articles attempt to analyze patterns of land use trajectories and discuss the implications of using different measures of 'intensification' to characterize these agricultural systems. In both cases, misinterpretation of their productive potential affects the economic identity and infrastructure support for these populations. Building upon the integration of survey, experimental, and multi-temporal remote sensing data, in the caboclo study we argue for a producer's identity 'from extractivist to forest farmer'

(Brondizio & Siqueira, 1997), while in the colonist case we show the importance of understanding colonists' land use within the context of generational cycles of lot formation and land allocation characterizing 'the colonist footprint' (Brondizio et al., 2002). This chapter attempts to bring these two pieces of research together to show that, apart from their socio-cultural differences, whether caboclo or colonist, small-scale producers in Amazônia share a condition of economic and social invisibility, at least in part fed by the ways we interpret (or misinterpret) their production systems. Consequently, the lack of political and basic infrastructure support for these areas leads to a sort of positive feedback process creating vicious cycles of economic failures and social poverty, further reinforcing their condition of invisibility and opinions that suggest their 'lack of entrepreneurial minds.'

Clarifying Terms and a Conceptual Framework for Smallholders in the Amazon

The current discussion relating to the use of the term caboclo leaves one close to a 'deadlock' situation on how to refer to and make generalizations about those we call caboclo and why we distinguish caboclos from other rural Amazonians. The use of the term caboclo to stress a historic-cultural group and/or to value an 'ethnographic other' is overwhelmed by implications of its biased social construction. Several attempts to illustrate the nuances and contradictions between popular and academic usage of the term exist, from Wagley's classic work (1955) on the use of the term caboclo to refer to distant subjects to Galvão's (1976), Moran's (1974), Parker's (1985), and Furtado's (1987) emphasis on a historically and geographically–situated social category, to more recent works by Lima (1992), Hiraoka (1992), Harris (1998), and Pace (1997) reviewing the term's ambiguity. In *Amazon Town*, a classic reference in caboclo studies, Wagley (1955) states that the term caboclo is used to refer to a person of lower class status than the speaker. Pace discusses the several 'typologies' used to define caboclos based on 'racial', 'ecological adaptation', and 'cultural' characteristics. A significant criticism was developed by Pace (1997) in analyzing the derogatory nature of the term and posing the question 'why do we insist on using the term [caboclo], particularly when it carries such pejorative connotations?' (1997:2). On the one hand, he suggests the academic use of the term is embedded in prejudice and reflects the views of the regional elite. On the other hand, there is a need across the research community to create an exotic, legitimated ethnic subject in order to validate one's ethnographic status, vis-à-vis others dedicated to 'true' Indigenous Amazonians. However, Pace goes further in proposing a series of alternative terms to replace caboclo. Examples include '*roceiros*' (small farmers), '*extratores*' (extractors of forest products), '*seringueiros*' (rubber tappers), '*ribeirinhos*' (river people), '*varzeiros*' (flood plain dwellers), and terms such as 'Euro-Native Amazonians,' or 'Afro-Euro-Native

Amazonians,' or regionally used terms like 'Amazonida' that refer, vaguely, to a regional cultural type (1997: 86).

Despite the importance of this discussion, one wonders whether we can change the regional pejorative views about caboclos by changing its denomination. Can we use the term caboclo in an analytical sense to criticize its own contradictions while stressing its 'otherness' by valuing their distinct contributions to food production and environmental management vis-à-vis the vast array of 'other' Amazonian peasantry? One could argue that the prejudice suffered by Amazonian caboclos, although presenting particularities, is historically rooted beyond Amazonian borders and common to a broader array of Brazilian peasants.

Caboclo populations have been a challenge not only to the conventional categorization of peasants in economics and sociology, but also to the conventional category of ethnicity in anthropology (Chibnik, 1991). Chibnik points out that the variation in the terminology for native non-Indigenous populations in Latin America is also a result of historic and demographic differences. He analyzes four 'ethnic' groups in Amazonia: *caboclo, cholo, ribereño, and camba.* He defines as ethnic groups those that have identified themselves and been identified by others as belonging to the same group, however, neither of which happens among caboclos. All those four groups are referred to as 'peasants,' but differences are enormous in terms of class, mode of production, and participation in the larger society. For instance, whereas ribereño is essentially rural and riverine, the term caboclo may incorporate non-riverine, town-dwellers, and urban riverine inhabitants. Ribereño is a category – while geographical – that involves different social classes, while caboclos are essentially lower class – reflecting Chibnik's use of the colloquial version of the term. Camba is a denomination encompassing most residents of the eastern lowlands of Bolivia, while Cholos, though detribalized, are not totally incorporated as part of a contemporary society.

Despite the inter and intra variability of these social groups, much of our understanding of these and other Latin American peasantry in general emerges from the conceptual construction present across rural development studies. An interesting parallel can be seen when we look at the threefold interpretation proposed by Cancian (1989) to categorize studies of peasant economic behavior: homogeneity, heterogeneity, and differentiation approaches. Homogeneity theorists emphasize the differences between peasants and other social groups. However, they tend to overlook internal differences inherent to peasant populations. There is a tendency to explain the poverty of the peasantry as a product of their resistance to social-economic integration within a wider society and to their avoidance of external and novel ideas. There is a strong emphasis on the role of history in characterizing the degree of relationship between peasants and the 'external' world. Heterogeneity theorists tend to deny both the internal homogeneity of peasant groups and their economic behavior as distinctive from that of others. Variability becomes an intrinsic characteristic of peasants, and their reaction to historical forces is seen as more dynamic rather than unilateral. Differentiation theorists stress the role of historical circumstances in shaping peasant behavior. There is an assumption that contemporary peasants have incorporated capitalist features into their economies. Although they maintain the ability to carry out strategies of self-sufficiency (e.g, dependence on land resources), they take advantage of outside opportunities (e.g., wage labor) to support their internal economies. Thus, there has been a step forward in the way we look at the traditional self-providing economy as isolated from the market. In the context of the literature reviewed in this chapter, not only caboclo, but colonist studies cut across these three analytical approaches, while presenting variation in agrarian history, social and institutional organization, and economic arrangements.

Roseberry (1993), in a comprehensive review of the Latin American peasantry, points out the negative assumptions behind the term *peasant* when used by the development 'establishment.' He does not neglect the use of the term as a general category, but recognizes the inviability of applying it as a concept in which economic development is framed. In this context, the construction of a new economic identity for caboclos, as well as for colonists, as small farmer producers necessarily requires a re-interpretation of their land use systems and recognition of their importance. This chapter argues that in both cases, these farmers are strongly engaged in the regional economy, responding to incentives, coping with limitations while using a variety of strategies and seeking economic and political representation against the odds that belittle smallholder production within the Amazon region and in Brazil as a whole.

In a broader context, much criticism has been devoted to the misinterpretation of swidden agriculture systems and their sophisticated agronomic nature, specially following up on the seminal work of Conklin (1957, 1961 and so forth). Another good example is the work of M. Dove (1983). He criticizes the so-called 'political economy of ignorance,' in the context of development projects involving swidden agriculture. These projects usually assume 'widely-accepted myths,' including the 'myth of communally owned land,' the 'myth of destruction and wastefulness,' and the 'myth of a subsistence focus system' (1983:85). In Amazônia, interesting examples include the volume edited by Beckerman (1983), Balee and Posey (1989), Smith, Falesi, Alvin, & Serrao (1996) and perhaps most important, Denevan and Padoch's (1987) re-interpretation of 'swidden agroforestry' systems in the light of intensification theory.[5] Revealing the diversity and complexity of indigenous agricultural practices, they have helped to reveal 'invisible' systems where annuals, perennials, and secondary species are intercropped and managed intensively, helping to debunk a still dominant view where the productive phase of a garden spans no more than 2 years (in a cycle of 20 to 30 years). Similarly, Pinedo-Vasquez, Zarin, Coffey, Padoch, & Rabelo (2001) and Padoch and Pinedo-Vasquez et al. (2006)

[5] A large literature not possible to be reviewed here has been developed around swidden agriculture in Amazônia.

interpretation of timber management practices on what they call 'invisible technologies' is an example that helps clarifying dimensions of regional resource management that defies the conventional interpretation of forest and productive land.

As previously developed by Brondizio and Siqueira (1997), this chapter uses the term caboclo in the sense of Netting's (1993) framework of 'smallholder.' While avoiding the use of the term 'peasants' – and for this matter we can apply it to caboclos and colonists, Netting's use of 'smallholder' (or 'small farmer' for that matter) contributes to a more positive socioeconomic identity of rural producers by calling attention to the important role that small scale farming systems play in the regional and national economies. Such an approach may help to redefine peasant societies in a more dynamic and dialectical way, as well as add to a more positive view of the role, peasant farming, plays and has played in the world of agriculture. On the other hand, the characteristics pointed out by Netting that typify smallholders are useful in order to understand the analytical concept of peasants. As rural and peri-urban inhabitants, they produce for themselves, but they also produce for markets; their economy depends on family labor, but they often employ themselves off their farms in a market economy and employ others when needed; they are not specific to any historical time nor geographic place, i.e., they existed before capitalism and probably will exist 'after' it and in different parts of the world. As a social category, they are not 'inexorably' doomed to disappear, nor are they a homogeneous group. The social, cultural, geographical, and historical diversities of this social category, as in the case of caboclos as well as colonists, must be recognized and taken into account. As rural producers, they are an important social category of our societies, and as such, they need to be recognized, especially by the political authorities establishing the regional economic and development policies.

A Framework for the Study of Land use Change in Rural Amazônia

While expressing linkages among social, economic, and environmental issues, land use can be looked at from different theoretical perspectives depending on the level in question. Rural studies in the Amazon,[6] particularly those on peasant economy, have typically focused on the articulation between factors mediating micro-macro levels, the organization of social groups, and historical conditions defining their relationship (of relevance to this chapter see for instance Nugent, 1993; Schmink & Wood, 1992). Most often, emphasis is on the internal structure of rural communities as it is subordinated to macro-level

[6] I use the term "rural studies" here to refer to various lines of research in anthropology, sociology, and geography concerned with rural development, socio-cultural change, land-use, and political ecology, among other themes.

external 'forces' characterized particularly by policy, market interests, and socio-cultural articulation between local communities and larger political structures. Highlighting the factors mediating these levels has contributed to our understanding of rural development problems, including commodity production and economic cycles, labor arrangements and control of capital, and feedback mechanisms underlying the economic and social behavior of householders and communities in relation to the 'outside' world.

In this context, my point of departure is that the study of land use and local production systems needs to integrate a larger array of variables. Intensification does not proceed linearly as dependent on one factor (e.g., population growth or market demand), nor it is a-historical (Balee, 1998). Instead, it occurs as a combination of these factors with other variables such as internal population dynamics and opportunistic advantages of external sources (e.g., incentives from development projects). Thus, it rather responds to multi-linear processes combining variables working at multiple scales that interconnect national, regional, local, household, and individual levels. For instance, external market demand for forest and agropastoral products has been historically one of the most significant elements underlying social and environmental change in Amazônia with strong implications for land use and livelihood strategies of rural populations. However, whereas long- and short-term market signals (e.g., price increases) may lead to intensification or extensification of land use activities in rural communities, this is a condition actually 'filtered' by household variables such as one's land tenure and access to resources, experience and available technology, and household labor availability, creating a diverse social response within a single community.

Using a framework that integrates a large array of variables, I try to show in this chapter that caboclo producers have properly perceived changes in agricultural market opportunities (short and long-term market trends) occurring within an increasingly urbanized and integrated Amazônia, and have acted to seize such opportunities by means of intensifying their production system by using their existing knowledge-base (of production techniques), as opposed to switching to exogenous production systems usually available by means of development projects and credit support. In the case of caboclos, they have taken advantage of market opportunities to intensify açaí palm fruit production through management of floodplain forest associated with agroforestry planting techniques. However, at the household level, I try to show that the ability to take advantage of market opportunities (in the short run) is largely constrained by the structure of land tenure. As we will try to show later in the paper, this case shows how Amazonian small farmers (caboclos) actively insert themselves into the regional economy in response to long- and short-term market opportunities rather than being passive actors of a regional labor force. Land tenure, however, remains a significant factor constraining caboclo's integration into the market (Brondizio, 2008; Brondizio & Siqueira, 1997; Brondizio, 1999; Brondizio, Safar, & Siqueira, 2003).

In addition, the focus on 'processes' of land use intensification (e.g., household feedback to external signals and the co-existence of land use strategies) becomes more relevant than characterization of 'stages' or patterns of intensification (i.e. Boserup's frequency of crop). A land use-based approach makes it possible to grasp the process of co-existence between intensification and de-intensification as related to temporal and spatial variation of economic strategies. For instance, increased intensification in one production zone coexists with transient de-intensification in another. This is the case of many rural populations in the Amazon estuary, which have virtually abandoned swidden agriculture in the upland forest in favor of açaí fruit production and trading. However, the thriving re-growth of fallows subjected to swidden agriculture allows manioc agriculture to be reconsidered at any time if the need arises. Within this framework, variability in land use intensification can be re-interpreted in terms of flexibility of economic and ecological strategies, rather than in terms of site-specific input/output ratios at one point in time. In this sense, another important aspect to be considered in a 'multilevel' analysis of intensification is related to the unit of observation and scale of analysis of land use systems. In order to place site specific measures into a regional perspective, one needs to scale up from a garden plot, to a farm, to a population, to a landscape, and finally to a regional context of intensification.

Study Cases and Data

Examples from the Amazon estuary (Ponta de Pedras, PA) and the Transamazon region (Altamira, Brasil Novo, Medicilândia, PA) are used in this chapter to represent caboclo and colonist populations, respectively. The data presented here are part of a larger data archive. In this context, the chapter takes advantage of a considerable number of publications spanning the last 14 years by a collaborative research network to support a series of detailed information too large to be expanded upon here. These works are cited below to refer to different aspects discussed in the text.

The Açaí Fruit Case

Brief Overview of the Study Area

The study area is located in the estuarine region of the Amazon, on Marajó island, in the municipality of Ponta de Pedras, state of Pará (Fig. 8.1). Special attention has been given to three populations representing different economic

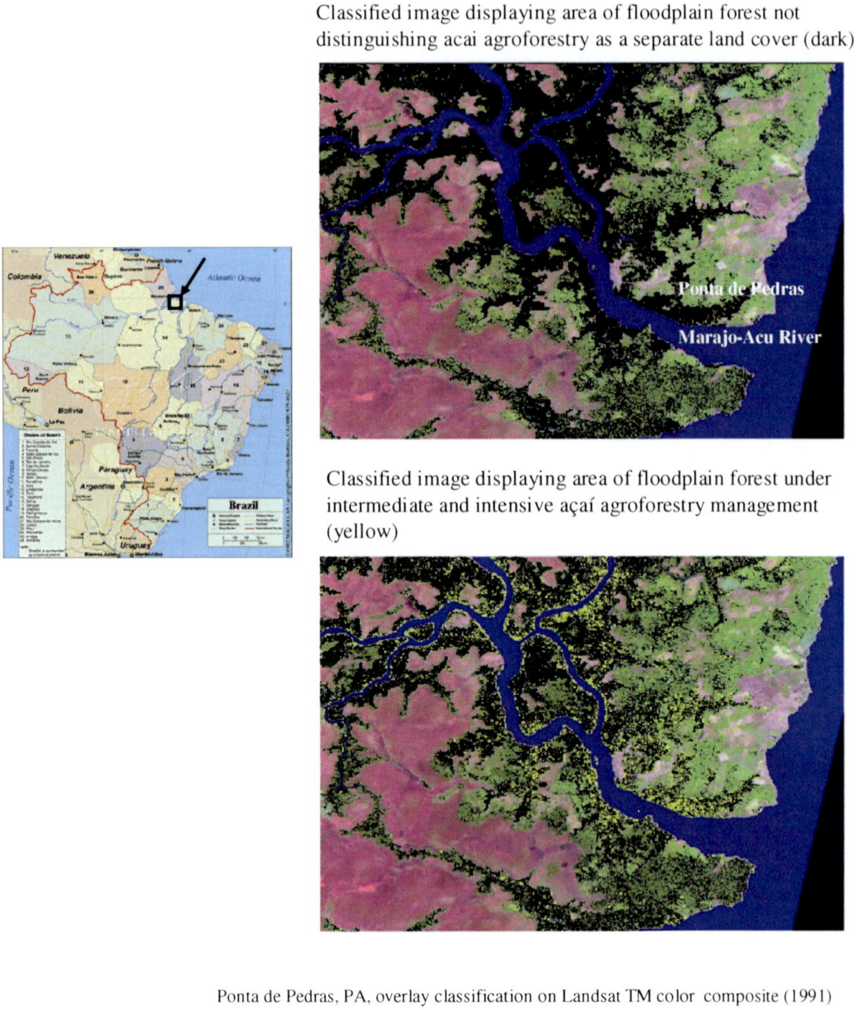

Fig. 8.1 Invisibility and visibility of açaí agroforestry, Ponta de Pedras, PA, Brazil

and land use patterns. Ethnographic, socio-economic, and ecological accounts concerning these populations can be found in Murrieta, Brondizio, Siqueira, & Moran (1989, 1992); Siqueira et al. (1993); Siqueira (1997); Neves (1992); Brondizio & Neves (1997); Brondizio et al. (1994) et al. (1994a); Murrieta (1994). Methodology and results concerning the use of satellite images to classify land cover classes, including unmanaged floodplain forest, açaí agroforestry, and three stages of forest succession can be reviewed in Brondizio, Moran, Mausel, & Yu (1993); Brondizio, Moran, Mausel, & Wu (1994a);

Brondizio et al. (1994b); Brondizio (1996, 1999); Moran, Mausel, & Brondizio (1994); Mausel, Wu, Li, Moran, & Brondizio (1993).[7]

Particularly relevant to this chapter is the data focusing on understanding the production system of açaí agroforestry. Vegetation stand inventories were developed for twelve fields characterizing different levels of management of açaí agroforestry. For fruit production experiments, four different producers, and eight different sites were selected to measure açaí production during the whole harvesting season of 1994–1995. A total of 20 production sites in different stages of management were studied. As part of the work, interviews were carried out with different segments of the açaí economy, such as small, medium, and large producers, and with different categories of share-croppers, market brokers, *carregadores* (porters), local and itinerant middleman, as well as processors and exporters.

Five local producers of açaí fruit have collaborated in this research with private data about daily production and prices of açaí fruit from 1984 to 1995. The price of açaí fruit and transportation (from Ponta de Pedras to Belém) was adjusted in relation to currency changes over a 10-year period (1984–1995). The Brazilian currency has changed five times during this period, thus requiring rectification of values before any price index could be derived. Two indices were developed: Açaí Price Index (API) and Açaí Freight Price Index (AFPI). One index was selected as the most relevant for comparison due to its regional characteristics: IPA-PA (Agricultural and Husbandry Price Index for the Pará State), published monthly by Fundação Getúlio Vargas (Conjuntura Economica, 1984–1995).[8] Data representing the distribution of the percentage of yield during each month of the season for each of the experimental sites allow us to derive figures respective to revenue/site/month[9] (Brondizio, 2008).

---

[7] Data representing Caboclo populations of the Amazon estuary (Ponta de Pedras, PA) include Landsat TM-based land use change analysis (1980s and 1990s), socio-demographic and land use surveys (86 households), experimental plots of açaí fruit production (season of 1994–1995), daily price variation and ethnographic material on agroforestry systems and resource management. Analysis includes measures of management intensity, productivity, labor allocation, market transactions, and economic return to compare this system to other regional land uses.

[8] This index is an official indicator of prices received by farmers in Pará State for agricultural and husbandry products. Its regional focus is especially important since it more closely reflects the economic context faced by açaí producers. The index is based on the price received by farmers for 24 agricultural products (including annual, biannual, and perennial crops) and seven husbandry products (including beef and poultry).

[9] To calculate the revenue on each of the experimental sites, the amount harvested each month was multiplied by the average monthly price of açaí fruit. The "net revenue" was calculated by discounting the cost of transportation and wage when applied. The production season stretched from September to February (data from market and experimental sites), although clear variations existed across the estuary.

Summary of Results: caboclo Case

Brief Overview on the Market Growth of Açaí Fruit

The market of açaí fruit has increased exponentially in the last 30 years (Fig. 8.2). A legacy of 'indigenous diet' and production technology, açaí fruit has been a top-ranked staple food and a key cultural symbol of estuarine life for a long time. Along with manioc flour, açaí fruit has continuously provided a caloric base for the rural diet throughout the different historical periods of the region, from floodplain chiefdoms to missionary occupation to the period of social transformation marked by directorate policies all the way to the boom and bust of the rubber economy (e.g., Wallace, 1853). In recent decades, açaí production continues to increase in order to meet the increasing 'staple food' demand prompted by low-income urban population growth after 1970 (Lopes, Souza, & Calvazara, 1982; Strudwick & Sobel, 1988) as well as an increasing external demand prompted by the emergence of a national and international 'fashion food' market that began in the early 1990s. For a detailed discussion of the social and economic history of açaí fruit expansion see Brondizio 2004.[10]

Açaí fruit looks like a blueberry, but only the appearance is similar. An *açaí* fruit is hard since it is a round seed covered with a thin mesocarp. The process of

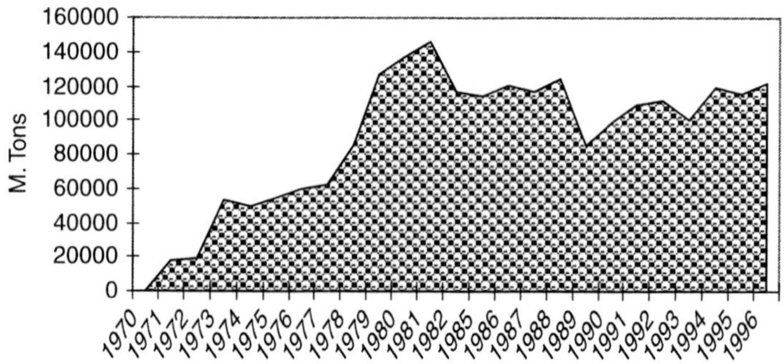

Fig. 8.2 National production of açaí palm Fruit (*Euterpe oleracea* Mart.) 1974–2000, FIBGE data. For an update see Brondizio 2008

[10] An important advance in the growth of the açaí economy during the early 1970s was the development and dissemination of electric machines used to process açaí pulp to make *vinho do açaí* (açaí juice). These machines replaced the *amassadeiras de açaí* (women who crush the fruit by hand), and hand processors made of wood. Nevertheless, the latter are still the main means of processing in rural households. Manual açaí processing requires hard labor and could not handle large quantities of fruits as required by the large urban market. Despite its dominant consumption by low-income populations, açaí is valued by other socioeconomic urban classes, not only in the form of staple food, but as a delicate dessert (ice cream, pudding, liquor, cake, among others). More recently, açaí juice has become popular throughout Brazil. For a detailed ethnographic account of açaí uses see Strudwick & Sobel (1988), Brondizio (2008), Brondizio 2004, Brondizio, Safar, & Siqueira 2003.

making the juice involves the removal and dilution of the mesocarp. Açaí juice is a purplish liquid of varied thickness, depending on how it is prepared.[11] The juice can be bought daily at numerous açaí stalls in urban areas in the region or prepared at home. Its popularity transcends social classes in the region, although it plays different roles in the diet of different groups. It is an important caloric source for the urban poor as well as for rural populations. Reports (e.g., Rogez, 2000, FIBGE, 1974–2000) indicate that consumption of açaí juice in Belém moved from 90,000 liters/day in the late 1980s to an estimated 400,000 liters/day in the late 1990s; this figure gives an estimated consumption of more than 60 liters/person/year, or, as noted by Rogez (2000), twice the amount of milk. Unknown until recently to most Brazilians, the expansion of açaí fruit consumption has been based on a myriad of new forms of preparation aiming at transforming the food in dietary and symbolic values. Recent examples of the expansion of the açaí fruit internationally include feature articles at the Gourmet Magazine (July 2002) and the celebrity-centered magazine InStyle (April 2002). One of the key distributors of açaí pulp in the United States, Sambazon, Inc., lists dozens of retailing shops across most American states and features in its website celebrity accounts and recipes on the health wonders of açaí juice. Distributors, although often facing importation constraints regarding hygienic safety of the product, have been able to grow by combining the 'health' and 'green' markets and focusing on the youth sectors such as surfers, skaters, conservationists, and those who are health conscious.

The so-called *açaízação*[12] of the estuary symbolizes the importance that açaí agroforestry has gained during the last 30 years in the region. The growth of the açaí economy is represented by two main industries, namely açaí fruit and heart of palm. Although sharing a common resource basis, *Euterpe oleracea* Mart., these industries have taken relatively independent paths and are based on socioeconomic structures not necessarily integrated.[13]

Today, açaí fruit is the most important income source next to government aid for a vast majority of riverine households. One can confirm this by looking at data from the regions of Ponta de Pedras (POEMA, 1994), Abaetetuba (Hiraoka, 1994), and the islands (e.g., *Ilha das Onças*) (Anderson & Ioris, 1992). In a Ponta de Pedras community for instance, açaí represents 64% of household income generated from agricultural products (including rice, beans,

[11] Rogez (2000) presents the most detailed study on açaí composition and processing.

[12] The term *açaízação* was used by Hiraoka (1994) to express the phenomenon of expansion of açaí agroforestry areas in the region.

[13] Despite other potential industrial uses, such as paper pulp (trunk), oil (fruit/pulp), animal food (fruit/seed), and ink (fruit/pulp) (Calzavara, 1972; Lopes et al., 1982; Strudwick & Sobel, 1988), there has been no significant commercial application of açaí besides heart of palm and fruit. However, it is important to consider the role açaí played during the 1960s in supplying fuel (use of stems as charcoal) to the brick (*olarias*) industries that prospered in the estuary during that decade, and even today in some areas, such as in Abaetetuba (Calzavara, 1972; Hiraoka, 1994).

and coconut). In Abaetetuba, açaí fruit is responsible for 50% of the household income of families involved in agroforestry, whereas in Ilha das Onças, açaí reportedly represents 63% of the income generated by commercial products (POEMA, 1994). The evolution of the açaí economy in the past thirty years has created a complex structure of production, distribution, commercialization and processing significantly specialized and ranked (for a detailed description, see Brondizio, 2008).

Management and the 'Invisibility' of a Production System

Açaí agroforestry management has been the focus of numerous works in the estuary (Calzavara, 1972; Anderson, Gely, Strudwick, Sobel, & Pinto, 1985; Jardim & Anderson, 1987; Anderson, 1988, 1990; Anderson & Jardim, 1989; Anderson & Ioris, 1992; Brondizio et al., 1993; Brondizio, Moran, Mausel, & Wu, 1996; Moran et al., 1994; Brondizio, 2008). Contrary to a system based on extractivism, management and planting of açaí agroforestry requires clear input of specialized agricultural and forestry labor in order to maintain and increase the stand crop productivity. Different management and planting strategies transform these areas into açaí agroforestry, locally called *açaizais*. The term encompasses different intensities of management (tree, sapling, seedling population densities and structure), and a diverse range of species composition. Despite encompassing a large range of management stages, the term *açaízal* is designated in this work as açaí agroforestry. The three main means of açaí agroforestry development are: (1) management of native stands; (2) planting of açaí stands following annual or biannual crops – that is, *roçado de varzea*; and (3) combined management and planting in native stands. In simple terms, management of açaí stands can be understood on two different levels: forest stand and plant levels. On the forest stand level, thinning and weeding techniques are used. On the plant level, management focuses on pruning techniques.

Stand thinning and *selection* control the density of individuals of all species competing with açaí and the balance between açaí basal area and other species. *Propagation* constitutes the planting and dispersion of seedlings and seeds of açaí, while simultaneously introducing other economic species to the stand. Finally, *pruning* controls the selection of productive clumps and stems. In the case of pure planted stands, i.e., roçado de várzea there is a need to include inter-cropping techniques between annual and perennial crops. These techniques demand intensive care of the crop site, including weeding, pest control and pruning of other crops. Despite the considerable modification of species composition, the managed areas largely retain the functional and structural characteristics of the floodplain forest – but with an overwhelming concentration of individuals of economic value.

The production pattern resulting from the experimental sites closely corresponds to the patterns found in relation to level of management at the sites were inventories were carried out. The three basic groups of açaí agroforestry distinguished by variation in stem/clump density can thus be related to fruit yield/

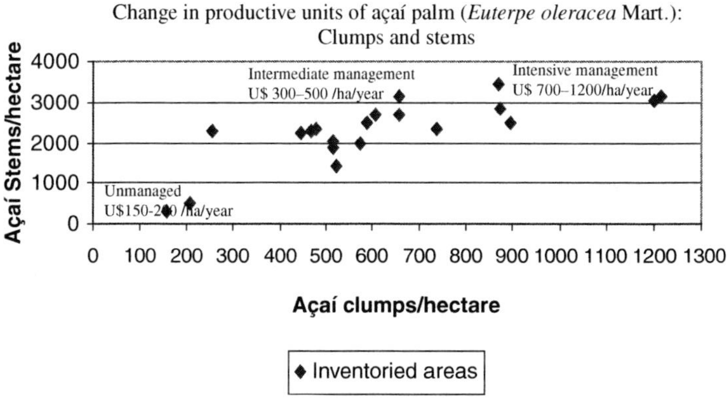

Fig. 8.3 Levels of management and range of economic return in açaí fruit production areas (adapted from Brondizio & Siqueira, 1997)

production (Fig. 8.3). Group 1, occurring in unmanaged sites evidences an average of 250 clumps/ha. In this group production output averages around 1,390 kg/ha/yr, that is, an average of 116 fruit baskets/ha. Group 2, occurring in initially and intermediately managed sites, has an average of 600 to 730 clumps/hectare. In this group, output production varies between 2,600 to 3,780 kg/ha/yr, i.e., an average of 269 fruit baskets/ha. Finally, group 3, characterized by more intensively managed sites, has an average between 890 and 1,200 clumps/ha. In this group, production varies more widely from 6,400 to 12,200 kg/ha/yr, an average of 760 fruit baskets/ha. Respectively, economic return in these groups range between US$ 150 – 200 / ha/year, US$ 300 – 500 / ha/year, and US$ 700 – 1,200 /ha/year. However, in all cases, the economic return depends upon harvesting schedules in relation to price fluctuations during the harvesting season. Based on our estimate integrating field inventories and Landsat TM data (Brondizio et al., 1996), the area under intensive açaí planting and management represents about 6% of the study area, surpassing in economic and spatial importance any other production system in the region. Whereas, if considered a mere 'extraction forest' (floodplain forest) this immense area under direct productive management tends to go unnoticed (see Fig. 8.1 to compare maps including and excluding açaí agroforestry as a separate land use class). These figures show the potential invisibility of this production system due to its forest characteristic and subtle differences between managed, planted, and unmanaged stands.

Responding to Price and Long Term Choice

Comparisons of the açaí price index (IPA) for the period between 1984 and 1995 are presented in Figs. 8.4 and 8.5. As the figures show, the pattern of increase is marked by seasonal variation of fruit production. It reflects a general pattern of

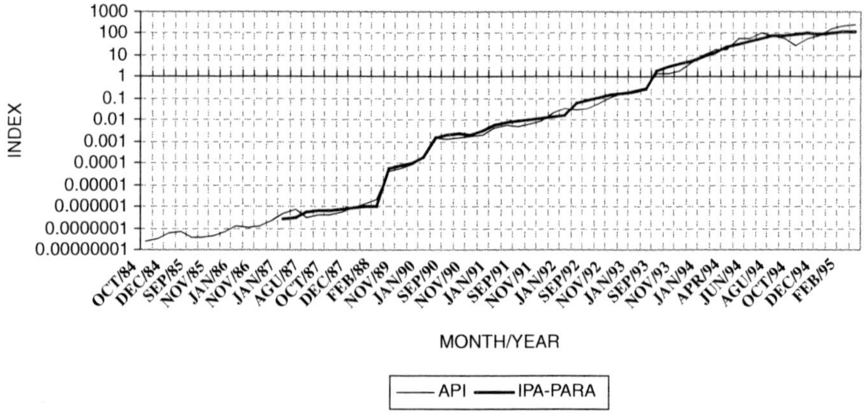

Fig. 8.4 Evolution of Açaí Fruit Prices 1984–1995: Açaí Fruit Price Index (API) & Para State Agro-pastoral Price Index (IPA) (adapted from Brondizio et al., 2003)

supply exceeding demand during the peak of the production season followed by the opposite trend towards the end. In order to put açaí prices into perspective, one can compare them with other indices, such as the Agricultural and Husbandry Index for the State of Pará (IPA-PARA), both shown in Fig. 8.4. This figure shows a similar growth of both indices. This is an important parameter in the success of the açaí economy over the ten-year period of study. Using a ratio between the two indices (Fig. 8.5) one observes that the açaí price index has followed and surpassed the inflation rates of the main rural products of the state (note price trajectories above value 1). Overall, açaí producers seem to have

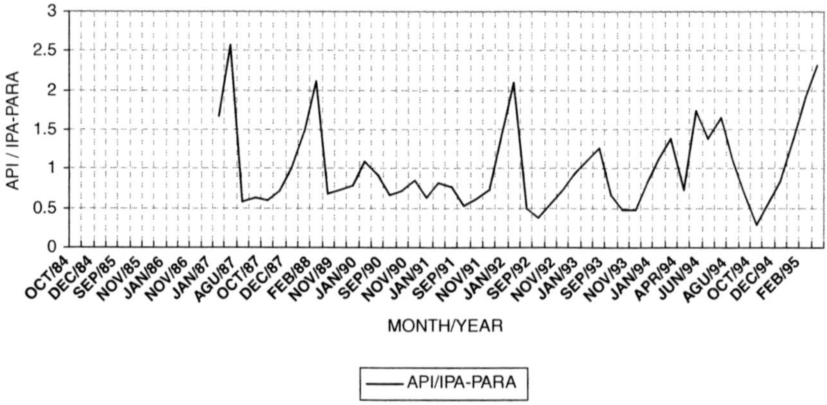

Fig. 8.5 Comparative performance 1984–1995: Açaí Fruit Price Index (API)/Para State Agricultural and Husbandry Index (IPA) (adapted from Brondizio et al., 2003)

received a better price than the average price of all agricultural and husbandry products of Pará. Analyzing the evolution of this ratio, we can see that açaí producers had an incentive to grow açaí, as its prices have followed those of other products, and during the end of the harvesting seasons, even surpassed them. We can, roughly speaking, assess the different opportunities caboclos face, and whether the choice of intensifying açaí production has an economic basis reflecting trends in regional markets.

As previously suggested, the main difference in economic return is related to the land tenure condition of the producer (Brondizio et al., 2003). In these experimental areas, the small owners spread out harvesting over the whole season, and thus had the chance to wait for higher prices, whereas the sharecropper producer was ordered to completely harvest his production area, mostly during November and December. During this period, açaí attains its lowest market price. In addition to low prices during the months of October and November, the cost of transportation during this period was higher in comparison to fruit prices.[14] The resulting variation in revenue is more related to harvesting period than to total area and intensity of production. Two main factors are working here. The first is the decision about the beginning and periodicity of harvesting, and the second is the decision concerning to whom they will sell their products. The decision regarding the harvesting period is directly related to market fluctuations and household needs. Since market supply is concentrated during the middle of the main harvesting season, it is likely that better prices can be demanded at the beginning or at the end of the season. Whereas owners have the autonomy of taking the risk of waiting for better prices, sharecroppers need to follow their landlord's schedule and decisions. Thus, sharecroppers may be subjected to selling all their production when the lowest prices are in place.

The evolution of açaí prices observed during this decade has shown a respectable performance, even when compared to all major crops and husbandry products in the state. Another important point is the consistent market for the product during the last decade, which shows signs of a well-structured production system. Production has increased five-fold during the past two decades based on management and planting, rather than extraction from untapped sources. The increases in production and price maintenance have been followed by the emergence of a socioeconomic organization around production, distribution, marketing, and processing, introducing a new class of producers and workers emerging from an extractivist economy but already functioning as a category of agricultural producers.

---

[14] Owner-sharecropper relationships during the harvesting season are typified by a number of informal and formal rules in relation to harvesting periodicity and schedule, price, and transportation costs. It has become more frequent for owners to organize a general meeting with the sharecroppers to decide on these issues. Owners usually decide on a starting date for harvesting that coincides with that of different sharecroppers working on the same property.

The Colonist Footprint Case

Brief Overview of Study Area

Around 1970, with financial loans from international banks and multinationals, Brazilian government started a new 'modernization' program of the country as a whole, and especially for the Amazonian region. The National Integration Plan (*Plano de Integração Nacional – PIN*) was created and aimed at interconnecting the various parts of the Amazon region internally and with the rest of the country, while inducing human occupation of the region through governmental programs of colonization (Moran, 1981; Mahar, 1979, 1988, among many others).

The colonization plan aimed to settle 100,000 families in 100-hectare lots along the Transamazon Highway in 5 years. According to the plan, small farmers would specialize in food crops in the first three years, and each year colonists would plant more of their land in permanent and cash crops such as coffee, sugar, black pepper and guaraná. The colonist should also leave 50% of his total area as a reserve of untouched forest. The bulk of candidates were landless people from the Northeast and from other parts of the country. Candidates from the South and Southeast regions, the most economically developed regions of Brazil, were considered essential as 'cultural brokers,' since government planners believed colonists from these regions could bring innovative technologies to the area and would help 'modernize' colonists from other parts of the country (Moran, 1981). However, the first three years of colonization were considered unsuccessful, and they were interpreted as a failure of the plan (Browder, 1988; Hecht, 1985; Ianni, 1979; Mahar, 1979, 1988; Velho 1972, among others). 'Blaming the victim' (Wood & Schmink, 1979) is probably the best expression to describe the end of the government-directed small farmers colonization projects in the Amazon. After 1974 the government changed its focus from small farm colonization and started a process of financing large enterprises, such as cattle ranchers, mining, lumber and large-scale agriculture for export (a process well described in Moran, 1981). Fewer and fewer resources were devoted to colonization projects. By 1980, INCRA (*Instituto Nacional de Colonização e Reforma Agrária*) – the governmental institution responsible for the colonization project – recognized that less than 8,500 families had been settled in the Amazon through their program (Miranda, 1990:41).

The Altamira region was one of most important *foci* of the government colonization program briefly described above. Altamira grew from a small riverine town based on rubber collection into a booming town of 85,000 due to agro-pastoral production stimulated by the Transamazon Highway built in 1971. It has one of the most significant patches of alfisols, or *terra roxa estruturada eutrofica*, in the Brazilian Amazon.

The study area is defined by a group of approximately 3,800 farm lots arranged according to different adjacent projects implemented by INCRA during the past 30 years. It cuts across the municipalities of Altamira, Brasil

Novo and Medicilândia, in the state of Pará (see Fig. 8.6) and encompasses an area of about 355,000 ha, stretching approximately from Km 18 to Km 140 of the Transamazon Highway west of the town of Altamira. By re-constructing the history of occupation of the study area through remote sensing data, we were

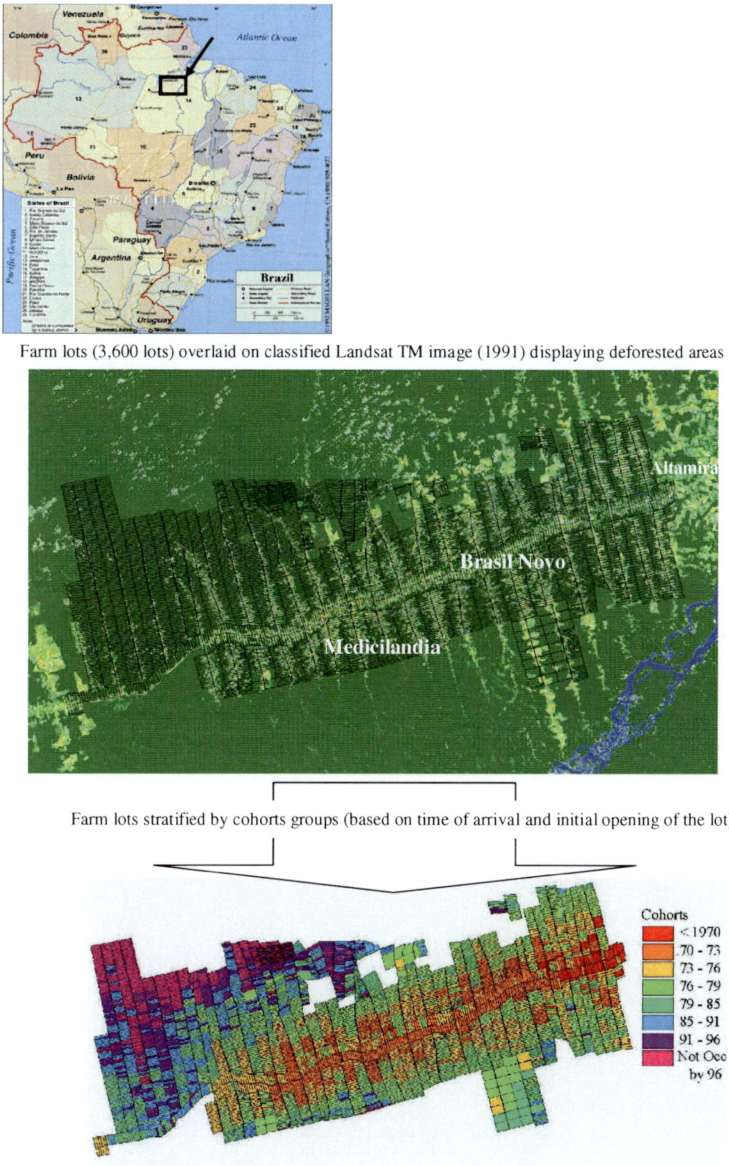

Fig. 8.6 Colonization settlements along the Transamazon Highway, Municipalities of Altamira, Brasil Novo, and Medicilândia

able to stratify farm lots by time of arrival and deforestation trajectory. The data presented in this chapter are part of a broader study, which addresses the relationships between household demography and socio-economic characteristics and the patterns of land use observed at the level of the farmer's individual plot.[15] Figure 8.7 highlights variation across neighboring farm lots. In this

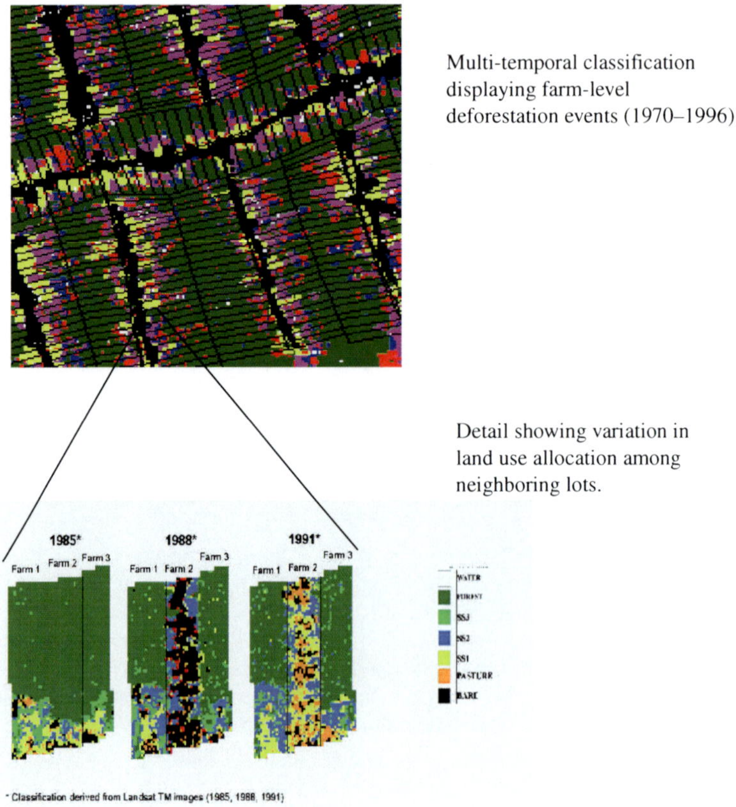

Fig. 8.7 Variation in farm-level land use allocation and land cover change. Examples from a stretch of the Transamazon Highway

[15] Data representing colonist populations of the Transamazon highway (Altamira, PA) include 402 detailed socio-demographic, economic, and land use questionnaires (household/farm lot level) sampled across colonization cohorts using a spatially-georeferenced sampling frame (3,800 farm lots). This includes remote sensing data capturing the entire period of frontier occupation (1970 to 1996) and a property grid that allows analysis at the level of settlement, cohort of farms (8), and household/farm lot (3,800 farm lots) level.

project, we highlight the need to study land use change on the frontier as resulting from both temporarily defined *period effects*, such as fluctuations in migration, different credit policies, inflation, etc.; *cohort effects* associated with the arrival and occupation of farm lots by groups of families, and; *age effects* associated with the transformation over time of households and their farms (McCracken et al., 1999; Moran, Brondizio, & McCracken, 2002; McCracken, Siqueira, Moran, & Brondizio, 2002; Brondizio et al., 2002; Siqueira, McCracken, Brondízio, & Moran, 2003).[16]

Summary of Results: Colonist Case

Cycles of Farm Lot Formation

Consolidating a farm in an Amazonian frontier puts the colonist in a paradox: having to open a rural property, consolidate its land use, and at the same time 'avoid' deforestation. This is an awkward position wherein they are either victims or aggressors depending on one's perspective. These issues raise questions about the role small colonist farmers are expected to play in frontier areas, the role of government policies, and the role of the scientific community in evaluating the causes and consequences of frontier occupation. Combining the need to develop a settlement, provide for their families, and cope with a new environment, a farmer's decisions about how much to deforest, what to plant, how to expand and consolidate a farm lot play a key role in their future success in the area (Brondizio et al., 2002; Futemma & Brondizio, 2003).

Figure 8.8 summarizes deforestation trajectories by taking into account average deforestation on farm lots across cohorts. Deforestation trajectories present a clear pattern across cohorts. Pulses of deforestation associated with crop and pasture development and secondary succession management mark these cycles of lot formation. Independent of cohort group, frontier farms show a developmental process associated with periods of establishment, expansion, and consolidation of land use activities. The magnitude of these pulses of deforestation relates to the interaction between farmers' decisions (in the household sense) and regional period effects ('signals' as used earlier), such as changes in economic, institutional, and infrastructure conditions motivating or inhibiting a particular land use behavior. We observe that intervals between pulses during stages of expansion and consolidation of a farm lot reflect processes of intensification and extensification, and relates to time of settlement, soil fertility

[16] In terms of economic and land use/agriculture trajectories, our study area can be divided into three main periods: 1972–1978-subsistence crops, such as rice, beans were dominant in the region; 1978–1988-highest production of perennial crops, such as cocoa and black pepper; 1988- to now, cattle ranching expansion and co-existence with other farming activities (Castellanet et al., 1994).

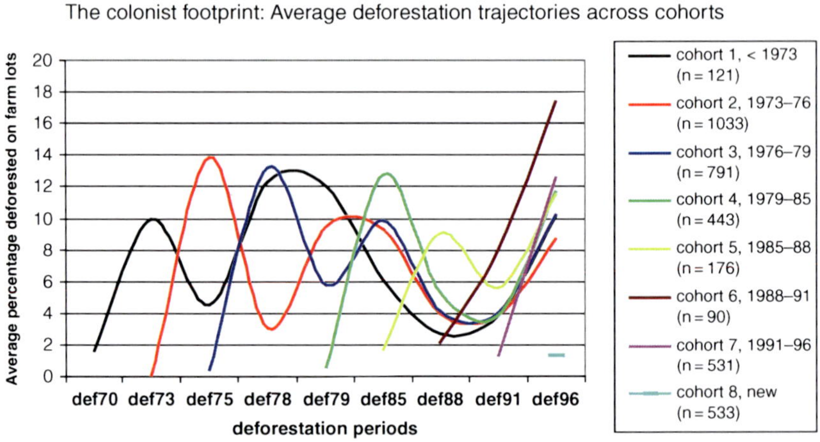

Fig. 8.8 Deforestation trajectories by colonization cohorts (distribution of deforestation events averaged by cohort of farm lots) (from Brondizio et al., 2002)

in the lot, available household labor, and opportunities created by credit and market – so-called period effects (McCracken et al., 2002).

Whereas positive significant correlation exists between time of settlement and deforestation, this is offset by the internal variability within cohorts, which is stronger than across cohorts (see Brondizio et al., 2002 for more detail). Such variability is even stronger in older cohorts suggesting variation in rate, extension, and direction of land use change probably associated with different trajectories in household economic strategies, composition, and in farm production potential. Decisions regarding deforestation may be taken to seize a 'period' opportunity, but not necessarily focus on long-term investment. This is the case, for instance, when a farmer allocates land to a particular crop in order to take advantage of a credit opportunity, then decides to discontinue the crop after the subsidy expires. As a result, large areas of secondary succession may appear. This reinforces the idea posed by the conceptual model presented by McCracken and colleagues (1999) that an initial period when farmers tend to deforest as much area as needed to establish their farm is followed by a consolidation period characterized by investment in perennial crop and secondary succession management.

The same conceptual framework used to explain the differential behavior of caboclos in relation to açaí fruit market signals applies here. Whereas signals of credit incentives and crop prices are regionally available, only farmers with necessary soil and labor endowment (as well as experience with the activity) are able to seize the opportunities. Figure 8.9 illustrates this point and the role of soil endowment upon land use decisions. Differences in soil quality explain much of the variance in crop choice and farmer persistence on rural properties. Upon arrival, most colonists did not recognize differences between alfisols and

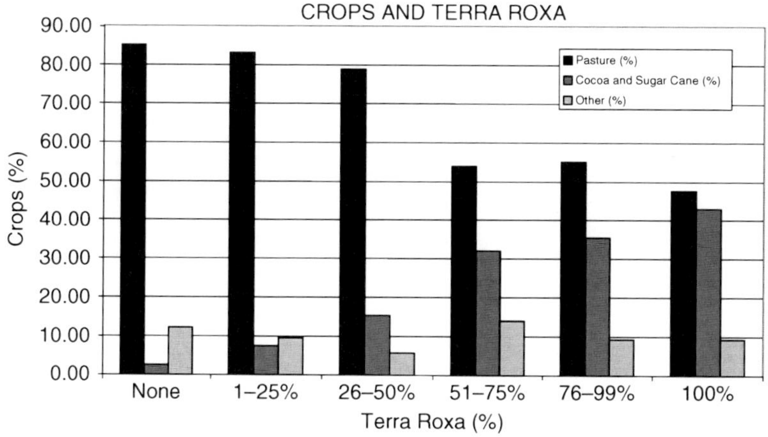

Fig. 8.9 Soil endowment (*% terra roxa*) and land use allocation among colonist farmers, Transamazon colonization area (adapted from Moran et al., 2002)

oxisols. However, over the past 25 years, colonists have learned the differences and today there is a clear association between the percentage of the property in alfisols and crop choice (see figure above on cacao and terra roxa). Crop choice is also constrained by the initial decision (and chance!) to locate on a specific property lot. Those who arrived early to the frontier acquired most of the plots with terra roxa– and these plots have not been resold as often as poor quality properties (see Moran et al., 2002).

Response to Credit Incentive

In many respects, it is difficult to quantify the effects of subsidized credit on the behavior of farmers in the Amazon, due to the lack of data availability and the difficulties in determining the effective use of the subsidized credit directed toward agriculture (Fiorini, Brondizio & McCracken, 2000). In the region of Altamira, more than half (56%) of the households interviewed received credit at least once. Figure 8.10 presents allocation of credit by type of land-use. Historical events clearly condition the variation and amount of credit allocated to different agricultural activities. Initial focus on annuals and perennials in the 1970s and early 1980s has changed to cattle ranching after 1991. During the 1990s, *Fundo Nacional do Norte* (FNO) has been almost the sole credit program available to small farmers; although often privileging cattle ranching, it mandates for the inclusion of a small area of some pre-defined perennials, which most of the time does not reflect the farmer's own crop choices (ibid.).

Credit used by the colonist farmer for equipment acquisition such as tractors has been generally unavailable and consistently low over the entire colonization period. This has been one of the main constraints to maintain opened areas in production as attested by most farmers we interviewed. In the area, the only

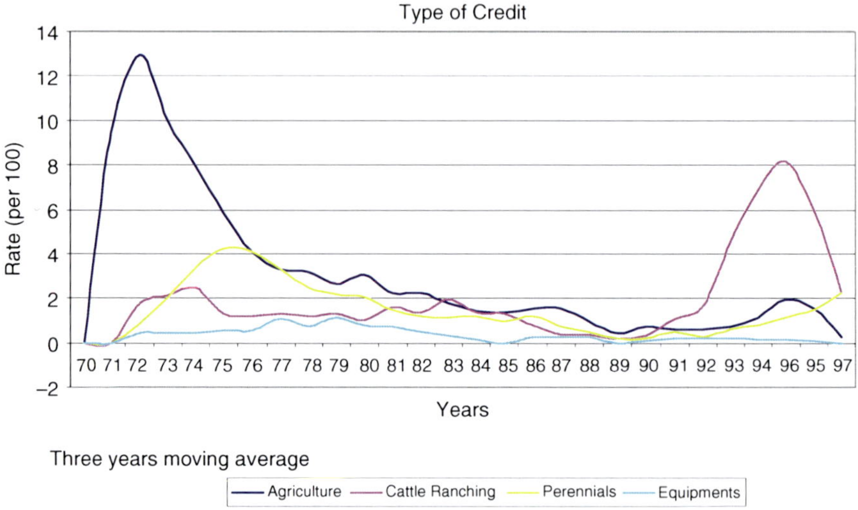

Fig. 8.10 Credit adoption by land use activity among colonist farmers, Trans-Amazon colonization area (adapted from Fiorini, Brondizio, & McCracken, 2000)

significant equipment is the chainsaw, owned by 77% of the farmers interviewed, followed by generators (owned by 22% of the farmers).

Data show that access to credit for agriculture in the area is not significantly correlated to variables like educational level or age of the household head, or to the economic conditions of the household at the time of settlement. On the other hand, Fiorini and colleagues (2002) observed a higher tendency to receive credit among farmers that arrived from the Northeast, Southeast and South of Brazil, or those born in the region, compared to farmers from other parts of Amazônia or the Midwest of Brazil. Previous experience with credit also was associated with higher levels of credit received. With regard to the characteristics of the lots, we observed a higher tendency to receive credit among farmers with *terra roxa* (62%) on their property compared to farmers without *terra roxa* (52%). The present study highlights clear links between 'frontier' farmers and regional and national credit policies. Credit represents a resource used by colonists to expand and/or consolidate a lot.

Discussion

Limitations of Intensification Theory to Capture Caboclo and Colonist Agrarian Systems

Application of conventional measures of intensification to Amazonian agriculture is challenged by numerous constraints. Nevertheless, as previously discussed, these limitations are not unique to the region, but rather, are common to the majority of small scale agriculture in the tropics. A primary problem is the focus on a single agriculture activity instead of land use systems in which an agriculture field fits as part of a larger economic strategy. This assertion recalls two concepts discussed above. The first concept is the importance of seeing agriculture within a spatial context, and the second is the co-existence of intensification and de-intensification of agriculture as part of a larger land use strategy (Futemma & Brondizio, 2003; Netting, 1993; Guillet, 1987). To these remarks one can merge a large body of literature on Amazonian floodplain populations showing more intensive use of the floodplain, associated with extensive swidden in the upland, both correlated to other economic activities, such as fishing, extractivism, hunting, cattle ranching, trading and off-farm jobs (Moran, 1989; Roosevelt, 1989; Denevan, 1984; Hiraoka, 1985; Padoch, 1989; Brondizio et al., 1994a).

The most problematic application of intensification models is related to agroforestry activities, especially in cases such as açaí agroforestry, where the distinction between agroforestry and native forest is not clear. Concerning the case of açaí agroforestry, the flaws of intensification measures to evaluate the production system can be explained (in summary) by the following five main reasons: (1) *technology* is based on indigenous management knowledge; (2) the *agroforestry structure* can fit into both extremes (intensive or extensive) of Boserup's frequency model; (3) *spatial dimensions* overlap areas of intensive, intermediate, and unmanaged areas; (4) the *multiple productive dimensions* produce a 'hidden harvest' within these areas, and; (5) *floodplain cycles* dictate cropping frequency more than fallow period (see detail discussion in Brondizio & Siqueira, 1997).

By the same token, colonisation areas challenge the application of conventional models of land use intensification based on fallow cycle and factors of production frequently used to explain the relation between agropastoral systems and deforestation in other areas. One of the most significant characteristics of these areas are the level of variability in deforestation and land use across farm lots of similar age and environmental conditions (example Fig. 8.7). The colonist footprint is characterized by the co-existence of extensification and intensification of production strategies marked by phases of expansion and consolidation of the farm operation. These processes, however, are characterized by high variation within farm cohorts resulting from differential rate, extent, and direction of land use change across farm lots. Understanding deforestation trajectories and the colonist footprint requires a combination of variables related to time of settlement (e.g., cohort and age effects), cohort and household dynamics (e.g., household labor composition, experience, origin), and period effects (e.g., credit, inflation), underlined by environmental, market, and infrastructural conditions. In these areas, agricultural systems combine activities aimed at increasing land value, consolidating tenure rights, and diversifying activities to minimize risks and to allow experimentation in a new environment.

Understanding these processes will help to put more attention on the improvement of existing infrastructure as well as value local experiences in order to help existing farmers maintain forest in their lots, increase agro-pastoral production, and improve the quality of life of their families – all of which are key elements for better policies aiming to decrease deforestation rates in the Brazilian Amazon.

Economic Rationality and Market Opportunities

Caboclos and small scale colonists are generally regarded as marginal actors of the regional market economy and are frequently questioned about the 'rationality' of their economic behavior, which is seen as backward and unfit to contemporary economic demands. A close look at the last twenty years of the açaí fruit economy, on one side, and the adoption of credit and land use systems by colonists on the other, shows signs to the contrary. Market demand and price changes have been the main motivation for the caboclos' decision to implement açaí agroforestry as their main agricultural activity.

The açaí agroforestry case helps us to re-think land use intensification in Amazônia from a market and sociocultural perspective. The caboclos' ability to participate in the açaí economy emerges from their aptitude to increase output production from an existing set of management techniques and knowledge, instead of adopting an exogenous system. This allows a progressive and flexible market insertion, which incurs fewer risks by combining subsistence and market outputs. However, whereas açaí production occurs across all property systems, the economic return of a producer is constrained by ones ability to decide when to best place one's yield on the market, a choice limited to sharecroppers. Thus, the açaí case reported here suggests that intensification in output production does not necessarily translate into improved return, but rather it depends on one's ability to take advantage of daily and weekly price fluctuations. In this case, land tenure, not a farmer's economic ability, explains the main differences between the marketing strategies of small owners, large owners, and sharecroppers.

Similarly, credit acquisition data show that colonists have responded to incentives, but in the case of most programs they were discontinued or mismanaged, leading to land use failures and abandonment of activities. Overall, most colonists have fulfilled their part (in planting and paying for credit), but were left without support (for instance during commercialization) or even roads to take advantage of their yields. The transportation and marketing infrastructure has historically been one of the main constraints of the regional economy. Distance, isolation, transportation means, and lack of capital have excluded a large number of Amazonian producers from taking part in the market without the reliance upon middlemen. In Amazônia, roads abandoned by discontinued policies are well known for letting successful crop yields rot on farmers' lots. This has created a complex structure of middlemen and a level of dependency on 'intermediate' markets for small farmers all over the region. This condition of unequal exchange further reinforces rural producer's invisibility in Amazônia. Colonists are eager to seize investment and market opportunities to consolidate their farm operations, but constantly face economic and infrastructure constraints.

Adaptation and Maintenance of Subsistence Basis

Worldwide, the economy of small-scale agriculturalists involves more than edible products (Ellen, 1982). The literature on non-timber products of tropical forests has flourished in the last few years, and the search for alternatives to deforestation and better knowledge of local economic strategies has taken place (Plotkin & Famolare, 1992; Nepstad & Schwartzman, 1992; Hecht, Anderson, & May, 1988). Studies have shown that forest products account for a considerable part of the local economy, and in some cases may exceed other activities, such as agriculture and ranching (Peters, Gentry, & Mendelsohn, 1989; Hecht, 1992). A good example is Hiraoka's work in the estuary (Abaetetuba) showing the importance of miriti palm (*Mauritia flexuosa*) to the household budget. The market of miriti fruit represents 13 to 15% of total household income (1994). In the estuary, açaí agroforestry provided staple food production and raw material. This can be contrasted for instance with local development projects that emphasized production of crops not consumed locally (Murrieta et al., 1992; Murrieta, 1994). In summary, access to a large portfolio of timber and non-timber products presented in agroforestry areas guarantees market independence in terms of raw materials, and an important part of both household and market economy that should be considered when accounting for agroforestry productivity.

In the case of colonists the focus on diversification of land use types reflects a strategy that helps to address market demand while providing for consumption. Planting annual crops (*lavoura branca* usually including rice, beans, corn, and manioc) followed by pasture formation allows one to address market and consumption needs while increasing property value by expanding the opened area. In most cases, areas of perennial crop are also present. By the same token, cattle ranching activities provide a stable source of income, less dependent upon transportation (easily sold on the property), storage, and price fluctuation (Hecht, 1993).

In both cases, we see a strategy that maintains intensive and extensive areas in production to attend both consumption and market needs. As previously noted, the co-existence of land use strategies can be re-interpreted in terms of flexibility and risk minimization by small farmers' households used to dealing with disadvantageous infrastructure and economic vulnerability.

Building upon Local Knowledge and the Need for Technological Support

It is common sense to say that caboclos have inherited the agricultural knowledge of pre-Colombian populations related to floodplain agriculture. However, as discussed earlier, the application of this knowledge has been shaped by both, historical factors, land tenure, and by available market opportunities. Although it appears technologically simple, this process involves specialized knowledge about the species and plant-soil interactions. Therefore, any account of technological input in these systems cannot rely simply on comparison with energy intensive agricultural technology, but should include considerations of accumulated management knowledge, specialized labor, and efficiency of production. The participation of caboclo farmers in the intensification of açaí fruit production reflects their technological background and ability to implement this knowledge. However, in the context of contemporary Amazonian economy, technological assistance for production and processing is a major need in these areas.

Colonists tend to reproduce systems that reflect their previous experience. However, most colonists are eager to experiment and develop new techniques to cope with the particularities of the local environment. Elsewhere, we have shown (Brondizio et al., 2002) that older colonists are able to keep open areas in production for longer periods. In part, this is due to their advantage in selecting the 'best' lots as they had a chance to interact with local caboclos and learn about soil selection criteria (Moran, 1981). It also reflects the trial and error experience they developed in the region. In areas such as the Transamazon, farmers have built up enough knowledge to inform at least better land use policies to foster regional development while attending to both market and household demand. Most colonists opt to increase the area deforested as a way to increase land value in the absence of technology and financial support in order to keep areas producing longer.

Conclusion

On one level, this chapter has tried to argue that the 'invisibility' of Amazonian caboclos is common to most smallholder rural populations in the region, despite their historical, cultural, geographic, and environmental differences. This is, at least in part, a result of historical social prejudice and misinterpretation of small-scale agriculture systems. However, at the same time, our data stress two important components that must be taken into consideration. First, particular to both cases, one finds an internal variability across and within these populations. Second, it becomes clear that sociocultural differences must be stressed and valued. In this sense, caboclos offer a unique contribution to the region based on their environmental knowledge, production techniques and management, and the historical context they bring, including a rich 'cultural inheritance' that characterizes Amazônia today. Even recent colonists, one may argue, bring a new blend of techniques that are transformed and translated into a new socio-economic and environmental reality that should not be dismissed. Today, colonists take the place of caboclos as the demographically dominant rural inhabitant.

In the context of creating a new economic development concept based on social justice and environmental grounds, it is important to modify our conception of caboclos and colonists as social categories as well. Improvement in their agricultural system should come with social-economic infrastructure and extension services that will help to target technological changes towards production systems without displacement of local resource bases. Estuarine and colonist populations focused upon in this chapter, as well as other rural populations of the region, have virtually no access to health and education services, nor to adequate agricultural credits. In such circumstances, underemployment in urban areas has been more attractive than farming. The lack of infrastructure in terms of energy, transportation, extension services, and cooperative organization lead rural producers to political isolation and continuous economic dependency on middleman and patronage. Products that have high market price, e.g. heart of palm (estuary) and lumber (Transamazon), have almost no economic value to the small producers. Small-scale transformation industries should be promoted as a way to aggregate value to local products, as well as increase employment and circulation of money within the region.

In summary, redefining the caboclos' identity as rural producers in the context presented by Netting's small farmer is a step towards overcoming the prejudices embodied in the term, as well as to overcome their extractivism background. This may help to change rural 'modernization' paradigms, which are so often presented on development projects in the region. Social and political recognition of caboclos as rural producers needs to be achieved and reinforced. Recognizing the role they play in regional agriculture may contribute to a shift that sees their production system no longer as extractivist but no less than forest farming (Brondizio & Siqueira, 1997). Similarly, Netting's small farmer framework may be applied to colonists. In lieu of nonexistent and inconsistent support, agrarian reform areas like along the Transamazon are turning to property aggregation, thereby repeating a vicious cycle of attraction and expulsion of small farmers.

At a regional level, an identity of Amazonian small farmers including a wide range of caboclos and colonists will surely help to increase political attention and foster more support. An important example is the rural union-based *Grito da Terra* initiative, which brings together all these categories under the identity of rural producers, furthermore linking them to a broader political and social movement in Brazil as a whole. Unfortunately, new development efforts, such as the *Avança Brasil* (1990s) and more recently PAC (2000s) programs are showing signs of repeating similar errors as its predecessor PIN. In various areas designated for agrarian reform, new incentives to large-scale soybean, logging, and ranching may continue to lead to high rates of lot aggregation, colonist out-migration and consequent swelling of urban areas. As a whole, we continue to reproduce a sense of shame, not pride, towards rural Brazilians, thus perpetuating a cloak of 'invisibility' for both caboclos and colonists alike. In this sense, the historical stigma of the term caboclo is just a reflection of a national, historic prejudice towards rural populations in general; as well as misinterpretation of agricultural intensification processes and the role of small-scale agricultural systems in the national economy.

Acknowledgments This paper was originally presented at the Workshop 'Sociedades Caboclas Amazonicas: Modernidade e Invisibilidade' (Amazon Caboclo Societies: Modernity and Invisibility) University of São Paulo, May 18–24, 2002. This paper benefits from the contribution of numerous projects and research efforts by a large number of colleagues from Indiana University, Museu Paraense Emilio Goeldi, University of São Paulo, Embrapa-Cpatu, and INPE, from 1989 to the present. I am especially thankful to my colleagues Emilio Moran, Walter A. Neves, Rui S. S. Murrieta, Andrea D. Siqueira, and Stephen McCracken. Research funds to colleagues and to the author include support from the NSF (PI Emilio Moran), CNPq (PI Walter Neves), NASA (PI E. Brondizio), NIGEC (PI E. Brondizio), McArthur (E. Brondizio), and NIH (PIs Emilio Moran and S. McCracken). Most important, it benefits from the collaboration of Amazonian farmers and people from Ponta de Pedras, Belém, Altamira, Brasil Novo, and Medicilândia as well as friends and local institutions throughout these places. At Indiana University, enormous support from colleagues, staff, and students at ACT, the Department of Anthropology, and CIPEC have been invaluable. I would like to thank the invitation and motivation to write this paper particularly from Cristina Adams and Walter A. Neves (for the workshop 'Sociedades Caboclas da Amazônia: Modernidade e Invisibilidade'), and comments and suggestions from Andrea D. Siqueira and Michael Sauer. Responsibility for the views expressed herein is the sole responsibility of the author, and does not represent those of the funding agencies, nor of other persons or institutions.

References

Anderson, A. B. (1988). Use and management of native forests dominated by Açaí palm (*Euterpe oleracea*) in the Amazon Estuary. *Advances in Economic Botany*, *6*, 144–154.

Anderson, A. B. (1990). *Alternatives to deforestation: Steps toward sustainable use of the Amazon rain forest*. New York: Columbia University Press.

Anderson, A., Gely, A., Strudwick, J., Sobel, G. L., & Pinto, M. G. C. (1985). Um sistema agroforestal na varzea do estuario Amazonico (Ilha das Oncas, Municipio de Barcarena, Estado do Para). *Acta Amazônica*, *15*(1–2 Supplementary issue, March/June), 195–224.

Anderson, A. B., & Ioris, E. (1992). The logic of extraction: resource management and income generation by extractive producers in the Amazon estuary. In R. Kent & C. Padoch (Eds.), *Conservation of neotropical forests* (pp. 175–199). New York: Columbia University Press.

Anderson, A. B., & Jardim, M. A. G. (1989). Costs and benefits of floodplain forest management by rural inhabitants in the Amazon estuary: A case study of Açaí palm production. In J. O. Browder (Ed.), *Fragile lands of latin America*. Boulder, CO: Westview Press.

Balee, W. (Ed.). (1998). *Advances in historical ecology*. New York: Columbia University Press.

Balee, W. & Posey, D. (Eds.). (1989). Resources management in Amazonia: Indigenous and folk strategies. *Advances in economic botany* (Vol 7). New York: New York Botanical Garden.

Beckerman, S. (1983). Does the Swidden Ape the jungle? *Human Ecology*, *11*, 1–12.

Boserup, E. (1965). *The conditions of agricultural growth: The economics of agrarian change under population pressure*. Chicago: Aldine.

Brondizio, E. S. (1999). Agroforestry intensification in the Amazon estuary. In T. Granfelt (Ed.), *Managing the globalized environment: Local strategies to secure livelihoods* (pp. 88–113). London: IT Publications.

Brondizio, E. S. (2004). From staple to fashion food: Shifting cycles, shifting opportunities in the development of the Açaí fruit (Euterpe oleracea Mart.) economy in the Amazon estuary. In D. Zarin et al. (Eds.), *Working forests in the American tropics: Conservation through sustainable management*? New York: Columbia University Press.

Brondizio, E. S. (2008). The Amazonian Caboclo: Forest farmers in the global market. New York: The New York Botanical Garden Press.

Brondizio, E. S., McCracken, S. D., Moran, E. F., Siqueira, A. D., Nelson, D. R., & Rodriguez-Pedraza, C. (2002). The colonist footprint: Toward a conceptual framework of deforestation trajectories among small farmers in frontier Amazônia. In C. Wood & R. Porro (Eds.), *Deforestation and land use in the Amazon* (pp. 133–161). Gainesville, FL: University Press of Florida.

Brondizio, E. S., Moran, E. F., Mausel, P., & Wu, Y. (1994a). Land use change in the Amazon estuary: Patterns of caboclo settlemnet and landscape management. *Human Ecology*, *22*, 249–278.

Brondizio E. S., Moran, E. F., Mausel, P., & Wu, Y. (1996). Changes in land cover in the Amazon estuary: Integration of thematic mapper with botanical and historical data. *Photogrammetric Engineering and Remote Sensing*, *62*, 921–929.

Brondizio E. S., Moran, E. F., Mausel, P., & Yu, W. (1993). Dinâmica da vegetação no estuario do Amazonas: Análise temporal do uso da terra integrando imagens Landsat TM, levantamento florístico e dados etnográficos. *Anais do 7°. Simpósio Brasileiro de Sensoriamento Remoto*, Vol.II, 38–46.

Brondizio E. S., Moran, E. F., Siqueira, A. D., Mausel, P., Wu, Y., & Li, Y. (1994b). Mapping anthropogenic forest: Using remote sensing in a multi-level approach to estimate production and distribution of managed palm forest (*Euterpe oleracea*) in the Amazon Estuary. *International Archives of Photogrammetry and Remote Sensing*, *30*, 184–191.

Brondizio, E. S., & Neves, W. A. (1997). A percepção do ambiente natural por parte de populações Caboclas do Estuário do Amazonas: Uma experiência piloto através do método de trilhas pré-fixadas. In C. Pavan (Ed.), *Uma estratégia Latino Americana para Amazônia* (pp. 167–182, Vol. I). São Paulo: Editora UNESP.

Brondizio, E. S., Safar, C. C. M., & Siqueira, A. D. (2003). The urban market of Açaí fruit (*Euterpe oleracea* Mart.) and rural land use change: Ethnographic insights into the role of price and land tenure constraining agricultural choices in the Amazon estuary. *Urban ecosystems* 6(1, 2): 67–98.

Brondizio, E. S., & Siqueira, A. D. (1997). From extractivists to forest farmers: Changing concepts of Caboclo agroforestry in the Amazon estuary. *Research in Economic Anthropology*, *18*, 234–279.

Browder, J. O. (1988). Public policy and deforestation in the brazilian amazon. In R. Repetto & M.Gillis (Eds.), *Public forest and the misuse of forest resources* (pp. 247–290). New York: Cambridge University Press.

Calzavara, B. B. G. (1972). As possibilidades do acaízeiro no estuário Amazônico. *Boletim Fundação de Ciências Agrárias do Pará*, *5*, 1–103.

Cancian, F. (1989). Economic behavior in peasant communities. In S. Plattner (Ed.) *Economic Anthropology*. Stanford: Standford University Press.

Castellanet, C., Simões, A., & Celestino Filho, P. (1994, October 20). *Diagnóstico preliminar da agricultura familiar na Transamazônica: Indicações para pesquisa-desenvolvimento*. Paper presented at Internal Seminar EMBRAPA/CPATU, Belém, Pará, Brazil.

Chibnik, M. (1991). Quasi-ethnic groups in Amazônia. *Ethnology*, *30*, 167–182.

Conjuntura Economica. (1983–1995). Rio de Janeiro: Fundação Getulio Vargas.

Conklin, H. C. (1957). *Hanunoo agriculture: A report of an integral system of shifting cultivation in the Phillippines*, FAO, Rome.
Conklin, H. C. (1961). The study of shifting cultivation. *Current Anthropology*, *2*, 27–61.
Denevan, W. (1984). Ecological heterogeneity and horizontal zonation of agriculture in the Amazon floodplain. In M. Schmink & C. H. Wood (Eds.), *Frontiers expansion in Amazonia*. Gainesville: University of Florida Press.
Denevan, W., & Padoch, C. (1987). Swidden-fallow agroforestry in the Peruvian Amazon. *Advances in economic botany* (Vol. 5). New York: The New York Botanical Garden.
Dove, M. (1983). Theories of swidden agriculture, and the political economy of ignorance. *Agroforestry Systems*, *1*, 85–99.
Ellen, R. (1982). *Environment subsistence and system*. Cambridge: Cambridge University Press.
FIBGE [Fundação Instituto Brasileiro de Geografia e Estatística] (1974–2000). *Produção da Extração vegetal e da silvicultura (Production of Vegetable Extract and of Silviculture)*. Rio de Janeiro: Departamento Agropecuário, Diretoria de Pesquisas.
Fiorini, S., Brondizio, E. S., & McCracken, S. (2000). Role of credit on the deforestation of an Amazonia frontier. *Poster presented at first international open science meeting of the large scale biosphere-atmosphere experiment in Amazônia* (LBA), CNPq, INPE, NASA, Belém, Para, Brazil, June 21–26, 2000.
Furtado, L. F. (1987). *Curralistas e Redeiros de Marudá: Pescadores do Litoral do Pará*. Belém: Museu Paraense Emilio Goeldi.
Futemma, C., & Brondizio, E. S. (2003). Land reform and land use changes in the lower Amazon: Implications to agricultural intensification. *Human Ecology*, *31*, 369–402.
Galvão, E., (1955). *Santos e Visagens-Um Estudo da Vida Religiosa de Itá, Amazonas*. São Paulo: Companhia Editora Nacional.
Guillet, D. (1987). Agricultural intensification and deintensification in lari, Colca Valley, southern Peru. *Research in Economic Anthropology*, *8*, 201–224.
Harris, M. (1998). What it means to be a Caboclo: Some critical notes on the construction of Amazonian Caboclo society as an anthropological object. *Critique of Anthropology*, *181*, 83–113.
Hecht, S. B. (1985). Environment, development and politics: Capital accumulation and the livestock sector in the eastern Amazon. *World Development*, *13*, 663–684.
Hecht, S. B. (1992). Valuing land uses in Amazonia: Colonist agriculture, cattle, and petty extraction in comparative perspective. In K. H. Redford & C. Padoch (Eds.) *Conservation of neotropical forests* (pp. 379–399). New York: Columbia University Press.
Hecht, S. B. (1993). The logic of livestock and deforestation in Amazonia. *BioScience*, *43*, 687–695.
Hecht, S. B., Anderson, A. B., & May, P. (1988). The subsidy from nature: Shifting cultivation, successional palm forests, and rural development. *Human Organization*, *47*, 25–35.
Hiraoka, M. (1985). Zonation of mestizo riverine farming systems in northeast Peru. *National Geographic Research*, *2*, 354–71.
Hiraoka, M. (1992). Caboclo and ribereño resource management in Amazonia: A review. In R. Redford & C. Padoch (Eds.), *Conservation of neotropical forests* (pp. 134–157). New York: Columbia University Press.
Hiraoka, M. (1994). The use and management of 'Miriti' (*Mauritia flexuosa*): Palms among the ribeirinhos along the Amazon estuary. *Paper presented at whitewater varzeas: Diversity, development and conservation of amazonian floodplain*, Macapa, Amapa, Brazil, December 12–14, 1994.
Ianni, O. (1978). *A Luta pela Terra*. Petrópolis: Vozes.
Jardim, M. & Anderson, A. (1987). Manejo de populações nativas de Açaízeiro no estuario Amazonico-resultados preliminares. *Boletim de Pesquisa Florestal*, *15*, 1–18.
Lima, D. M. (1992). *History, social organization, identity and outsiders' social classification of the rural population of Amazonian region*. Unpublished PhD. dissertation, King's College, Cambridge.

Lopes, A. V. F., Souza, J. M. S., & Calvazara, B. B. G. (1982). *Aspectos econômicos do Açaízeiro*. Belém: Sudam.

Mahar, D. (1979). *Frontier development in the Brazilian Amazon: A study of Amazônia*. New York: Praeger.

Mahar, D. (1988). *Government policies and deforestation in the Brazilian Amazon region*. Washington, D.C.: The World Bank.

Mausel, P., Wu, Y., Li, Y., Moran, E. F., & Brondizio, E. S. (1993). Spectral identification of successional stages following deforestation in the Amazon. *Geocarto International, 8* 61–72.

McCracken, S., Brondizio, E. S., Nelson, D., Moran, E. F., Siqueira, A. D., & Rodriguez-Peraza, C. (1999). Remote sensing and GIS at farm property level: demography and deforestation in the Brazilian Amazon. *Photogrammetric Engineering and Remote Sensing, 65*, 1311–1320.

McCracken, S., Siqueira, A. D., Moran, E. F., & Brondizio, E. S. (2002). Land use patterns on an agricultural frontier in Brazil; Insights and examples from a demographic perspective. In C. Wood & R. Porro (Eds.), *Deforestation and land use in the Amazon* (pp. 162–192). Gainesville: University Press of Florida.

Miranda, M. (1990). Colonização Oficial na Amazônia: O Caso de Altamira. In B. Becker, M. Miranda & L. Machado (Eds.), *Questões Sobre a Gestão do Território* (pp.35–46). Brasília: UNB and UFRJ.

Moran, E. F. (1974). The adaptive system of the Amazonian caboclo. In C. Wagley, (Ed.), *Man in the Amazon* (pp. 136–159). Gainesville: The University of Florida Press.

Moran, E. F. (1981). *Developing the Amazon*. Bloomington: Indiana University Press.

Moran, E. F. (1989). Models of native and folk adaptation in the Amazonia. In D. Posey & W. Balee (Eds.), *Resources management in Amazonia: Indigenous and folk strategies. Advances in economic botany* (Vol. 7, pp. 22–29). Bronx: The New York Botanical Garden.

Moran E. F., & Brondizio, E. S. (2001). Human ecology from space: Ecological anthropology engages the study of global environmental change. In M. Lambek & E. Messer (Eds.), *Ecology and the sacred: engaging the anthropology of royal A. Rappaport* (pp. 64–86). Ann Arbor: University of Michigan Press.

Moran, E. F., Brondizio, E. S., & McCracken, S. D. (2002). Trajectories of land use: Soils, succession, and crop choice. In C. Wood & R. Porro (Eds.), *Deforestation and land use in the Amazon* (pp. 193–217). Gainesville: University Press of Florida.

Moran E. F., Mausel, P., & Brondizio, E. S. (1994). Secondary succession and land use in the Amazon. *National Geographic Research and Exploration, 10*, 456–476.

Murrieta, R. S. (1994). *Diet and subsistence: Changes in three caboclo populations on Marajo Island, Amazonia, Brazil*. Unpublished Master Dissertation, University of Colorado, Boulder.

Murrieta, R. S. S., Brondizio, E. S., Siqueira, A. D., & Moran, E. F. (1989). Estratégias de subsistência de Uma população ribeirinha da Ilha de Marajó, Brasil. *Boletim do Museu Paraense Emilio Goeldi. Serie Antropologia, 5*, 147–63.

Murrieta, R. S. S., Brondizio, E. S., Siqueira, A. D., & Moran, E. F. (1992). Estratégias de subsistência da comunidade de Praia Grande, Ilha de Marajó, Brasil. *Boletim do Museu Paraense Emilio Goeldi. Serie Antropologia, 8*, 185–201.

Nepstad, D. C., & Schwartzman, S. (1992). Non-timber products from tropical forests – Evaluation of a conservation and development strategy. *Advances in economic botany* (Vol. 9). New York: New York Botanical Garden.

Netting, R. M. (1993). *Smallholder, householders – Farm families and the ecology of intensive, sustainable agriculture*. Stanford: Stanford University Press.

Neves, W. A. (1992). *Antropologia Ecológica de Populações Ribeirinhas do Estuário do Amazonas: Subsistência e Adapatação*. Relatorio Científico, CNPq, Brazil.

Nugent, S. (1993). *Amazonian Caboclo Society. An Essay on Invisibility and Peasant Economy*. Province, Oxford: BERG.

Pace, R. (1997). The Amazon caboclo: What's in a name? *Luso-Brazilian Review*, *34*, 81–89.
Padoch, C. (1989). Production and profit in agroforestry practices of native and ribereno farmers in the lowland Peruvian Amazon. In J. O. Browder (Ed.), *Fragile lands of Latin America*. Boulder: Westview Press.
Padoch, C., & Pinedo-Vasquez, M. (2006). Concurrent activities and invisible technologies: An example of timber management in Amazonia. In D. A. Posey (Ed.), *Human impacts on the Amazon: The role of traditional ecological knowledge in conservation and development*. Oxford: Oxford University Press.
Parker, E. (1985). Caboclization: The Transformation of the Amerindian in Amazonia, 1615–1800. *Studies in Third World Societies*, *29*, xvii–li.
Peters, C., Gentry, A., & Mendelsohn, R. (1989). Valuation of an Amazon rainforest. *Nature*, *339*, 656–657.
Pinedo-Vasquez, M., Zarin, D. J., Coffey, K., Padoch, C., & Rabelo, F. (2001). Post-boom logging in Amazonia. *Human Ecology*, *29*, 219–239.
Plotkin, M., & Famolare, L. (1992). *Sustainable harvest and marketing of rain forest products*. Washington, D.C.: Island Press.
POEMA [Pobreza e Meio Ambiente na Amazônia] (1994) *Amazônia: Alianças em Defesa da Vida*. Belém: Universidade Federal do Pará.
Rogez, H. (2000). *Açaí: Preparo, composição e melhoramento da conservação*. Belém: Editora da Universidade Federal do Pará.
Roosevelt, A. C. (1989). Resource management in Amazônia before the conquest: Beyond ethnographic projection. *Advances in Economic Botany*, *7*, 30–62.
Roseberry, W. (1993). Beyond the Agrarian question in Latin America. In F. Cooper, A. F. Isaacman, F. E. Mallon, W. Roseberry, & S. J. Stern (Eds.), *Confronting historical paradigms*. Madison: The University of Wisconsin Press.
Schmink, M., & C. Wood. (1992). *Contested frontier in Amazon*. New York: Columbia University Press.
Siqueira, A. D. (1997). *The ecology of food and nutrition: Patterns of land use and nutritional status among caboclo populations on Marajó Island, Pará, Brazil*. Ph.D. Dissertation. Indiana University, Department of Anthropology.
Siqueira, A. D., Brondizio, E. S., Murrieta, R. S. S., Silva, H. P., Neves, W. A., & Vietler, R. B. (1993). Estratégias de subsistência da população ribeirinha do Igarapé do Paricatuba,Ilha de Marajó, Brasil. *Boletim do Museu Paraense Emilio Goeldi, Serie Antropologia*, *9*, 153–170.
Siqueira, A.D., McCracken, S. D., Brondízio, E. S., & Moran, E. F. (2003). Women in a Brazilian agricultural frontier. In G. Clark (Ed.), *Gender at work in economic life*. Society for Economic Anthropology Monograph Series, No. 20. Walnut Creek, California: Alta-Mira Press.
Smith, N. J. H., Falesi, I. C., Alvin, P. T., & Serrao, E. A. S. (1996). Agroforestry trajectories among smallholders in the Brazilian Amazon: Innovations and resiliency in pioneer and older settled areas. *Ecological Economics*, *18*, 15–27.
Strudwick, J., & Sobel, G. L. (1988). Uses of *Euterpe Oleracea* in the Amazon estuary, Brazil. *Advances in Economic Botany*, *6*, 225–253.
Velho, O. (1972). *Frentes de expansão e estrutura agrária*. Rio de Janeiro: Zahar.
Wagley, C. (1955). *Amazon town*. New York: Macmillan.
Wallace, A. R. (1853). *Palm trees of the Amazon and their uses*. London: J. Van Voorst.
Wood, C., & Schmink, M. (1979). Blaming the victim: small farmer production an Amazon colonization project. *Studies in Third World Societies*, *7*, 77–93.

Chapter 9
The Use of and Access to Forest Resources: The Caboclos of the Lower Amazon and Their Socio-Cultural Attributes

Célia Futemma

Abstract This chapter discusses the issue of natural resource use by riverine populations, presenting what the author has designated as an institutional analysis of the access to and use of forest resources by a section of Patos riverine community, in the Lower Amazon. Futemma pays special attention to the key role that the social networks, above all those woven around kinship and neighbourhood, play as they propitiate the access of local residents to the *várzea* and dry land resources. Another central aspect in the analysis carried out by the author is the flexible role played by formal land property (not shared by all of the community's households), in the sense of rendering viable access of all – depending on social relations – to the resources that are vital for the material survival of Patos residents. The author concludes that, over the state's formal rules (which has promoted and designed the agrarian land reform implanted in the area), a system of local rules and informal access to forest resources is superimposed, which tries to accommodate the social diversity and landscape, minimising inequalities among the small rural Patos producers.

Keywords Economy · Kinship · Floodplain · *Terra firme* · Land access · Land ownership

Introduction

The initiative of the conference and the subsequent publication of this book offer us the chance to recover and perhaps even create a space lacking to the *caboclo* societies in academia and public policy. I use the plural in order to show that there is not just about one caboclo society, as given the sheer cultural, geographical and historical/formative diversity, the Amazon is home to various caboclos. In this volume, caboclos from different Amazonian spaces are

C. Futemma (✉)
Federal University of São Carlos, São Paulo, Brazil
e-mail: cfutemma@uol.com.br

C. Adams et al. (eds.), *Amazon Peasant Societies in a Changing Environment*,
DOI 10.1007/978-1-4020-9283-1_10, © Springer Science+Business Media B.V. 2009

approached through different lenses, demonstrating the complexity and historical trajectories of these populations either forgotten by or insufficiently portrayed in the literature, never mind in public policy.

The importance of the Amazon as a storehouse of biomass and vegetal and animal biodiversity, not to mention its aquiferous magnitude, conspires to diminish somewhat the importance of the human beings who live there and depend on that same stock not only for their survival, but their social reproduction, as Nugent makes clear (in this volume). Among Amerindians, ranchers, loggers and settlers, caboclos exist and persist. The network of social relations constructed by these people has enabled them to abide through time and establish their own organizations and strategies of adaptation to different identities, institutions and natural systems.

The caboclos present very particular characteristics that reveal their mixed origin. Miscegenation understood in the non-pejorative of sense the process through which the caboclos acquired customs, technologies, knowledge, and the values of various peoples that enabled them to form their own flexible cultural – and even cosmopolitan – profile (Moran, 1995; Castro, in this volume). On one hand, they know and practice the Indian agricultural techniques, fishing and hunting, while, on the other, they have worked their way into the market and developed successful economic strategies (Brondízio & Siqueira, 1997). Their European roots saw Christianity blended into the Indian religious rituals (Galvão, 1951), forming a whole network of interrelations (Wagley, 1953; Futemma, 1995). Like their Indigenous antecedents, they dwell on the floodplains, uplands and inter-fluvial tracts of the Amazon (Castro, 2000).

The socio-cultural aspects of the caboclos are fundamental to understanding how they use and gain access to natural resources. In other words, the relationships of kith and kin that determine access to resources that, in turn, help maintain the integrity of the community – 'kinship communities', as Lima calls them (in this volume). This chapter aims to understand this relationship by analyzing a caboclo community that experienced agrarian reform, nestled between the floodplain and the uplands, where it can avail of the resources of both ecosystems. Using traditional instruments of analysis, this chapter will examine in which conditions the access to and use of natural resources occur among the caboclos of the Lower Amazon, demonstrating the key role played by social networks, particularly bonds of kinship, in this arrangement.

Studies on the use of and access to environmental goods and services are part of an analysis of the institutions on their various levels: local, municipal, state, national and even international. These institutions – a body of rules and norms – can be either formal or informal (North, 1990). It is through rules that individuals define their access to and use, management and monitoring of these goods and services. However, drawing up rules is no easy task, as it requires a great deal of time and involves a tentative and error-intensive process (Ostrom, 1990).

In order to be efficient, an institutional arrangement must have clear rules concerning the problems of provision and appropriation, and the instruments to comply with the law (Ostrom, Gardner, & Walker, 1994). The rules of provision determine which human and material resources the individual needs in order to replace the extracted resource. An individual may extract without contributing to the replenishment of the system, but this might lead to its destruction. For example, if an individual extracts specimens of tree from a forested area, he must replace them, that is, he has to manage the forest with a view to maintaining stocks of the extracted resource; otherwise the forest is doomed to depletion. Rules of appropriation determine when, how and to what extent the individual can extract a resource without threatening its extinction (Ostrom et al., 1994). Ostrom et al. propose an analytical tool – Institutional and Developmental Analysis – that makes it possible to analyze institutional arrangements as per their cultural aspects, biophysical conditions and the rules in use[1] at the fora where decisions are made on how these arrangements should be defined and executed.

In the present case study, the cultural attributes of caboclo societies, considering both the community and the household (HD),[2] are key to understanding the patterns of forest and land use. The decision-making process inside the HD concerning natural resources is affected by various socio-cultural and ecological factors. The agrarian reform enacted in 1987 endowed each family in the study area with a plot of roughly 50 hectares; these families acquired land use and tenure rights that amounted to private appropriations. Today, some residents use plots belonging to third parties, for three main reasons: (1) the residents have no land of their own; (2) some lots are privileged by the spatial distribution of the resources; and (3) social relations between the community residents. Ten HDs, mostly the married children of plot owners, do not have land of their own on which to supply their household needs, and these are the *sem terra* (landless).[3] The distribution of the forest resources across the lots is not homogeneous, in other words, some resources are clustered whilst others are scattered. Each lot therefore varies in terms of types of fruit, game, woods and soil. The residents are

---

[1] Ostrom and Schlager (1996: 131) define rules thusly: 'Rules are prescriptions that define which actions are required, prohibited and permitted, and which sanctions are authorized if the rules are not respected'. Ostrom et al. (1994: 38) add: 'Rules provide information on the actions the individual must perform (obligations), must not perform (prohibitions) or may perform (permissions) if that individual wants to avoid sanctions'.

[2] The household is the social unit in various rural villages in Amazonia, and these have been the subject of studies by various researchers (Chibnik, 1994; Futemma, 1995; Lima-Ayres, 1992; Siqueira, 1997). The household represents the hub of transmission of information and material goods, and the production, sustenance, and the social and biological reproduction of the individual (Netting, 1993; Netting, Wilk, & Arnould, 1994). Analysis of the household allows the researcher to observe the choices made by the individuals and how decisions are taken in this microcosm of the social universe.

[3] The designation 'sem terra' is not used by the local population, but is merely the term adopted in this chapter to refer to families in the Patos community that have no lots of their own.

connected by tightly-knit social connections, whether of kith or of kin, creating a lasting social network.

This study has three main goals: (1) to analyze if the rights of use and access to forest resources and lands differ from HD to HD within the same rural community; (2) analyze whether the differences between HDs are related to their economic dependence upon (a) given resource(s); and (3) analyze whether the differences between HDs stem from the social bonds between them. With these ends in mind, the location chosen as the study site was Patos, a caboclo community in the Lower Amazon (municipality of Santarém, Pará), where the forest area had been divided into private lots by agrarian reform.

Environmental and Institutional Attributes

The study area is located 55 km east of Santarém (Pará State, Brazil), the most important commercial and urban centre in the Lower Amazon. Transport to and from the town of Santarém is by bus (a six-hour journey) or by motor boat (three hours). The region is characterized by high but variable annual precipitation, ranging from 1,000 to 3,000 mm of rainfall, mostly spread over the months of March (358 mm) and April (362 mm), and an average annual temperature of 26°C (Junk, 1984; Radambrasil, 1976). The river level fluctuates by roughly five metres between the peaks of the dry season (July to December) and the rainy season (January to June).

The community of Patos is situated between two ecosystems: the terra firme (uplands) and várzea (floodplains). Given the objectives of this analysis, focus will be placed on the upland ecosystem (for more information on the white water floodplains see Futemma, 2000; and Futemma, Castro, Silva-Forsberg, & Ostrom, 2002). The upland ecosystem can be divided into two main ecological zones according to historical land use – the flank and plateau.

The chosen community consists partly of rural settlements established by the agrarian reform of 1987 (Ituqui), which covers some 16,589 hectares. Ituqui consists of seven communities. According to the oral history, the current residents of the area have been living there since the early 20th Century – roughly 1920. They had free access to the forests up to the 1960s, when a large cattle company threatened to expel them from the region. The company bought the land with a view to extracting and selling the timber in order to turn it to pasture for cattle-raising. After two decades fighting for the land, the National Institute of Colonization and Agrarian Reform (Incra) – the federal organization responsible for agrarian reform – expropriated the land and created settlements in the region[4] destined for the traditional residents of Ituqui

[4] The land was expropriated by decree 94.169 and the settlement Project (Gleba do Ituqui) was created by Incra decree 806/87, on September 19th, 1987. The settlement consists of seven communities, including Patos (Serviço Público Federal, 1994).

(Leroy, 1992). With the agrarian reform, the area located in the terra firme ecosystem was divided into 28 roughly 50-hectare lots and each family or single man over eighteen years of age received tenure rights. However, according to Incra policy, while the owners have the right to occupy and use the forest lot, they cannot sell it or transfer it, as their rights extend to Authorized Occupation, but not definitive deeds of ownership.

After the privatization, the HDs without lots – the landless – began to emerge in virtue of two main factors. Firstly, children who were under 18 in 1987 did not receive allotments and tenure rights. These children are now adults with families of their own. As a result, they use their parents' land to meet their subsistence needs. Secondly, in the mid to late 90s, families from other localities moved to Patos, where they did not own land. It must be clarified that these moves generally took place because one of the members of these families married a Patos resident. As such, the privatisation created two distinct groups of HDs (1) the landowners (LO) and (2) the landless (LL). There are currently 23 LOs and 10 LLs.

The Terra Firme Ecosystem

The terra firme (upland) ecosystem is humid tropical forest. The flank is predominantly covered by secondary vegetation, the legacy of its long history of land use. The flank is a 600-hectare area, 1,200 m wide, that runs along the river adjacent to the floodplains, where the residents have their houses. The plateau covers an area of 1,700 hectares covered by mostly mature forest housing countless species of noble woods (Pitt, 1969; Radambrasil, 1976). The plateau includes some sparse areas where forest was recently cleared to make way for arable land. The slope from plateau to flank runs to roughly 200 m at a 50° gradient. The soil is predominantly yellow oxisol, in other words, highly acidic and nutrient-poor (Radambrasil, 1976). The plateau is interspersed with fragments of anthropogenic soil (the Indians' black soil), which is highly fertile (Balée & Posey, 1989).

The flank zone is characterized by predominantly secondary forest, with no timber-grade woods, no wild fruits (like the *piquiá* [*Caryocar villosum*]), and few game animals. The HDs that live on the flank have to obtain these resources from some plot on the plateau, where there is mature forest cover.

Resources of the Terra Firme

The terra firme presents an array of forest resources and soil types, with variations from the flank to the plateau. The flank zone has had a long history of agricultural use (for annual crops and pasture), unlike the plateau zone, where usage is more recent. Since the privatization process, the local residents have used the plateau much more intensively for agricultural purposes than the flank, with its impoverished soils and slopes dotted with houses (see Futemma & Brondízio, 2003). In fact, the residents tend to use the flank area more for hunting, logging, gathering assorted raw materials for domestic use (for fences, firewood, palm leaves for building, etc.) and for picking wild fruits (piquiá [Caryocar villosum] and uxi [Endopleura uxi]).

Soil for Tilling

Despite its sandy, nutrient-poor soil, in the past the residents used the flank intensively to grow crops. The flank is environmentally privileged, as being located near the riverbank means there is a ready supply of water and easier access to urban markets. Though patches of black earth are scattered about the plateau, the main environmental obstacle the farmer faces here is lack of water, probably the locals' biggest impediment to using the land on a larger scale. The farmers mainly plant manioc (*Manihot esculenta*) and corn (*Zea mays*), followed by beans (*Phaseolus spp*) and rice (*Oriza sativa*). Although perennial cultivars are not very common among the farmers of Patos, six HDs do plant these species: pineapple, orange, cocoa and banana. These fruits are grown on a small scale for domestic consumption, though some are occasionally sold on the local market. The locals practice traditional shifting, slash-and-burn agriculture and use both mature and secondary (*capoeira*) forest areas for their farming/ranching purposes. Most of these farmers (71%) clear plots in the more advanced brush (locally referred to as *grossa* [gross/thick], compared with 18% in mature forest and 11% in the younger brushwood [*finas*].

Soil for Cultivated Pasture

Cultivated pastures were introduced to the region after the tax exemptions of the 1970s, a federal government incentive to cattle raising. Other authors have also observed the introduction of cattle to the region (McGrath, Castro, Futemma, Amaral, & Calábria, 1993; Castro, in this volume). The first pasture was opened in the community of Patos between 1992 and 1993, on the flank, due to water access (Futemma et al., 2002; Futemma & Brondízio, 2003). The cattle graze the terra firme pasture for six months during the rainy season (January to July). The pasture is then left to stand for the next six months, all through the dry season (June to December), when the cattle are put to pasture on the floodplains. This annual rotation allows the grass on both the floodplain and the uplands to recover over a six-month rest period. There are two systems for raising cattle: beasts of burden (used for subsistence), and slaughter cattle (for the market).

Wood

Commercial-grade woods were depleted on the flank a long time ago. The plateau has more recently begun to be prospected and still retains a stock of commercially valuable species. These species have made logging a relevant source of income for the regional and local economies since the 1960s (Barros & Uhl, 1996). The owners of the land in Patos simply negotiate the timber, as the logging companies do all the work of extracting the wood and transporting it to the saw mills. Felling usually occurs during the dry season (July to December), when the roads are in better conditions. These companies often do not pay for the trees in cash, but in a variety of other forms: helping the residents clear tree cover to prepare tillable land, supplying water, providing medical assistance, issuing cash loans and, particularly, by opening local roads. In the early 90s, various landowners in Patos told how they sold wood in return for the opening of roads into the community's forested area, making it easier for the residents to move about and transport their produce.

Logging is therefore a special case that does not directly involve local manpower. Even so, with the introduction of the chainsaw into the community, the residents of Patos started to hire saw-owners to fell trees for them. This extraction largely goes toward meeting the domestic needs. The logging companies are always responsible for all commercial felling and log transport. In general, the seller (in this case, the land owner) does not monitor the work of the logging company, but will count the stumps once the work is complete to ascertain the number of trees that have been removed.

Another change that came with the introduction of the chainsaw was the increased availability of wood for construction. The type of house the resident lives in is an indication of the family's social status, or at least of some economic improvement in its conditions. As such, HDs with wooden houses are considered better-off than those living in thatched and mud-wall huts. In Patos, there are 22 wooden houses, against seven of thatch and four of lath-and-plaster.

Non-Timber Products

Game is another much sought-after resource for home consumption. Hunting is done year-round on the flank and plateau areas. The number and size of the species hunted tend to be larger on the plateau than on the flank, where the vegetation is secondary. Hunting dogs are used to track game and a pellet gun for the killing. Game and fish are the residents' main sources of protein (Futemma, 2000).

Other products appreciated locally, and currently found only on the plateau, are the wild fruits piquiá and uxi. The residents consume and sell a lot of these fruits, and their value as a forest resource is attested to by the fact that these trees are spared on farmland, even after slash and burn clearing. The piquiá tree bears fruit from January to March and the uxi from March to April. The ripe fruits are simply collected from the ground.

Despite the introduction of allopathic (industrialized) medicines, the people of Patos still largely depend on household remedies – from plants and animals – to cure ailments. In addition, various types of vine are used to make rope, baskets and other artefacts. Palm leaves are gathered from the flank to cover walls and to roof houses, but there are other plants and animals found only on the plateau. This kind of raw material is available all year round.

Socio-Cultural Attributes

There are 33 households (HDs) in the caboclo community of Patos and roughly 200 inhabitants. The caboclos to whom I refer are native Amazonian populations with strong indigenous features mixed with European and African traits that inhabit the riverine areas and depend conspicuously upon natural resources, though they do also have some insertion on the local and even regional markets (both state and nation). One of the characteristics of caboclo social structure in the Amazon region is their tightly-knit bonds of kinship (Futemma, 1995; Lima-Ayres, 1992; Wagley, 1953; Lima, in this volume), a trait also encountered in other peasant populations throughout Brazil, as demonstrated by Woortmann (1995). In Patos, relations of kinship are established by blood or affinity. Blood relations are the biological connections between parents and children, siblings, nephews/nieces and uncles/aunts, grandparents and grandchildren. Relations of affinity include endogamic and exogamic marriages, including those amongst cousins. There is a third type of relation that may or may not involve blood ties and which occurs through the adoption of a child – designated locally as a son or daughter *de criação* (adopted) (Futemma, 1995; Gentil, 1988; Wagley, 1953).

In Patos, blood relations are bilateral, in other words, members on both the father's and mother's sides are considered relatives, and, as we shall see later, the degree of kinship affects the local institutional structure when it comes to natural resources. Endogamic and exogamic marriages are common amongst the caboclo populations of the Amazon. In the case of exogamic marriages – those between residents of different communities – the man and woman will move to one community or the other, and, in many cases, an entire family may move with the son or daughter who gets married. There are no set rules regarding residence – patrilocal/virilocal, matriclocal/uxorilocal or avunculocal – but the predominant form is neolocal (the newly-weds move into a new house, sometimes on the plot of either the bride or groom's parents). Exogamic marriage represents 77% of the HDs in Patos, and endogamic – between residents of the same community – 33%.

Bringing up adopted children, referred to locally as *filho-de-criação* (adopted boy) or *filha-de-criação* (adopted girl), is a characteristic common to most if not all caboclo communities. Adopted children contribute immensely to maintaining the household. The adopted child is often a grandchild or niece or nephew, or some other blood relation, whose biological parents are unable to rear the child. A mutual commitment holds between the adopted child and the adoptive parents. On one hand, adopted children may look after the adoptive parents in old age, while, on the other, the adoptive parents may confer material assets upon the adopted child, such as a piece of land or some heads of cattle. In the absence of biological offspring, land can be inherited by adoptees, something Woortmann (1995) also observed among German immigrant settlers in the Brazilian south. Another interesting aspect of this relationship, different to legal adoption, is that the biological parents of the child may actually live in the same community and be in constant contact with the child/children. In Patos, seven HDs have adopted children and 12 of these cases are of children adopted by grandparents. The total number of adoptees is 24, though some households adopted more than one child.

Another important aspect of caboclo societies is the religious syncretism between Indigenous and Christian beliefs. For a long time, Catholicism was the main religion in caboclo communities, albeit having mixed with Indigenous values and beliefs, many of Tupi origin, since its original introduction by the Portuguese (Parker, 1985). More recently, however, evangelism reached the caboclo communities, and eight families from the present case study converted in the early 90s.

One aspect of Catholicism that plays a fundamental role in the caboclo economy and kinship system is the practice of godparentage. Godparentage was introduced to the Americas through Iberian culture and it is a defining social characteristic of the peasantry throughout Brazil and Latin America, allowing for the extension of ties beyond the family sphere (Woortmann, 1995). Godparentage is an important social trait among the caboclos and its influence extends over both economic and political life (Wagley, 1953: 152). The parents of a child invite a man and a woman to serve as godparents at their child's baptism.[5] This system was very important during the rubber (Wagley, 1953; Weinstein, 1985) and jute cycles (Gentil, 1988).

Though a purely fictional relation, godparentage ensures that the caboclos maintain ties and mutual reciprocity (Lima, in this volume). As in other caboclo populations, in Patos, bonds of godparentage facilitate economic transactions and political relationships. Godfathers extend political and economic favours to their godchildren (Futemma, 1995; Wagley, 1953) and godmothers help produce the manioc flour (*farinha*), look after the children and prepare large

[5] This man and woman become the child's godfather and godmother through this ceremony, thus establishing a strong social bond between the parents of the child and the chosen godparents, who are henceforth referred to as *comadres* (female: com + *madre* = mother) and *compadres* (male: com + *padre* = father).

Table 9.1 Distribution of households according to number of kinship bonds shared with other units in the community of Patos

Number of social bonds with other households	Intensity of social bonds	Frequency	(%)
0	**None**	1	3.1
1–2	**Weak**	10	31.3
3–4	**Moderate**	5	15.6
>10	**Strong**	16	50.0
TOTAL		32	100.0

Note: N = 32 households.

meals. Bonds of godparentage are also exercised when it comes to lending land for tilling or tending to the sick.

In order to make a quantitative evaluation of the degree of social interpenetration amongst the HDs in Patos, an index was created based on the simple calculation of the number of HDs with which any given HD has some social connection. It was possible to identify four categories of social connectedness for the HDs depending on the number of ties: None, Weak, Moderate and Strong (Table 9.1). The degree of social connectedness is considered 'weak' when the HD has kinship bonds with only one or two other HDs. Those that have social ties with three or four HDs are considered 'moderate', while integration is rated as 'strong' when a family has connections with ten HDs or more.

In Patos, only one HD had no bonds whatsoever with any of the other HDs in the community. More than two-thirds had moderate to strong bonds of kinship. On the other hand, 31.3% revealed weak integration, with family ties to only one or two other units. Table 9.1 shows that the majority of HDs in Patos (96.9) are connected through some degree of kinship. Knowledge of family ties is necessary to understanding how the private forest lots are managed in Patos, as we shall see later.

In short, strong social bonds involve various spheres of life within the group. The caboclos traditionally exchange products, share resources (e.g., they may divide a piece of land for agricultural purposes) and reallocate manpower so as to reduce risk and pool forces in order to overcome local difficulties and ensure family subsistence (see Lima in this volume).

Collective Work

In agriculture, there is a tradition of collective work known as *puxirum*. This collective effort illustrates how the social relations work within the kinship system. In the past, at a time when nobody permanently appropriated land, the puxirum helped mitigate the lack of manpower and make a more equitable distribution of the land. Before privatization, tenure rights were claimed on a 'first come, first served' basis. This right to use is a kind of informal contract between the HDs which involves the reciprocal exchange of manpower (Castro, 2000; Lima, in this volume). This group effort is called upon when the task in hand is too big for the human resources of any given HD to handle. Despite the fact that the members of more than one household may work together to carry out an agricultural task (whether through the *puxirim* or normal daily exchanges), the HDs generally do not share the produce. Each reaps what it sows.

Crop share is also common practice in Patos. As a third of the HDs do not have land of their own – though many have parents who own plots – they need to plant on a third-party's land. Each landowner allows more than one other HD to farm his lot. Given the temporary nature of these usage rights, many crop-share partnerships involve just one annual yield of manioc and corn, and the owner will control the use and access to his forest lot.

Besides the large pitch-ins, even today it is not unusual for two or three HDs to come together to carry out certain activities. In general, these HDs will share some form of family relation and refer to themselves as kinsfolk (Futemma, 1995; Murrieta, 1998). The women gather to process the manioc flour and to prepare other manioc derivatives (pancakes, tapioca, tapioca flour, and others). They also exchange tips on gardening, medicinal herbs (see Murrieta in this volume) and may even tend to the plots together. The men generally fish and hunt in groups. A band of three or four men hunting or fishing together is a common sight, and contrary to what happens in agriculture, the catch and game is shared amongst the participants.

Activities that involve the sharing of manpower, foodstuffs or other products help sediment certain forms of mutual commitment amongst community members. However, such collaboration most frequently occurs with small rather than large work groups. From an early age, the caboclo learns that some tasks are best done collectively rather than individually, as the workload can be shared. The transmission of information is a key aspect of cooperative behaviour among HDs. Neighbours generally exchange information on new technologies, better agricultural techniques, ways to increase crop yields, new planting strategies (e.g. the 'green plantation', see Futemma & Brondízio, 2003) or even on new agrarian and environmental policies. An individual may attempt to freeload and not contribute to the benefit of the HD or community, in which case he or she will be harangued by the rest of the group, either through gossip or verbal admonishment. As such, group work and the existence of mechanisms for curtailing opportunistic behaviour contribute to developing a sense of commitment and reciprocity that, in turn, helps reduce the costs of social transactions and facilitate team work. In this sense, living in close proximity over a long period of time and sharing interactions in various spheres of social and economic life have helped the local population find various ways to overcome the problems inherent to collective activities. Thus embarrassing situations, resentments and even severe punishments can be avoided, preventing a loss of community feeling and a straining of social relations.

Methodology

Data Collection

In order to collate data on patterns of land use and systems for regulating the use of forest resources on private lots, interviews were conducted at all 33 households in the community of Patos. These structured interviews were designed to ascertain how each HD reaches decisions on the use of and access to the forest lots and how many users work on them, as well as to gather information on the transmission of material goods and the economic roles these resources play in sustaining the household. When possible, the interview was conducted in the presence of the husband and wife and elder children (over 18 years of age). In addition to the individual interviews with each HD, we also held collective interviews in order to confirm the information gathered and supplement it with details concerning the collective activities. A forest inventory was also carried out in order to provide a snapshot of mature humid tropical forest and the occurrence of treewood species (for more information, see Futemma, 2000).

Data Analysis: Socio-Cultural and Economic Criteria

Two factors affect decision-making on systems of usage and access rights to forest resources and land: degree of social connectedness and level of dependency on the resource.

The intensity of the social connections between one unit and the others varies within the community. Variations in kinship, godparentage and friendship affect the patterns of natural resource usage and the systems that regulate it. Members with some bonds of kinship can be divided into two groups: (1) first degree blood relatives (between parents and children, and between siblings); and (2) second-degree blood relatives (uncles/aunts, nephews and nieces, grandparents and grandchildren). In relation to blood relatives, HDs with first-degree ties are more closely bound than those with second-degree ties. However, HDs connected by little more than acquaintance may occasionally pool manpower resources, exchange products or come to each other's aid in an emergency. Bonds through godparentage are weaker than blood ties, but stronger than friendship or acquaintance. Even so, non-family connections do strengthen relations between HDs.

The residents depend on upland forest lots for their subsistence and livelihoods. As the HDs use the land primarily for subsistence and secondarily for commercial purposes, the rules for access vary in accordance with the level of dependence on a specific resource. Of the array of resources available in the upland forest of Patos, as seen earlier, four products – some exclusively for subsistence and others almost entirely market-bound – were selected for examination: piquiá (native fruit), timber, arable land and pastureland.

Though piquiá is greatly appreciated locally, it is not essential to subsistence and has little commercial value. The second product, timber, unlike piquiá, is highly valued on the market and the fact that the logging companies fell and remove the trees is a bonus to the lot-owners when it comes to selling their yields. Timber is also used to meet domestic needs, such as house-building. Other products derived from tree trunks – firewood and fencewood – are also consumed domestically. The third product, agricultural produce, is an importance source of staple foods, such as manioc flour. The sale of flour also supplements household income. The fourth product is use of pastureland. Cattle are strongly connected to the market, where heads command high values, but livestock is not essential to family subsistence. One important domestic use of cattle is as beats of burden and yoke animals – 'bull cars' as they are called locally. Each HD will therefore have one or two heads of cattle grazing on the grounds of the house or in the yard, wherever there is available grass cover, thus dispensing with the need for pasture. However, if a family's stock exceeds three or more head, then pasture will be required.

In summary, this analysis focused on four natural resources – piquiá, timber, arable land and pastureland – in order to determine how rights are established (at least up to the study date – 2000), and how these rights are affected by degree of social ties among the users and by the level of dependence upon a given resource. Social ties are broken down into four categories: 1st degree – first degree blood relations; 2nd degree – second degree blood relations; 3rd degree – godparentage; and 4th degree – neighbours (friends and acquaintances). Dependence on a resource is classified in three categories as per the purpose of use: subsistence, market or subsistence & market. The category 'subsistence' refers to HDs that use the resource exclusively for consumption. 'Market' refers to HDs that use the resource for commercial purposes and 'subsistence & market' is reserved for HDs that both consume and sell a given resource.

Results

Managing a Private Forest Lot

Despite the recent history of agrarian change, the HDs that own lots have been trying to create informal, verbal rules to regulate access to different forest products and soils. These arrangements have emerged in response to experiences indicating the need for norms or codes of conduct to be followed by the residents in relation to the use of and access to certain natural resources.

Use of Private Forest Lots

In a bid to regulate their own private forest lots, the owners of land have determined rights for different resources. Though each owner sets his own rules, a discernible system of rights has developed amongst the land owners (LOs), who have tended to adopt the same criteria in determining conditions for access and use.

In order to understand the self-management of forest resources on the uplands, I have opted for the definitions proposed by Ostrom & Schlager (1996: 131–2): access rights, extraction rights, exclusion rights, management rights and alienation rights (Box 9.1).

Box 9.1 Definitions proposed by Ostrom & Schlager (1996: 131–2):

Access Rights: Right to enter a determined physical area and avail of its non-deductable benefits.

Extraction Rights: Right to obtain items of a resource or products from a resource.

Management Rights: Right to regulate patterns of internal use and transform the resource through improvements.

Exclusion Rights: Right to decide who can and cannot have access to these rights and the conditions under which they can be transferred.

Alienation Rights: Right to sell and/or confer one or more of the above-mentioned rights.

Rights to access and extraction specify what an individual can and cannot do within the operational environment. Within the scope of collective-choice, an individual has the right to determine and/or change the forms of usage pertaining to a given resource (management rights) and can decide who is permitted (or not) to enter and extract that resource (exclusion rights). Finally, alienation rights allow the individual to transfer ownership to others within the scope of collective-choice, though these rights can also be determined constitutionally (Ostrom, 1990; Ostrom 1992; Ostrom et al., 1994).

The LOs have defined who can enter the forest (access rights) and how much each user can extract of a given product (extraction rights). They have also determined the forms of resource management to be adopted (management rights). There have been cases in which LL children partake of the decisions made by their land-owning parents, helping them determine who can and cannot use a given forest resource or stretch of land. These groups can also specify who is allowed to sell, rent or transfer a given resource (alienation rights).

In order to restrict the activities of particular individuals, the LOs set verbal rules granting access to forest products or soil by third-parties. Requests for permission, even if informal and verbal, are nonetheless important signals of one individual's regard for the other. This type of permission is part of the local moral code and is being incorporated into the consuetudinary system governing natural products, thereby distinguishing the rights a specific HD is granted to a given product.

In order to understand how the LOs in Patos create and enforce rights to forest products and use of land, the following section will focus on rights to four consumable resources, considering the social ties and economic dependency involved.

Rules of Access to Private Forest Lots

In general, the LOs and LLs gather piquiá for private consumption and for sale. Family members (1st and 2nd degrees) can pick fruits without needing permission, which can be extended to any resident. Non-relatives (3rd and 4th degrees), however, need to request permission from the owner of the lot (Table 9.2).

Table 9.2 Rights of access and extraction among households in the community of Patos according to the degree of social bonds and the economic use of products from forest lots

Forest resources	Degree of social relations[1]	Use for subsistence	Commercial use
Piquiá (edible fruit)	1st degree	Free Access	Free Access
	2nd degree	Free Access	Free Access
	3rd degree	With Permission[2]	With Permission
	4th degree	With Permission	With Permission
Timber	1st degree	With Permission	With Permission
	2nd degree	With Permission	Prohibited
	3rd degree	Purchase/w. Permission	Prohibited
	4th degree	Purchase[3]	Prohibited
Arable Land	1st degree	With Permission	With Permission
	2nd degree	With Permission	With Permission
	3rd degree	With Permission	With Permission
	4th degree	With Permission	With Permission
Pastureland	1st degree	With Permission	With Permission
	2nd degree	Partnership[4]	Partnership
	3rd degree	Partnership	Partnership
	4th degree	Land-lease[5]	Land-lease

[1] 1st Degree: first degree blood relatives (parents and children, siblings); 2nd Degree: second degree blood relatives (nephews/nieces, uncles/aunts, grandparents); 3rd Degree: godparentage (between godparents and godchildren and with the parents of the godchild); 4th Degree: friends/acquaintances.
[2] Verbal permission (informal and unwritten).
[3] 4th degree users can buy or try to negotiate access with the landowner.
[4] Users from the 2nd and 3rd degrees can form a partnership with the LO to gain access to areas of pasture.
[5] 4th degree users can only gain access and usage rights to pastureland through a leasing agreement or by negotiating a partnership with the LO.
Note: N = 33 Households. Unlike table 1, this Table features 33 HDs as opposed to 32, as it includes the additional HD of a farmer who moved into the area in the mid 90s, and it was important to show how this 33rd HD relates with the other HDs in Patos.

Given its low commercial value and the difficulties inherent to monitoring its extraction, piquiá is used more for consumption than for sale and can be gathered freely by anyone. When asked from which lot they picked piquiá, 64% of HDs said they gathered the fruit from their own lots while 36% said they collected it from other lots (those of parents or friends) (Table 9.3). HDs that picked piquiá on a friend's land did so for their own consumption. However, those who gathered fruit on their parents' land did so for two ends: consumption and sale. Table 9.3 would indicate that regulations on the gathering of piquiá are less restrictive (more flexible).

If wood is needed for subsistence, such as for house or fence building, 1st and 2nd-degree relatives can collect wood so long as they ask permission first (regardless of kinship bond). Access to wood resources becomes more restrictive in terms of bonds of godparentage (3rd degree) and friendship/acquaintance (4th degree). In fact, these will often have to buy or barter for wood, and the cost will be higher for 4th-degree ties. When asked about where they get their wood from, 81% of the HDs said they harvested timber from their own lots, while 19% said they gathered wood from the lots of parents, children or

Table 9.3 Distribution of households according to usage patterns for four consumable products, considering the economic use to which each resource is put and the lot from which it is extracted (N = 33 households)

Source of resource (private upland forest lot)	Economic use of resource	Relative frequency of households			
		Piquiá fruit	Wood	Arable land	Pasture-land
Own Lot	Subsistence	47	–	12	20
Own Lot	Market	–	–	–	40
Own Lot	Subsistence & Market	17	81	55	7
1st-degree relative's lot	Subsistence	10	14	–	–
1st-degree relative's lot	Subsistence & Market	3	5	12	–
2nd -degree relative's lot	Subsistence	–	–	–	–
2nd -degree relative's lot	Subsistence & Market	–	–	6	–
2nd -degree relative's lot	Subsistence & Market (Partnership)	–	–	–	20
3rd -degree compadre's lot	Subsistence	–	–	3	–
3rd -degree compadre's lot	Subsistence & Market	–	–	3	–
4th -degree friend's lot	Subsistence	23	–	3	–
4th -degree friend's lot	Subsistence & Market	–	–	6	–
4th -degree friend's lot	Subsistence & Market (Leasing)	–	–	–	13
TOTAL		100%	100%	100%	100%

siblings (1st degree) (Table 9.3). We can see that wood, as a resource of high commercial but relatively low subsistence value, is subject to more stringent (less flexible) rights of access than piquiá.

Given the sheer weight of agricultural produce in a HDs subsistence, such as flour production, for example, there are fewer restrictions on access to arable land (more flexibility). All HDs, regardless of degree of social bond, must request permission to clear or till an area on a lot (Table 9.3). While parents, children and siblings (1st degree) can cultivate such perennial species as fruits and coffee, all other units (2nd, 3rd, and 4th degrees) can only grow annual species – manioc and corn. In this latter case, permission is provisional. One way or another, everyone can grow and sell manioc to obtain their livelihood. In many cases, LLs work in crop-share partnership with LOs. When asked what lot(s) they used for planting, 67% of HDs said they used their own lots while 33% used the land of third-parties – relatives (1st and 2nd degrees), *compadres* (3rd degree) or friends/acquaintances (4th degree). The LLs usually paid their leases with a share of the harvest (e.g. manioc flour) rather than in cash. The use of the forest lot as arable land demonstrates the importance of agricultural produce to family sustenance. As such, rights to till and plant are less restrictive or exclusive than those for extracting wood or opening pasture.

Concerning cattle, only 1st degree blood relatives can clear areas for pasture, but even these have to request permission (Table 9.2). Second degree relatives must also ask permission or enter into partnerships with the owners of the land. Partnerships are also quite common among people with ties of godparentage (3rd degree). Acquaintances (4th-degree social ties) encounter the heaviest restrictions, as these are only allowed access to land through commercial transactions – leases.

When asked where they cleared ground for pasture, 12 units (36%) said that they had cleared areas of mature forest. Of these 12 (67%) used their own lots to raise slaughter cattle (for commercial purposes), while 3 (20%) created partnerships to use the land of 2nd-degree relatives. The remaining 13% took out leases with friends or acquaintances. Although cattle raising is more of a commercial than subsistence activity, it is a medium to long-term investment. Hence the use of a forest lot for pasture is extremely restrictive (Table 9.3).

One case that deserves special attention is that of a medium-sized farmer who lives in the Patos community and has a private lot on the flank (Futemma et al., 2002). This farmer has been trying to establish closer ties with the community residents through partnerships and even by leasing (crop-share). He belongs to the acquaintances category (4th degree bonds), as he has no blood relatives in the area and only recently moved to Patos. At the time of the study, he was the owner of the largest herd of cattle in the area, with some 50 head.

Decisions on forms of using or managing the resources on private lots – management rights – are taken only by members of the same HD. However, as mentioned earlier, there are cases in which 1st-degree relatives help land-owning parents to reach decisions on such rights. These cases tend to involve married children who look after the lots of aged parents (especially when the head of the

HD is an elderly woman). Under these circumstances, the children assume the role of 'owners of the lot', participating in all decisions regarding its management, or at least playing an active role in the decisions their parents take. As such, children who are LL or LO ensure exclusion rights for their parents' lots, deciding who may or may not enter, how much they can extract of each forest product, or what use they can make of the soil.

The fifth type is related to rights to transfer a private lot to other members of the same HD or to other HDs (alienation rights). The transfer of material assets, including private lots, is similar for men and women; in other words, male or female children may inherit any material asset. The criteria adopted to establish land rights for heirs are practically unanimous among the HDs. The land goes to children who work it (in this case, either through cattle raising or agriculture). In other words, if a married son/daughter works the land, then he/she will inherit the lot partially or fully. Land inheritance implies usage rights: whoever uses the land has the rights over it. Even adopted children (not direct blood relatives) can acquire this right to land or to any other material asset or benefit. In Patos, there have been two cases in which filhos-de-criação, (despite having no blood ties) inherited forest lots from adoptive parents.

In summary, this analysis allows us to identify two rules that hold between the HDs of Patos: boundary rules and authority rules (Ostrom et al., 1994: 41–2) (see Box 9.2).

Box 9.2 Definition of two land use rules that hold between the HDs of Patos community, Pará.

Boundary rules: These determine how participants assume or abdicate positions within the group. In other words, these rules define who can access and use a resource and the qualifications individuals must possess in order to be considered eligible for such use.

Authority rules: These determine which actions will be delegated to each participant. These rules define how much of a resource each user can extract, when and how (technology).

(Ostrom et al., 1994: 41–2)

Considering the social ties between HDs, the land owners and their landless married children can determine who can have access to and use their forest resources (boundary rules). The quantity of a given resource that can be extracted varies depending on whether it is for household consumption or commercial markets (authority rules).

Table 9.4 shows that the boundaries of the group of users[6] are more permeable (less exclusive) when it comes to subsistence produce (1st, 2nd, 3rd and 4th

[6] By group of users I mean the group of individuals with equal rights to a given resource, whether private, public or communitarian. In this case, the users refer to individuals who enjoy the same rights over the resources of private forest lots.

Table 9.4 Economic Use of Forest Lot Resources

Degree of Social Bonds

Subsistence 1st degree* HIGHLY PERMEABLE BOUNDARIES	Market 1st degree PERMEABLE BOUNDARIES
Subsistence 2nd degree PERMEABLE BOUNDARIES	Market 2nd degree LESS PERMEABLE BOUNDARIES
Subsistence 3rd degree LESS PERMEABLE BOUNDARIES	Market 3rd degree RESTRICTIVE BOUNDARIES
Subsistence 4th degree LESS PERMEABLE BOUNDARIES	Market 4th degree HIGHLY RESTRICTIVE BOUNDARIES

*1st degree: 1st-degree blood relatives (parents and children, siblings). 2nd degree: 2nd-degree blood relatives (grandparents/grandchildren, uncles/aunts, nieces/nephews). 3rd degree: bonds of godparentage. 4th degree: friends/acquaintances.

degrees), but much less permeable (more exclusive) in relation to market-bound resources, where the group of users is restricted to members of the same household and, in some cases, other kinsfolk (1st-degree and occasionally 2nd-degree relatives, but never bonds of types 3 and 4).

Consuetudinary System and Social Structure of the Caboclos

The existence of informal rules among HDs helped fill the lacunas left by federal, state and municipal environmental legislation. These 'de facto' rules are often more adequate to the local ecological and socio-cultural contexts than the laws.[7] In other words, they prove more diversified than the de jure rules

[7] In Brazil, two federal agencies are responsible for drafting and implementing environmental and agrarian policy: Ibama (Brazilian Institute for Renewable Natural Resources and the Environment) and Incra (National Institute for Colonization and Agrarian Reform), respectively. Brazilian environmental policy underwent massive change with the new Federal Constitution of 1988 (see Machado, 1989). In relation to non-timber forest products, such as vines, roots and fruits, there is no law restricting use for consumption, much less for commercialization, which means that no license, authorization or permit from Ibama is needed to handle these resources. However, the extraction of woods is subject to legal restrictions and those wishing to sell woods can only do so under license from Ibama (decrees 302/88, 011/89 and 732/91). However, extraction for domestic purposes or for arts and crafts is totally exempted from these decrees. There are therefore more restrictions over wood in both the formal institutional and informal consuetudinary arrangements. In order to fell 20 hectares or more of any species of tree in the Amazonian region (categorized as deforestation by Ibama), the individual must have legal authorization (decree 449/87). The federal government exercises legal control over all forms of deforestation, but the legal framework does not distinguish between cuts on the grounds of purpose – whether for agriculture, cattle raising or agroforestry – and each cut represents different forms and durations of occupation. Pasture, for example, implies medium to long-term occupation and involves the complete removal of tree cover (see Ibama, 2002).

established by the government insofar as they attempt to accommodate the social and ecological diversities (Agrawal, 1996; Ostrom et al., 1994; Tang, 1992). Moreover, the local institutional arrangements are more flexible to adjustments in the event of changes in local conditions.

In the Amazon, small rural producers depend upon multiple natural resources that come from different natural systems – terra firme, the floodplains and other water systems, and the rules created by the users to regulate these systems reflect the ecological and cultural diversity and multiplicity of the resources (Castro, 2000). Under consuetudinal systems, the different resource types, patterns of land use and the local demographic conditions affect the decisions made as to how this natural produce should be appropriated.

Patos presents a combination of forms for appropriating the various environments: the floodplains have been used collectively since the mid 1990s (Futemma et al., 2002) while the uplands is divided into private lots. This differentiation in appropriation depending on type of resource has been observed in other geographic regions by many other authors: the Swiss Alps (Netting, 1993); Peruvian communitarian lakes (Leviel & Orlove, 1990); the African floodplains (Thomas, 1996); in Mexico (Bellon, 1996); in the Andes (Guillet, 1987); and in the Philippines (Wiber, 1993). Though the upland lots of Patos are private, the social relations are permeable enough to allow for agreed access to various resources among the HDs, both for subsistence and commercial needs. This permeability of the group of users, as witnessed in Patos (Table 9.4), affords some form of land and forest use to relatives and non-relatives alike. Behind these relations lies a tradition of reciprocity not exclusive to caboclo societies (Castro, 2000; Lima, in this volume), but inherent to all peasant societies throughout Brazil (Woortmann, 1995). The social network developed by caboclo societies translates into an accumulation of social capital that ensures their sustenance and social reproduction.

Even if the Amazonian caboclos are considered the product of a history of exploitation that stretches back to the times of European colonization and to the cocoa and rubber cycles (Ross, 1978), these populations have managed to abide and adapt to the variety of natural environments they now inhabit (Brondízio, 1996; Castro, 2000; Futemma, 1995; Lima-Ayres, 1992; Moran, 1995; Siqueira, 1997). This close relationship with the natural environment and the empirical knowledge that comes of it have been passed down through the generations, and this has only been possible thanks to a sedimented social structure – bonds of kinship, godparentage and friendship. 'Kinship communities', as Lima calls them, transform private property into an almost collective appropriation for the group of users included. Otherwise put, the private lots belonging to individual households allow for the inclusion of other residents from the community. Likewise, collective labour through the day-to-day sharing of manpower or through the puxirum rally-rounds ensures continued reciprocity among households, as reflected in this normative arrangement.

Conclusion

The results presented in this chapter indicate that the development of local rules occurs gradually by trial and error. These local and informal rules seek to accommodate social and biophysical diversities and minimize disparities in access to natural resources among the rural smallholders. The rules for access to and use of forest resources and soils vary depending on the economic characteristics of the resource (for subsistence or commercialization) and the intensity of social relations between the households (first or second degree blood relations, godparentage or friendship/acquaintance), including the relationship between land owners and the landless.

The households that share strong bonds of kinship (blood relations) give and receive more access to resources than would occur with lesser bonds (non-blood relations). As such, first-degree relatives can use the land for agriculture, ranching and logging activities of high commercial value. The importance of subsistence farming for these smallholders raised an interesting aspect concerning these lots. All households, whether related to each other or not, allow access to their lands for the planting of manioc, even for purposes of sale, as this is the basic form of obtaining income to sustain the members of a household.

The consuetudinary system proves complementary to the legal norms, but the specific characteristics of each resource poses a real challenge to the generalized form of privatization that has been occurring across the Legal Amazon through agrarian reform. The existence of a local institutional arrangement, albeit informal and verbal, underscores the importance of recognising the capacity for organization and coordination that these residents possess and their ability to manage their own resources, despite the economic, political and environmental difficulties. The local normative framework reveals a more inclusive network of users that enables non-owners to draw from the sources of nourishment and income (the produce of the land). It is the construction of such a sedimented and normalized social network that makes the social reproduction of caboclo societies possible.

Acknowledgments The author would like to thank the American National Science Foundation (NSF) for the financial support (SBR9521918) extended through the Centre for the Study of Institutions, Population and Environmental Change (Cipec) of the Indiana University, the World Wildlife Fund (WWF) CSR-107-98 for its support for her doctoral research and, lastly, the CNPq (process 20.0685/95-3) for its international doctoral scholarship. The author is also deeply grateful for the logistical support provided by the Várzea Project (IPAM – Amazon Research Institute). Thanks also to Fábio de Castro, Rui Murrieta and Andréa Siqueira for their revisions. Finally, huge thanks to all the residents of the community of Patos and the entire region of Ituqui for their support and collaboration, without which the collection of this data would have been impossible. The author assumes full responsibility for the content presented in this chapter.

References

Agrawal, A. (1996). Group size and successful collective action: a case study of forest management institutions in the Indian Himalayas. In C. Gibson, M. A. Mckean, & E. Ostrom (Eds.). *Forest resources and institutions* (Working Paper n. 3, pp. 49–74). Indiana: Workshop in Political Theory and Policy Analysis.

Balée, W., & Posey, D. (1989). *Natural resource management by indigenous and folk societies of Amazonia*. New York: New York Botanical Garden.

Barros, A. C., & Uhl, C. (1996). Padrões, problemas e potencial da extração madeireira ao longo do rio Amazonas e do seu e Estuário. In A. C. Barros., & A. Veríssimo (Eds.), *A expansão da atividade madeireira na Amazônia: Impactos e perspectivas para o desenvolvimento do setor florestal no Pará* (pp. 109–39). Belém: Imazon.

Bellon, M. R. (1996). Landholding fragmentation: are folk soil taxonomy and equity important? A case study from Mexico. *Human Ecology*, *24*, 373–93.

Brondízio, E. S. (1996). *Forest farmers: Human landscape ecology of caboclo populations in the Amazon Estuary*. PhD Dissertation, Indiana University, Bloomington.

Brondízio, E. S., & Siqueira, A. D. (1997). From extractivism to forest farmers: changing concepts of caboclo agroforestry in the Amazon Estuary. *Research in Economic Anthropology*, *8*, 233–79.

Castro, F. (2000). *Fishing accords: The political ecology of fishing intensification in the Amazon*. PhD Dissertation, School of Public and Environmental Affairs, Indiana University, Bloomington.

Chibnik, M. (1994). *Risky rivers: The economic and politics of floodplain farming in Amazonia*. Tucson: The University of Arizona Press.

Futemma, C. (1995). *Agriculture and caboclo household organization in the Lower Amazon Basin: Case studies*. M. Sc. Dissetation,Tulane University, New Orleans.

Futemma, C. (2000). *Collective action and assurance of property rights to natural resources: a case study from the Lower Amazon Region, Santarém, Brazil*. PhD Dissetation, School of Public and Environmental Affairs, Indiana University, Bloomington.

Futemma, C., Castro, F., Silva-Forsberg, M. C., & Ostrom, E. (2002). The emergence and outcomes of collective actions: an institutional and ecosystem approach. *Society and Natural Resources*, *15*, 503–22.

Futemma, C., & Brondízio, E. (2003). Land reform and land-use changes in the Lower Amazon: implications to agriculture intensification. *Human Ecology*, *3*, 369–402.

Galvão, E. (1951). Panema: uma crença do caboclo amazônico. *Revista do Museu Paulista*, *5*, 221–5.

Gentil, J. (1988). A Juta na agricultura de várzea na área de Santarém – Médio Amazonas. *Boletim do Museu Paraense Emílio Goeldi, Série Antropologia*, *5*, 118–205.

Guillet, D. (1987). Land tenure, ecological zone, and agricultural regime in the Central Andes. *American Ethnologist*, *5*, 139–56.

Ibama (Instituto Brasileiro de Meio Ambiente e Recurso Natural Renováveis). (2002). Disponível em http:/ www.ibama.gov.br/notic.htm. [On-line data].

Junk, W. J. (1984). Ecology of the várzea, floodplain of Amazonia white-water rivers. In H. Sioli (Ed.), *The Amazon. Limnology and landscape ecology of a mighty tropical river and its basin* (pp. 215–43). Boston: Dr. W. Junk Publishers.

Leroy, J. P. (1992). *Uma chama na Amazônia*. Rio de Janeiro: Editora Vozes.

Leviel, D. P., & Orlove, B. (1990). Local control of aquatic resources: community and ecology in Lake Titicaca, Peru. *American Anthropologist*, *5*, 62–82.

Lima-Ayres, D. M. (1992). *The social category caboclo: history, social organization of the rural population of an Amazonian Region (The Middle Solimões)*. PhD dissertation, King's College, University of Cambridge, Cambridge.

Machado, P. A. L. (1989). *Direito ambiental brasileiro*. São Paulo: Revista dos Tribunais.

McGrath, D., Castro, F., Futemma, C., Amaral, B., & Calábria, J. (1993). Fisheries and the evolution of resource management on the Lower Amazon floodplain. *Human Ecology*, *21*,167–95.

Moran, E. (1995). Disaggregating Amazonia: A strategy for understanding biological and cultural diversity. In L. E. Sponsel (Ed.), *Indigenous peoples and the future of Amazonia: an ecological anthropology of an endangered world* (pp. 72–95). Tucson: The University of Arizona.

Murrieta, R. (1998). O Dilema do papa-chibé: consumo alimentar, nutrição e práticas de intervenção na Ilha de Ituqui, Baixo Amazonas, Pará. *Revista de Antropologia, 41*, 97–150.

Netting, R. (1993). *Smallholders, householders: farm families and the ecology of intensive, sustainable agriculture*. California: Stanford University.

Netting, R., Wilk, R., & Arnould, E. (Eds.). (1994). *Households: comparative and historical studies of domestic group*. Berkeley: University of California.

North, D. C. (1990). *Institutions, institutional change and economic perfomance*. Cambridge: Cambridge University Press.

Ostrom, E. (1990). *Governing the commons: the evolution of institutions for collective action*. Cambridge: Cambridge University.

Ostrom, E. (1992). *Crafting institutions for self-governing irrigation systems*. San Francisco: Institute for Contemporary Studies.

Ostrom, E., & Schlager, E. (1996). The formation of property rights. In S. Hanna, C. Folke, & Maler, Karl-Goren (Eds.), *Rights to nature: Ecological, economic, cultural, and political principles of institutions for the environment* (pp. 179–203). Washington: The Island.

Ostrom, E., Gardner, R., & Walker, J. (Eds.). (1994). *Rules, games, & common-pool resources*. Michigan: The University of Michigan.

Parker, E. (1985) The Amazon caboclo: An introduction and overview. In E. Parker (Ed.), *The Amazon caboclo: historical and contemporary perspectives* (pp. xvii–1). Williamsburg: College of William and Mary.

Pitt, J. (1969). *Relatório ao governo do Brasil sobre aplicação de métodos silviculturais a algumas florestas da Amazônia*. Belém: Sudam/FAO.

Radambrasil. (1976). *Projeto radambrasil (Programa de integração nacional)*. Levantamento dos recursos naturais. Vol. 10. Folha SA-21-Santarém. Rio de Janeiro: Ministério das Minas e Energias/Departamento Nacional de Produção Mineral/Projeto Radambrasil.

Ross, E. B. (1978). The evolution of the Amazon peasantry. *Lat. American Studies, 10*, 193–218.

Siqueira, A. D. (1997). *The ecology of food and nutrition: patterns of land use and nutritional status among caboclo populations on Marajó Island, Pará, Brazil*. PhD Dissertation, Department of Anthropology, Indiana University, Bloomington.

Tang, S. Y. (1992). *Institutions and collective action: self-governance in irrigation*. San Francisco: ICS Press.

Thomas, D. H. L. (1996). Fisheries and tenure in an African floodplain village and the implications for management. *Human Ecology, 24*, 287–313.

Wagley, C. (1953). *Amazon town: a study of man in the tropics*. London: Oxford University.

Weinstein, B. (1985). Persistence of *caboclo* culture in the Amazon: the impact of the rubber trade, 1850–1920. In E. Parker (Ed.), *The Amazon caboclo: historical and contemporary perspectives* (pp. 89–114). Williamsburg: College of William and Mary.

Wiber, M. G. (1993). *Politics, property and law in the Philippine uplands*. Canadá: Wilfrid Laurier University.

Woortmann, E. F. (1995). *Herdeiros, parentes e compadres*. Brasília: Edunb/Hucitec.

Part IV
Gender and Daily Life

Chapter 10
Women, Gender Relations and Decision-Making in Caboclo Households in the Amazon Estuary

Andrea Siqueira

Abstract This contribution presents an ethnographic report of the social relationships between men and women in five households situated in the region of Ponta de Pedras, Marajó Island, Pará, in the beginning of the 1990s. From these reports, Siqueira concludes that the women's power degrees in influencing household decision making is highly affected by the following factors: type of rights they hold regarding land property, effective economic contribution and education, the latter not necessarily formal. In these arrangements, the decision making process can be centralised on the masculine figure (the 'head of the household') or else shared, in a negotiated way, between the spouses. Siqueira stresses that to understand and to valorise woman's role, as well as understanding the households' dynamics in detail, is fundamental to the eventual success of projects aiming the improvement of these families' material life quality.

Keywords Women · Men · Resources · Economy · Work · Domestic life · Household organisation

One of the particularities most often mentioned in relation to the rural and peri-urban 'caboclo communities and populations' of the Amazon is the diversity of their economic activities and the character of their adaptation to the local physical and socio-cultural environment (Moran, 1974, 1990; Parker, 1985; Hiraoka, 1992; Padoch et al., 2008). As with other groups of rural smallholders, these activities are highly dependent on family labour, in which men, women and children all actively participate. However, despite the fact that they are responsible for a large portion of the region's food production, the contribution of the rural Amazonian caboclo population and their agricultural and agro-forest systems remains largely invisible economically, socio-culturally and

A. Siqueira (✉)
Center for Latin American and Caribbean Studies, Indiana University, Bloomington, IN, USA
e-mail: asigueir@indiana.edu

C. Adams et al. (eds.), *Amazon Peasant Societies in a Changing Environment*,
DOI 10.1007/978-1-4020-9283-1_11, © Springer Science+Business Media B.V. 2009

politically, as demonstrated by many of the articles in this volume. 'More invisible still' is the role of women, as argued in what little academic literature has been written on the issue (see Alvares & D'Incao, 1995).[1]

This article aims to discuss the participation of women in the caboclo family economy and the role they play in domestic decision-making, with emphasis on the diversity of such processes among the rural households of the Amazon Estuary.[2] The article is divided into three parts, followed by final considerations. The first part briefly presents the main analytical models for households within the scope of the present study, while the second part presents the arrangements and dynamics of caboclo households. Lastly, part three will present five 'cases' of different domestic arrangements and gender relations that serve to illustrate the variations on the staple decision-making processes, referred to here as 'shared/negotiated' and 'centralized'. By 'shared/negotiated decisions', I mean those referring to economic activities and the maintenance of the household that are taken and negotiated by both the man and woman of the household, and sometimes even their children. By 'centralized decisions' I mean those taken by the male 'head' of the family, normally without the consensus of his wife and children. The cases presented here illustrate a spectrum of 'shared' and 'centralized' decisions.

By approaching these themes, this article aims to help garner greater visibility for the 'muted presence' (Simonian, 1995) of women in the regional Amazonian context, especially in terms of women and work.

Households: 'All for One' or 'Each to His/Her Own'?

Households and their socio-economic dynamics have been studied by various areas of the social sciences. Anthropology in particular has been very keen to define what constitutes a household in different socio-cultural contexts

---

[1] Few but important works have focused on the issue of gender relations and/or the economic participation of women in the Amazonian rural and riverine context. See, for example Motta-Maués (1993), Maneschy (1994), Simonian (1995) and Wolff (1999).

[2] The data presented herein was collated through participative observation, interviews, censuses and surveys conducted on various field trips during the period 1989–1994 to rural caboclo households in the municipality of Ponta de Pedras on Marajó Island. Some of this data was gathered by a multidisciplinary team between 1989 and 1991. E. Brondízio and the author continued these studies in the area in subsequent years. The first visit to the area, in 1989, was made possible by the Wenner-Green Foundation through funding granted to Emilio Moran (Indiana University). Between 1990 and 1991, the project was financed by the CNPq (PI. Dr. Walter Neves, 1992), which also awarded individual scientific top-up and doctoral scholarships. Some of the resulting works include: Murrieta, Brondízio, Siqueira, & Moran (1989, 1992), Siqueira et al. (1993), Murrieta (1994), Brondízio (1996), Siqueira (1997), Brondízio, Moran, Mausel, & Wu (1994), Brondízio & Siqueira (1997), Brondízio & Neves (1992), Murrieta, Dufour, & Siqueira (1999), and Brondízio, Safar, & Siqueira (2002), among others.

(see Barlett, 1989; Folbre, 1988; Fricke, 1986; Mencher, 1998; Netting, 1993; Netting, Wilk, & Arnould, 1989; Roldán, 1998; Wilk, 1984, 1989, 1992, among others). The definition of household based on residence – 'all those living under the same roof or eating from the same pot' (Peterson, 1994: 92) – tends to establish a rigid division in terms of what truly constitutes a social group and rarely expresses the social, cultural and economic diversity of the existing domestic arrangements and relations (Wilk, 1992; Yanagisako, 1979). In addition, this definition also tends to reinforce a harmonic image of domestic relations, predicated on the intrinsic cooperation of the household members, where all contribute and benefit equally from the collective activities (moral economy). According to this reference, households are understood as indivisible units, where individual attitudes are guided by altruism in the name of the common good, in this case, the maintenance and survival of the domestic structure. This approach ignores the relations of gender, age and power – and the conflicts inherent to them – between the members of a household, such that 'households end up being studied as objects rather than as activities and relations' (Wilk, 1989: 25).

An opposite reference to the moral economy is the political economy. According to this perspective, the household is understood as the locus of constant conflicts and negotiations among members, albeit within an ideology of harmony (Cheal, 1989). In general, this model of household would be characteristic of industrial capitalist societies, while the cooperation model would relate more to pre-industrial societies.

However, far from being exclusive, these two models are analytical instruments that can be employed in analyzing domestic dynamics in rural and urban contexts, as well as within agricultural and industrial societies.

Caboclo Households: Internal Dynamics and Arrangements

The three communities studied in this research (Marajó-Açu, Paricatuba and Praia Grande) are located in the municipality of Ponta de Pedras, on Marajó Island in the state of Pará. From the macro-environmental perspective, this region is situated within the Amazon River estuary. In these communities, households consist of individuals that share various aspects of domestic production and consumption. The members present biological or fictitious bonds of kinship and tend to live and sleep under the same roof throughout most of the year. However, they may also include absent members who continue to contribute to the household economy by sending money, food or gifts and or through physical work, and who have rights of inheritance over the productive unit (plot, forest, watercourse, plantation, mill, etc.). Relations between members may be of cooperation and/or conflict; that is, among the households studied, we identified cases of moral economy, political economy and, very often, a blend of both. The power of negotiation the members have/present varies depending on gender, age and rights of inheritance.

In 1991, the Marajó-Açu community consisted of 46 households mostly living as sharecroppers along the banks of the river of the same name. The main economic activities of these households were the cultivation and extraction of açai palm fruit (*Euterpe oleracea* Mart.), prawn fishing and *marretagem*[3] (see Murrieta et al., 1989). At the time of the study, the Igarapé do Paricatuba community consisted of 19 households, mostly smallholders whose economies were based on the harvesting and cultivation of açai, fishing, marretagem and slash-and-burn agriculture (Siqueira et al., 1993). Unlike the two above mentioned communities – both riverside settlements – the 21-unit Praia Grande community is located on terra firme along the shores of Marajó Bay. Since the late 1960s, with the help of the local Catholic Church and national and international organizations, the residents of Praia Grande switched from sharecropping for large absentee landowners to being members of an agricultural cooperative, a shift that has guaranteed their collective land tenure. Since then, countless agro-pastoral projects have been implemented aiming to promote the residents' permanence on the land, economic autonomy and self-sustainability (Reymão, D'aguiar, & Duarte, 1986). Despite their unquestionable social merit, none of these projects obtained economic success – at least not by the end of our study. By economic success I mean a self-sufficient system that does not depend on outside subsidies (Murrieta et al., 1992; Siqueira, 1997). Nevertheless, years of community projects led to the generation of massive social capital among the households, unlike anything seen among households in the municipality. In 1994, faced with dwindling funding from the Church, but thanks to the community's organization and leaders, Praia Grande went in search of new financers for its community projects (POEMA, 1994).[4]

Despite their differences in land tenure, social organization and prime economic activities, the households in the three communities presented numerous similarities. All of the units studied carried on a variety of production activities destined for domestic consumption and for the market, and these were almost

---

[3] 'Marretagem' is the local name given to the river trade involving the purchase and sale of products, especially açai fruit, fish and manioc flour. Men and boys buy these products at remote locations and sell them at the markets in Ponta de Pedras or Belém. These trips may take up to 15 days and are made more often during the local açai intercrop period. This is an economic strategy adopted by many riverine families, especially sharecroppers.

[4] Wind power and the implementation of a Bahia coconut fibre processing plant producing car seat upholstery for Mercedes-Benz were some of the projects the community was able to set up with funding from the Federal University of Pará and the German car manufacturer. By 2005 the project had failed and the plant was closed. In the 1990s, many community members were active participants in the Rural Workers' Union and in organizing the municipal branch of the Partido dos Trabalhadores (Worker's Party), through which a member of the community was appointed Municipal Secretary for Agriculture for a two-year period (1995–96), which was an important landmark in local history, as never before had a smallholder held an administrative post (Siqueira, 1997).

exclusively dependent on family labour, and the source of the highest annual income in all three communities was the cultivation and commercialization of açai fruit (Brondízio, 1996; POEMA, 1994).

In the study communities, women and men marry young, usually at the age of 18 and early 20s, respectively. In general, the union is a process rather than an event and is rarely formalized, although they may be confirmed some years later in a religious ceremony (normally Catholic). Ideally, such a union signifies the birth of a new, independent household. However, this is rarely the case, as the newly-weds continue living with and working for and/or with their parents or in-laws until they can arrange a plot of their own or a sharecropping arrangement with a landowner.[5] Migration to the local urban centers of Ponta de Pedras or Belém can be a temporary or even permanent alternative for some young couples. Setting up an economically independent household may prove an unattainable goal, and the young couple and their children may have to live with parents or in-laws forever. For many women, living with parents is preferable to living with in-laws. Expressions like 'I wouldn't be treated like a doormat if I was living with my mother' not only express the conflicts that hold between daughters and mothers in-law, but also between husband and wife. Despite being rarely observed or mentioned, many women feel they would be better protected from physical or mental aggression if they were living with or near their own parents and siblings.

Children of either sex are hoped for and considered essential to the maintenance of households. They generally start helping their parents at a young age, with girls taking on chores earlier than boys. At six years of age, girls are already looking after their younger siblings and helping their mothers with the house chores, while boys of the same age spend most of their time playing, with few or no domestic duties. However, by the age of eight, both boys and girls can be seen working with their parents on the plot (planting, raking, reaping), in the açai-palm stands (planting pruning, and harvesting), in the manioc flour houses, fishing for prawn (usually preparing bait – *poquecas*) and hunting (preparing traps). It is more common for boys to go fishing, hunting and trading at the local markets with their parents than for girls.

Use of modern contraceptive methods (pill, injection or condoms) is rare among the study sample of female family heads aged between 15 and 49, though in 1994 a quarter of those surveyed had been sterilized (tubal ligation). Most of the surgeries had been financed by candidates for political posts, who literally exchanged sutured tubes for votes, with the procedures carried out in Belém. No contraceptive methods or reproductive health information are provided at the local hospital (the only health service available), though most women know that reproductive alternatives existe. Some women mentioned cost as an

[5] In the study region, the market for the sale or purchase of rural smallholdings (5–15 hectares) is practically inexistent. Of the households surveyed (n = 106) not one had acquired land through purchase, but only through inheritance or the local Catholic church-run agricultural cooperative projects.

impediment, especially in relation to the pill. Others sees the use of pills or injections as 'undermining the strength of a woman's body'. For most women, condoms and coital interruption are convenient solutions, though largely unacceptable to their husbands or partners. When asked about the practice of abortion, none of the women surveyed acknowledged having had one, though many claimed to know someone who had tried to abort using herbs or potions, though these are considered dubiously effective methods.

The level of schooling among female family heads is low, at an average of 2.7 years of completed study. However, the education of male and female children is considered an important goal by both men and women, although girls are thought to be more 'cut out' for study than boys, who are often labeled 'lazy' or 'dumb' when it comes to the books. Although girls are generally better educated than their brothers, the households studied did not show higher retention rates for male workers than female, unlike frequently observed elsewhere in the Amazon and Brazil. In many regions, girls tend to migrate to the urban centers for work or study more often than boys.[6] Normally, the first four years of schooling are at the local (multi-series) school, after which pupils have to row, walk or take a town-provided motor boat to the municipal seat. Some families even move to the town in order to facilitate their children's schooling. Many women would like to see their children finish secondary school so they could take jobs as teachers, which would give them a monthly wage and a respectable social position, and perhaps even enable them to help out financially in the future. However, the reality is that few children manage to complete even primary school, much less secondary level, and repeats and drop outs are very common, especially during the harvesting of açai fruit.[7]

A couple or individual, usually a widow or widower, will live alone only so long as he or she is physically capable of carrying out all of the tasks that go with the rural estuarine life, such as rowing, fetching water from the river or well, fishing, planting and harvesting açai, making manioc flour (*farinha*), cooking etc. When they are no longer capable of doing these activities, arrangements are made within the family structure. For example, elderly or younger couples with small children receive visits from relatives and friends at harvest time, especially during the açai production period. The age of the helpers varies from 8 to 50 and the 'help' can be repaid in money, presents, or part of the yield. It is also common to see young girls – nieces, granddaughters or goddaughters – helping out with household chores. However, when such arrangements cannot be made or kept, the elderly will usually go to live with one of his/her children. The presence of elderly relatives in the home appears to be appreciated in virtue of

---

[6] See Siqueira, McCracken, Brondízio, & Moran (2003) and McCracken, Siqueira, Moran & Brondízio (2002) for more on this subject among migrant households of the Transamazon Highway. See also Camarano & Abromovay (1999) for discussion of this matter throughout Brazil.

[7] In 1994, in a bid to retain students, the local town council adopted a school calendar organized around the açai harvest.

the added help with the housework or extra financial contribution to the family income through retirement pensions, which are often the only fixed incomes in a household.[8]

Theoretically, in terms of gender, the man is considered the family provider, while the woman is responsible for the housework and childcare. However, the division of labour and sharing of responsibilities with regard to production and reproduction in the household is more dynamic than that. Men, women and children actively participate in the productive activities (agriculture, agroforestry, fishing), but few men help out in the kitchen or with the cleaning and washing. This double workload for women is rarely acknowledged and frequently contested. It is quite common to hear the women themselves remark, after hours of toil on the plantation or milling manioc flour, that she was simply 'helping her husband' as opposed to working. Nevertheless, women do tend to be less involved in the sale or negotiation of commercial products. For example, the women of the study area were never seen negotiating the sale or purchase of açai fruit – the most important and lucrative regional product – and they were also less present during the sale of prawn at the local market. One exception was in the sale of products to Avon, in which the women were very much involved. The proceeds of these sales may have been small, but they were constant.[9]

Failure to provide shelter and nourishment for the family is one socially acceptable reason for a woman to leave her husband. In reality, however, infidelity is generally the main cause of marital separations. In these cases, the tension caused overspills the limits of the household and spreads throughout the community, with everyone taking sides. While socially condemned, man's infidelity tends to be tolerated much more than woman's.

In summary, the study households present a diversified economy and require complementarity in terms of gender and age to ensure the continuity of their production and reproduction. In many aspects of economic activity and consumption, these households run on a moral economy. Access to health and education services is precarious for all, regardless of sex or age, and no alimentary consumption privileges were observed on grounds of gender (Siqueira, 1997; Adams et al., in this volume). However, one observes that, despite working in various productive activities and being responsible for all daily domestic affairs, women were allotted very little space in economic transactions or control over the use of family resources. These restrictions tend to create a relationship of subordination to husbands or partners when it comes to decision-making. Far from homogeneous or constant, there is a variety of possible household arrangements, in which woman's role and bargaining power can vary over time and from household to household.

---

[8] With their pensions, the elderly can keep slates or accounts at the local shops, enabling them to buy foodstuffs and durable goods for various other members of the household(s).

[9] The economic and socio-cultural importance of the sale of products to Avon and Natura among caboclo women warrants further study.

Between 'Centralizing and Negotiating': Some Examples of Gender Dynamics and Relations in Household Decision-making

Case 1

The first case presented here, under the title 'negotiated decisions', concerns a household that typifies a mixture of moral and political economy. In 1991, this household consisted of a large smallholder family with eight members – two adults (Maria and João),[10] one elderly woman (Maria's mother), three adolescents and two children. The household's main economic activities were the cultivation of açai, fishing (fish and prawn), slash-and-burn agriculture and temporary wage-earning labour. The pension received by Maria's mother contributed toward the monthly domestic budget, especially through the purchase of foodstuffs.

Maria owned the riverine plot on which the family lived. João is her second husband and the father of her youngest child (a one-year old daughter). Her older children (aged 17, 14, 13 and 10) are from her first marriage, which ended when her youngest son was only a year old. With the break-up of this union, Maria became the sole provider for the family, with the help of her mother, who came to live with her. Seven years later, João, a single man and long-term resident in the community, moved in with Maria and her family. João's authority over Maria's children was always negotiated. They all worked together on the manioc plantation and in the processing of flour, most of which was used for domestic consumption. However, the stewardship, harvesting and sale of açai fruit were done separately. The açai stands were considered the sole property of Maria and were managed with the help of her eldest sons. The proceeds from the sale of this fruit were shared exclusively with them. João managed and collected açai on an area near his own plot and sometimes assumed the role of middle-man, buying and selling the yields of his wife, stepsons and neighbors. The income from these sales was spent collectively and individually: collectively to buy food and medicine, individually to purchase durable goods, such as nets, a television, watch or canoe, as well as toiletries like perfumes, shampoos and creams. The black and white television in the house is a good example of an individual asset used collectively. It belonged to Maria's eldest daughter and was only turned on for a few hours each day, usually at noon and in time for the seven o'clock soap-opera. The television ran on a battery that lasted roughly three days, after which it had to be recharged in town. The cost of this service was split between Maria, João and the children.

In 1994 they were building a new wooden house to replace the old. All of the household members were contributing to the construction with cash and/or work. However, Maria's children only agreed to collaborate after negotiating with the couple. They wanted the new house to have a large sitting room where

[10] The names of all those featured in the cases that follow have been changed.

they could hold parties. In October that year, with the house newly built, Maria's eldest son hired a sound system, generator and DJ and threw a huge party with an admission fee and drinks for sale. João helped organize the party and Maria and her mother sold food during the event. The party, as is typical in the area, lasted twelve hours and was considered a success. The profits from the event were divided between João and the eldest son, while Maria and her mother kept the proceeds from the sale of food.

These events illustrate the fact that in this household, despite some productive work done collectively, other activities are executed independently, and the decisions on the allocation of resources (including work) are negotiated by Maria, João and Maria's children. The fact that Maria had already inherited land and had grown-up children when João went to live with them most likely explains why she retained much of her former role as family 'head', with an active interest in the management of her açaí palm stands and the sale of their yield, and therefore also a certain financial independence from João. In this case, land ownership and the cycle of the household (grown-up children) would seem to be the key factors behind the Maria's domestic decision-making power.

Case 2

In 1991, this household was a nuclear family consisting of Laura and Roberto – the owners of a small riverside plot that had belonged to her family – and their six children (aged 9, 7, 4, 3, 2 and 1). The family's main sources of income were prawn fishing and the cultivation and sale of açai. Laura took an active role in fishing for prawn and selling the daily catch at the local market. In 1993 they moved to a house they owned in the town of Ponta de Pedras. One of the reasons for this move was to make it easier for the kids to attend school. Some of the family's economic activities changed with their relocation to the town, but the main source of income remained prawn fishing and the sale of açaí fruit. On their riverside plot, the couple kept small animals (chickens, ducks and pigs). In town, Laura rented a stall at the municipal market, where she sold fruit and vegetables bought from local producers. She also sold Avon cosmetic products and made ice-pops and pastels for her older children to sell at the market and school gate. The proceeds from the sale of these goods were hers to spend as she saw fit, without needing her husband's consent. She generally used this income to cover household and school expenses, such as groceries, water and electricity bills and school materials, but also to buy her own toiletries (shampoo, lipstick, nail varnish and creams). According to Laura, Roberto had started drinking too much since moving into town, so, toward the end of 1994, she moved back to the riverine property, where she felt it was easier to keep the home and family together. Food was easier to come by on the farm, and it was less expensive to keep the kids. In town, the children needed more clothes and shoes and were always asking her to buy this or that. Laura also felt that she had less authority over the children in the town (or 'on the street', as the locals commonly call it), as they spent more time playing unsupervised.

In this case, owning the land and participating in the economic dealings of the household would seem to be what sustained Laura's authority and decision-making power in relation to the allocation of household resources. The move into town did not interrupt the family's agroforest and fishing activities, which remained important sources of income. Like a minority of the local women, Laura actively participated in the prawn fishing and daily sale, and this ensured her autonomy in allocating domestic income. Though the move into town presented Laura with new work opportunities, it does not seem to have increased her decision-making power within the household. In fact, she felt overburdened and concerned that she was losing authority over her children. At the same time, she also felt that living in town exacerbated her husband's drinking habits, leaving her to shoulder most of the household responsibility alone. Returning to the farm was one way she found to maintain her active domestic role without exonerating her husband of his responsibilities to the family.

Case 3

In this case, the household in 1991 consisted of a nuclear family. Rita and Claudio had been living and working as sharecroppers on a riverine plot for about 20 years. They had seven children (aged 15, 14, 12, 10, 7, 4 and 2). Claudio and the eldest sons worked in the açaí stand and fished for prawns, selling both products at the local market. The proceeds from the açaí sales were split with the landowner, who lived in Belém. Between crops, and mainly during rainy season (winter), Claudio traveled the estuary to buy açaí, fish and manioc flour from other producers for re-sale (marretagem). Normally, Rita and her elder daughter tended to the domestic chores, but when Claudio and the elder sons were traveling, she took over the fishing and sale of the fish and prawn hauls at the market. According to Rita, family investments were decided by consensus between she and Claudio.

In 1993, the area they worked was sold by the owner and they used the compensation money to buy a small wooden house in Ponta de Pedras and an electrical açaí processor. However, as there was no electricity where they lived, they could not set up a stall for selling açaí juice as they had planned. As soon as they moved into town, Rita started making coffee and savory snacks to sell at the market, with the help of their younger kids. Her eldest son got a job at a bicycle repair shop, where he received a fixed wage, half of which Rita demanded to go toward the household expenses. Claudio and the older sons continued fishing and river trading and, by the end of 1994, had also managed to lease an açaí stand to work the following year's yield.

Unlike Case 2, Claudio and Rita's move into town was not their own choice, but was forced upon them by the precariousness of sharecropping. As sharecroppers, their economic life-line could be cut at a moment's notice, creating financial and alimentary insecurity. However, while trying their hand at other, mostly informal activities in town, the family managed to maintain some of their core economic activities (açaí, fishing and river trading). In this case, Rita's decision-making power seemed to stem from her economic participation, especially during her husband's absence up-river. Furthermore, the move into town required domestic re-arrangements, such as increased monetary input from Rita and her children. Unlike the previous case (Laura, case 2), Rita did not feel overburdened by her heavier economic contribution to the domestic finances, nor did she feel she had acquired more decision-making power.

Case 4

In 1991, the next household to be presented was composed by Gessy, Diomar, their three children (aged 8, 6 and 4) and Gessy's mother. Unlike the above-mentioned cases, this household was part of a communitarian cooperative established by the Catholic Church in the 1970s. Alongside other 21 other families, they had collective access to coconut stands and pastureland, plus their own private plot, vegetable garden and açaí stand. The main activities of this household were the cultivation and sale of açaí, Bahia coconut and beans, as well as some occasional heads of cattle. Gessy and Diomar were both actively involved in the community.

The economy and organization of this community were heavily influenced by the local diocese, which had long placed emphasis on social organization and the formation of local leaders, both male and female. Another important characteristic of this program was that it registered the couple, and not just the man, as members of the cooperative, and therefore as owners of the land.[11]

In this household, the work was done by Gessy, Diomar and their two elder sons and the decisions were taken by the couple. Gessy's mother helped with the housework and child-minding. In addition to their community work, Gessy and Diomar were both deeply engaged in the organization of the Rural Workers' Union and local directorate of the Workers' Party. However, in 1994, Gessy was considering cutting down on her community and political activities because she had too little time to work on the family's private lots and in the home. In this case, despite the social capital she had earned through her communitarian and political experience, she felt it was her duty rather than Diomar's to reduce

[11] Also see Murrieta et al. (1992), Murrieta (1994), Brondízio (1996) and Siqueira (1997) for more detailed discussion on this issue. Another example of the social and political role of the Catholic Church in the Amazon is given in Futemma's contribution to this volume.

her public participation, thus following the general pattern of gender relations that sees women as chiefly responsible for looking after the home and the kids.

In terms of domestic dynamic, Gessy's decision-making power would seem to have been based on her active economic participation, her membership of the agricultural coop (which granted her tenure over collective and individual plots) and, particularly, the 'informal education' she had acquired from her countless communitarian projects and leadership development courses.

Case 5

In 1991, the household about to be described consisted of a large family – a couple (Bete and Otávio), their seven children (aged 13, 12, 8, 6, 5, 3 and 1) and Otávio's elderly mother. As owners of a riverine property, their main sources of income were the cultivation and sale of açaí fruit, slash-and-burn farming and fishing, as well as two pensions. However, in this case, all decisions on the allocation and management of resources were taken by Otávio, who owned the land on which they lived. Bete had no involvement in the agricultural production or sale of the yields. Though born and raised in the area, Bete rarely left the grounds of the home and claimed to know nothing about the *mato* (wilderness). However, her mother-in-law, two eldest daughters (8 and 6) and her sons all worked with Otávio on the plantations and açaí groves.

Bete used to say that the durable assets in the home (radio, clock, TV, motor oat and pellet gun) belonged to the couple, but that Otávio decided what to do with them and what investments to make. In 1994, Otávio sold their black and white television without consulting her, thus robbing her of what she called 'her only entertainment'. Another source of discord was Otávio's construction of a motor boat, which Bete said had already cost a lot of money and was not worth the investment. However, while she complained about his centralized decisions ('no point in arguing, what he decides is decided'), Bete recognized that Otávio was a good husband and provider, attributes highly praised by the neighbors, who considered him 'hard-working' and 'the best hunter' in the area.

Tension between Bete and her mother-in-law were also observed and referred to openly. Bete claimed she interfered in her relationship with her husband and children.

This household, though aggregating resources (labour and money) to meet its production and maintenance needs, presented very little consensus between the couple as to how investments should be made and resources allocated. All of the decisions were taken by Otávio and Bete had no bargaining power whatsoever. In this case, Bete's less active participation in the economic activities of the household (agriculture, fishing and agroforestry) and the fact that she was not the heir to the land on which they lived significantly reduced her negotiating power and leverage in domestic decision-making.

Final Considerations

This article has sought to show some of the possible arrangements and dynamics within the rural caboclo households of the Amazon estuary. More specifically, it attempted to enumerate the factors and processes that affect the participation of women in households' decision-making. Although women play a significant part in the local and regional economies, they are nonetheless affected by customary and legal systems that tend to favor the male gender. The cases presented here illustrate the various different roles women have in the economic maintenance of their families and in the local economy. However, their participation in the taking of domestic decisions varies according to their land ownership status, economic contribution and level of education. The examples also illustrate the connections between the rural and urban settings in the economic strategies of many households, as well as the importance the Catholic Church has had in forming local leaders, females included.

Ideally, the rural caboclo units require complementarity of gender and age, as children and adults of both sexes are necessary to the production and reproduction of rural life. However, complementarity does not necessarily mean equality. From an early age, boys and girls help their parents in productive activities, and they are expected to look after their parents when they age and can no longer fend for themselves. In return, the elderly are expected to contribute to the family income with their retirement pension. Far from being homogeneous or constant over time, domestic arrangements and the negotiations among household members tend to be dynamic, with a blend of 'moral' and 'political' economic systems, in which the decision-making processes can be centralized on the figure of the male (the 'family head') and/or be shared/negotiated between the couple, and between the couple and their children.

Land ownership – whether through acquisition, inheritance or concession – and economic contribution to the household seem to be the most important factors determining a woman's active participation in the taking of domestic decisions. The same can be seen in other rural Brazilian settings, where access to land is the pivot in the struggle for gender equality (Abramovay & Castro, 1998). Women living or working on land that belongs to them or to their families, or to a cooperative arrangement, have a more active voice in the decisions reached by the family than do those living on land taken as belonging to the husband alone. Women with some form of land tenure also participate more actively in management decisions concerning crops and cultivars, though still have less involvement in the commercialization of the agroforest and fishing yields. Not a single woman among all of the households surveyed was responsible for selling açaí directly to the market (currently the most lucrative product for sale) or practiced river trading, while few were involved in selling the daily prawn catch at the local market. However, whenever possible, many women sold Avon cosmetics or produced homemade foodstuffs (ice-pops, snacks, coffee) to be sold by their younger children, activities that generated small but constant income.

A move into town can often be a temporary or permanent event for these households, but does not necessarily signify an end to their rural activities. In these cases, there seems to be more of a continuum between rural and urban life than a dichotomy. While migration may not result in any significant changes for some women, for others it can offer new opportunities or greater responsibility as a provider, which can be perceived as either a good or a bad thing. This research also shows that moving into town requires new strategies to increase family income in order to cover the higher costs of urban living, but in the two such cases described in this article, fishing and agroforest activities continued to play an important role in the household economy. However, these two cases reveal very different situations, in which land ownership was a clear factor in determining when and why to move. The sharecroppers were obliged to move when the owner of their land demanded it and had no way of returning to rural living should life in the town not work out.

Finally, it is important to stress the importance of social projects designed to form male and female local leaders, such as those implemented by the Catholic Church. These projects have been very effective in preparing local leaders and pioneering in ensuring land tenure for the couple as opposed to the 'family head', i.e., the man. Understanding and valuing of the role of women and having a clear notion of the internal dynamics within households are fundamental to the success of any intervention project hoping to improve the well-being of these populations. Access to land, credit and education should be directed to and guaranteed for both men and women, otherwise such projects will only serve to perpetuate the existing sexual inequality.[12]

References

Abramovay, M., & Castro, M. G.(1998). *Engendrando um novo feminismo: Mulheres líderes de base*. Brasília: Unesco.

Alvares, M. L. M., & D'Incao, M. A. (Eds.). (1995). *A mulher existe? Uma contribuição ao estudo da mulher e gênero na Amazônia*. Belém: Gepem/Goeldi.

---

[12] It is important to note that many of the challenges and problems faced by the caboclo women of the Amazon estuary, such as access to land, health services and education, are shared by other working women throughout rural Brazil. Though these are broadly class-related issues, they also reveal a gender component. It must be remembered that women only won the legal right to be beneficiaries of governmental agrarian reform and settlement programmes and to receive full rural pensions in the Constitution of 1988 (Deere & León, 1999), and that, to this day, only three of the eighteen million female rural workers have registered professions, such is the bureaucratic red–tape they have to face (Abramovay & Castro, 1998: 73).

Barlett, P. E. (1989). Introduction: Dimensions and dilemmas of house holding. In R. Wilk (Ed.), *The household economy*. Boulder: Westview.

Brondízio, E. S. (1996). *Forest farmers of the Amazon: Human and landscape ecology of caboclo populations of the Amazon Estuary*. PhD Dissertation, Indiana University, Bloomington.

Brondízio, E. S., & Neves, W. A. (1992). Populações caboclas do Estuário do Amazonas: A percepção do ambiente natural. In C. Pavan (ed.), *Uma estratégia latino-americana para a Amazônia* (pp. 167–82). São Paulo: Mmarhal/Memorial/Unesp.

Brondízio, E. S., Moran, E., Mausel, P., & Wu, Y. (1994). Land use change in the Amazon estuary: Patterns of caboclo settlement and landscape management. *Human Ecology, 22*, 249–78.

Brondízio, E. S., & Siqueira, A. D. (1997). From extractivists to forest farmers: changing concepts of caboclo agroforestry in the Amazon estuary. *Research in Economic Anthropology, 18*, 233–79.

Brondízio, E. S., Safar, C., & Siqueira, A. D. (2002). The urban market of açai fruit (Euterpe oleracea) and rural land use change: Ethnographic insights into the role of price and land tenure constraining agricultural choices in the Amazon estuary. *Urban Ecosystem, 6*, 67–97.

Camarano, A., & Abramovay, R. (1999) *Êxodo rural, envelhecimento e masculinização no Brasil: Panorama nos últimos 50 anos*. Rio de Janeiro: IPEA. (Texto para Discussão n. 621).

Cheal, D. (1989). Strategies of resource management in household economies: Moral economy or political economy? In R. Wilk (Ed.), *The household economy*. Boulder: Westview.

Deere, C. D., & León, M. (1999). *Towards a gendered analysis of the Brazilian agrarian reform*. Occasional Papers. Latin American Studies, Consortium of New England. 54p.

Folbre, N. (1988). The black four of hearts: Toward a new paradigm of household economics. In D. Dwyer, & J. Bruce (Eds.), *A home divided*. Stanford: Stanford University.

Fricke, T. (1986). *Himalayan households: Tamang demography and domestic processes*. Ann Arbor: UMI Publications.

Hiraoka, M. (1992). Caboclo and Ribereño resource management in Amazônia: A review. In K. H. Redford, C. Padoch (Eds.), *Conservation of neotropical forests* (pp. 134–57). New York: Columbia University Press.

McCracken, S., Siqueira, A., Moran, E., & Brondízio, E. (2002). Land use patterns on an agricultural frontier in Brazil: Insights and examples from a demographic perspective. In C. Wood, R. Porro (Eds.), *Patterns and processes of land use and forest change in the Amazon* (pp. 162–192). Gainsville: University of Florida.

Maneschy, M. C. (1994). Uma presença discreta: A mulher na pesca. In M. A. D'incao, I. M. Silveira (Ed.), *A Amazônia e a crise da modernização* (pp. 251–8). Belém: MPEG.

Mencher, J. (1988). Women's work and poverty: Women's contributions to household-mantainance in South India. In D. Dwyer, J. Bruce (Eds.), *A home divided: Women and income in the Third World* (pp. 99–119). Stanford: Stanford University Press.

Moran, E. (1974). The adaptive system of the Amazonian caboclo. In C. Wagley (Ed.), *Man in the Amazon* (pp. 136–59). Gainsville: University of Florida Press.

Moran, E. (1990). *A ecologia humana das populações da Amazônia*. Petrópolis: Ed. Vozes.

Motta-Maués, M. A. (1993). *'Trabalhadeiras' e 'Camarados': Relações de gênero, ritualização e simbolismo numa comunidade amazônica*. Belém: UFPa.

Murrieta, R. S. (1994). *Diet and subsistence: Changes in three caboclo populations on Marajó Island, Amazônia, Brazil*. M.A. Dissertation, Department of Anthropology, University of Colorado, Boulder.

Murrieta, R., Brondízio, E., Siqueira, A., & Moran, E. (1989). Estratégias de subsistência de uma população ribeirinha do rio Marajó-Açu, Ilha do Marajó, Brasil. *Boletim do Museu Paraense Emilio Goeldi, 5*, 147–63, 1989.

Murrieta, R., Brondízio, E., Siqueira, A., & Moran, E. (1992). Estratégias de subsistência da comunidade de Praia Grande, Ilha de Marajó, Brasil. *Boletim do Museu Paraense Emilio Goeldi, 8*, 185–201.

Murrieta, R. S., Dufour, D. L., & Siqueira A. D. (1999). Food consumption and subsistence in three populations on Marajó Island, Amazon, Brazil. *Human Ecology, 27*, 455–75.

Netting, R. (1993). *Smallholders, householders: Farm families and the ecology of intensive, sustainable agriculture*. Stanford: Stanford University Press.

Netting, R., Wilk, R., & Arnould, E. (Eds.). (1989). *Households: Comparative and historical studies of the domestic group*. Berkeley: University of California Press.

Neves, W. A. (1992) *Antropologia ecológica de populações ribeirinhas do Estuário do Amazonas: Subsistência e adaptação*. Relatório Científico apresentado ao CNPq.

Padoch, C., Brondízio, E., Costa, S., Pinedo-Vasquez, M., Sears, R. R., & Siqueira, A. (2008). Urban forest and rural cities: Multi-sited households, consumption patterns and forest resources in Amazonia. Ecology and Society 13(2). [on-line] URL http;// www.ecologyandsociety.org/vol13/iss2/art2.

Parker, E. P. (Ed.). (1985). The Amazon caboclo: Historical and contemporary perspectives. *Studies in Third World Societies*. (Virginia, Publication n. 32).

Peterson, J. T. (1994). Anthropological approaches to the household. In, R. Borooah, K. Cloud, J. T. Peterson, S. Seshadri, T. S. Saraswathi, & A.Verma (Eds.), *Capturing complexity*. New Delhi: Sage Publications.

POEMA (Programa Pobreza e Meio Ambiente na Amazônia). (1994). *Amazônia: Alianças em defesa da vida*. Belém, UFPA.

Reymão, M. E .G., D'aguiar, A. M. M., & Duarte, M. T. S. (1986). *Organização popular e mudança*. São Paulo Ed. Loyola.

Roldán, M. (1988). Renegotiating the marital contract: Intra-household patterns of money allocation and women's subordination among domestic outworkers in Mexico City. In D. Dwyer, & J. Bruce (Eds.), *A home divided: Women and income in the Third World* (pp. 229–77). Stanford: Stanford University Press.

Simonian, L. T. (1995). Mulheres e seringueiras na Amazônia brasileira: Uma vida de trabalho silenciado. In M. L. M. Alvares, & M. A. D'incao (Eds.), *A mulher existe? Uma contribuição ao estudo da mulher e gênero na Amazônia* (pp. 97–106). Belém: Gepem/ Goeldi.

Siqueira, A. D. (1997). *The ecology of food and nutrition: Patterns of land use and nutritional status among caboclo populations on Marajó Island, Pará, Brazil.* PhD Dissertation, Department of Anthropology, Indiana University, Bloomington.

Siqueira, A. D., Brondízio, E., Murrieta, R., Silva, H., Neves, W., & Vietler, R. B. (1993). Estratégias de subsistência da população ribeirinha do Igarapé do Paricatuba, Ilha do Marajó, Brasil. *Boletim do Museu Paraense Emilio Goeldi, Série Antropologia, 9*, 153–70.

Siqueira, A. D., McCracken, S., Brondízio, E., & Moran, E. (2003). Women in an Amazonian agriculture frontier. In G. Clark (Ed.), *Gender at work in economic life* (pp. 243–67). Walnut Creek: Altamira Press.

Wilk, R. (1984). Households in process: Agricultural change and domestic transformation among the Kekchi Maya of Belize. In R. Netting, R. Wilk, E. Arnould (Eds.), *Households: Comparative and historical studies of the domestic group*. Berkeley: University of California Press.

Wilk, R. (1989). Decision making and resource flows within the household: Beyond the black box. In R. Wilk (Ed.), *The household economy: Reconsidering the domestic mode of production* (pp. 23–52). Boulder: Westview Press.

Wilk, R. (1992, November). *The census as an ethnocentric grid: Methodological problems in understanding the Caribbean household.* Paper presented at Population Institute for Research and Training, Working Paper n. 92-15, Indiana University, Bloomington.

Wolff, C. S. (1999). *Mulheres da floresta: Uma história do Alto Juruá, Acre (1890–1945)*. São Paulo: Hucitec.

Yanagisako, S. J. (1979). Family and household: The analysis of domestic groups. *Annual Review of Anthropology, 8*, 161–205.

Chapter 11
'I Love Flowers': Home Gardens, Aesthetics and Gender Roles in a Riverine Caboclo Community in the Lower Amazon, Brazil

Rui Murrieta and Antoinette WinklerPrins

Abstract This chapter brings a detailed ethnographic analysis of the motivations at play in the cultivation of gardens and backyards by a group of kinship-related women from the São Benedito Community, in Ituqui island, Lower Amazon. The authors initially discuss the dichotomy still present in the studies about the Amazonian *caboclo* societies, and specially their cultivation practices, represented by the studies of socio-cultural emphasis, on the one hand, and agro-forest and ecological, on the other. By means of a personalised and constantly multivocal narrative, Murrieta and Winkler-Prins demonstrate how the cultivation practices of these women are an intricate product of social, ecological, economic and emotional motivations. Among herbs, trees and flowers, social roles are contested, affectivities and alliances ratified, agricultural practices are experimented, thus creating a great stage for the women's expression of their subjectivity. With this study, the authors seek to show how the wider and concomitant knowledge of the social and material contexts, as well as the personal and subjective ones, and their practical implications, is central for the understanding of the motivating elements present in São Benedito women's actions. Consequently, one needs a greater integration between the studies of both economic-ecological and socio-cultural emphasis in order to reach such level of understanding.

Keywords Gender · Agriculture · Floodplain · Homegardens · Women

R. Murrieta (✉)
Department of Genetics and Evolutionary Biology, University of São Paulo, São Paulo, Brazil
e-mail: murrietabr@yahoo.com.br

C. Adams et al. (eds.), *Amazon Peasant Societies in a Changing Environment*,
DOI 10.1007/978-1-4020-9283-1_12, © Springer Science+Business Media B.V. 2009

Introduction

The role of Amazonia's historical peasantries, the so-called caboclos,[1] in the academic and political discourse has shifted from extremes such as the degenerated outcomes of Amerindian detribalization to legitimate heirs of the vanished knowledge of Amazonian native societies (Murrieta, 2000, 1998; Nugent, 1993; Pace, 1997; Parker, 1985a, 1985b). The latter view has had increasing sympathy from academics and the general public, thus making caboclos truly 'people of the forest.' Such an increasing interest in caboclo societies brought to light forms of resource management and repertoires of ecological knowledge much more extensive and complex than expected (some of the best examples are Anderson & Ioris, 1992; Brondízio & Neves, 1996; Brondízio & Siqueira, 1997; Castro, 2000; Frechione, Posey, & Silva, 1989; Furtado, 1993; Futemma, 2000; Futemma, Castro, Silva-Forsberg, & Ostrom, 2002; Hiraoka, 1985, 1992, 1999; Murrieta et al, 1999; McGrath, Castro, Futemma, Amaral, & Calabria, 1993, Moran, 1981, 1990, 1993; WinklerPrins, 1999, 2001). Despite the undeniable merit of these studies and the valuable information they produced, most of them[2] have given very little emphasis to the intricate, and frequently tense, dynamics between utilitarian aspects of environmental knowledge and management, and the multiple layers of the construction of social meaning and everyday practices.

On the path of such a trend, even less has been achieved regarding the role of gender differences in this process and the contradictions implicit in it. Some recent works have exercised a reflection upon this gap (see Alvarez & D'Incao, 1995). However, it is still Motta-Maués' research in the village of Itapuá on the Amazonian coast that stands out as the most influential ethnography on gender issues among caboclo populations (Motta-Maués, 1993). The author describes in detail many aspects of female activities and social roles in Itapuá. She principally analyses the forms of food restrictions and body rituals, which permeated the local symbolic representation of women. Despite its pioneering value and elegant analysis of social structure, Motta-Maués' ethnography is not so attentive to the nuances and contradictions of everyday practices of female lives and their relationship with subsistence activities and the environment, and consequently, its implication for household and community dynamics.

[1] 'Caboclos are the largest non-Indian population in the Amazon region and are the mainstream of the Amazonian peasantry. They have developed a diversified economy based on fishing, hunting, slash-and-burn and intensive agriculture, and the extraction and commercialization of forest products. Although the term Caboclo has been widely used, it should be considered as an analytical tool without implying any quality or social identity' (Murrieta, Dufour, & Siqueira, 1999: 456).

[2] Although the works of Brondízio and Siqueira (1997), Castro (2000), Futema (2000), Castro (2000) are primarily focused on environmental issues, they already present concerns which include aspects of identity constructions and community institutional history and dynamics. Castro and Futema do an excellent work relating resource control to local institutional dynamics.

By contrast, the literature on homegardens has stood out by its greater attention to their functional/utilitarian aspects. Homegardens are part of a continuum of cultivation areas and take on many forms and functions throughout the world (e.g. Niñez, 1984; Landauer & Brazil, 1990, Kumar & Nair, 2004). Often they grade into other types of cultivation areas (cropping, forest management, etc.). The intensity and diversity of cultivation varies considerably in different parts of the world, depending on the region's historical development and social and economic context (Kumar & Nair, 2004). Amazonia's homegardens have not been acknowledged substantially in the literature as much research effort in the region has focused on colonist cropping systems and biodiversity research (WinklerPrins, 2002). Gardens are 'in-between' spaces that have often gone unnoticed, especially since it is typically the domain of women, and not considered to be 'real' agriculture (Rocheleau & Edmunds, 1997).

Although well documented throughout Latin America, much of what has been written about homegardens has emphasized the ecological structure and the material gain of those gardens in economic terms as well as their possible contribution to sustainable development.[3] Examples for the Amazon include works by Guillaumet et al. (1988), Padoch and de Jong (1991), Smith (1996), Slinger (2000), Lima and Saragoussi (2000), WiklerPrins (2002) and WinklerPrins and Souza (2005). But as mentioned, their view of gardening is very much from an agroforestry and economic perspective, with little attention to the social dynamics and multiple layers of meanings behind these practices.

Based on the ideas and concerns discussed above, we want to propose a more multi-leveled approach to understanding how some of the caboclo women farmers on Ituqui Island, in Lower Amazon, Brazil, relate to their physical environment, especially their homegardens, and the implications for their lives as well as for their families and immediate community. For this, we will introduce aspects of the social, economic and environmental interactions of Ituqui women and their activities in homegardens. Thus, we intend to complement existing studies demonstrating the complexity of the process that produces social action as well as the implications of those actions for everyday life.

The Setting

Ecology and Landscape

Ituqui is a 21,000 hectare island in the Amazon River floodplain, located in the geographic region called the Lower Amazon (Fig. 11.1). The main canal of the Amazon River passes on the north side of the island, while a side-channel called

[3] Exceptions are the works of Lerch (1999), Thomasson (1994), Greenberg (1996) and Keys (1999), which focus on social networking, identity, and educational functions respectively.

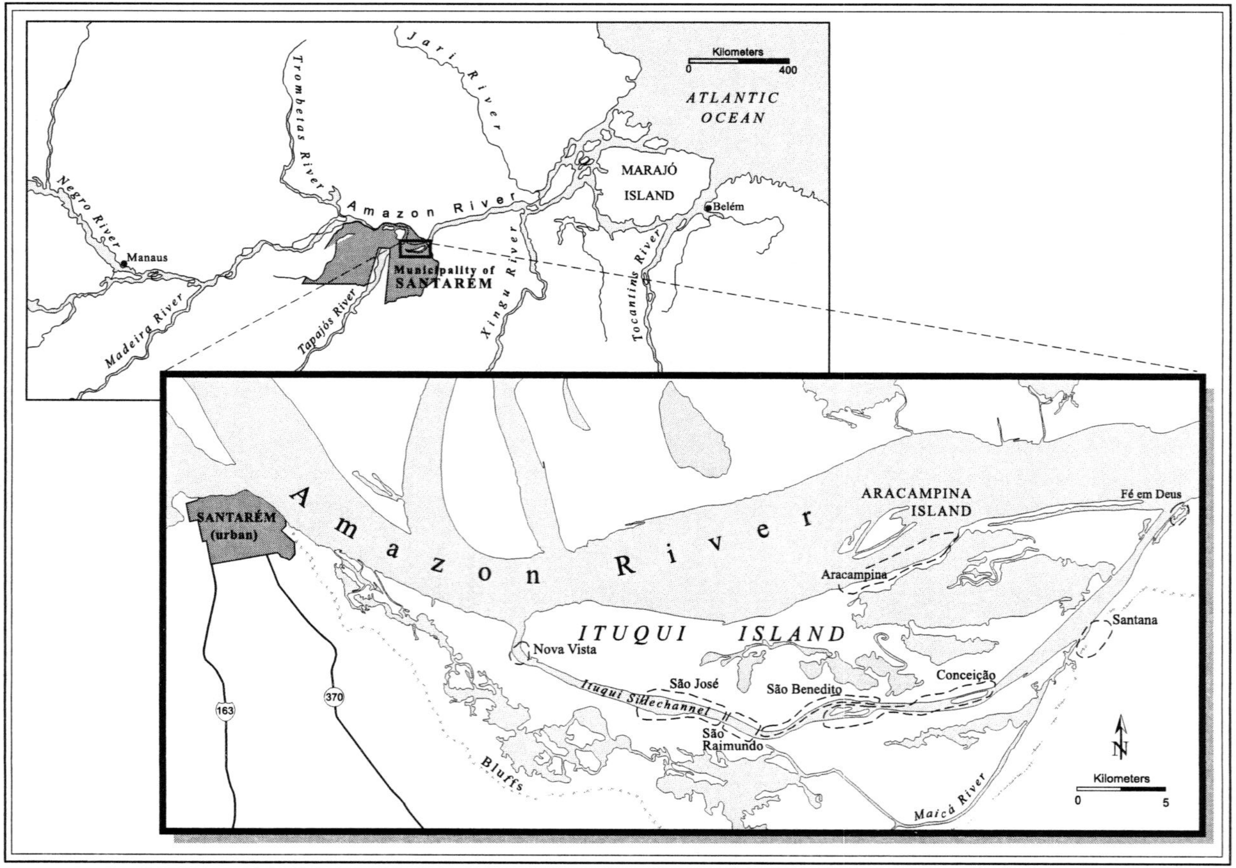

Fig. 11.1 Map of Santarem, Para and Ituqui Island. (Cartography by Ellen White, Michigan State University)

Paraná do Ituqui flows past the island on the south side. There are seven communities on the island. They are: São José, São Raimundo, São Benedito, Fé em Deus, Conceição, Boa Vista and Aracampina. We focused most of our research on the community of São Benedito (Fig. 11.1).

São Benedito is located on the south side of Ituqui Island. It is spread along the levee of the Paraná do Ituqui (Fig. 11.1). In 1994, the community consisted of 37 households with a total of about 184 individuals. Due to the peculiarities of local geomorphology, São Benedito inhabitants held control over one of the largest areas available for agriculture and cattle pasture on the island and have access to two of the most productive lakes of Ituqui, Munguba and Munguba Grande. From the river shore to the inland stream that dissects most of the island, the *Igarapé do Santíssimo*, a mosaic of floodplain microenvironments unfolds before the observer's eyes. Little of the floodplain forest is left close to the Paraná do Ituqui shore; it has been replaced by pastures of native and exotic grass, agricultural fields, and secondary vegetation in different stages of succession (Murrieta, 2000; WinklerPrins, 1999). Ituqui Island, like most of the Amazonian floodplain, is an ever-changing landscape that presents sharp differences between the seasons. Water level is the main factor in the seasonal change. During the dry season a mosaic of pastures, grasslands and secondary forest (*capoeira*) dominated the landscape immediately behind household fences. Differences in topography, the ubiquity of cattle, and subtle human manipulations of the landscape are the main shaping factors of the landscape (WinklerPrins, 1999; Raffles & WinklerPrins, 2003). During the rainy/flood[4] season, water level rises (flux is 7–10 meters), most of the vegetation becomes flooded, enlarging the size of lakes. It becomes difficult for unfamiliar eyes to distinguish one lake from another (Murrieta, 2000).

A mixture of private and community control exists in the smallholder areas of Ituqui Island (McGrath, 2000). By law, all the coast and river shore land up to 50 meters from the water is owned by the Brazilian navy, and can only be used (through usufruct rights) by local inhabitants (Santos Vieira, 1992). Despite the legal impediment of 'owning' floodplain areas, most people in São Benedito and elsewhere in the Ituqui area saw private control as a legitimate system. Levee land was privately controlled, while lakes and pastures (these are the same: lakes in the rainy season dry out and become pasture in the dry season) were communally held, although sometimes even these resources were seen as privately controlled. Only the permanent watercourses and the perennial lakes were treated as completely communal.

---

[4] The annual Amazon River flood occurs is off-set from the onset of the rainy season by only about a month in this region. Hence the season is best described as the 'rainy/flood season.'

The People of São Benedito

Like most island inhabitants the dwellers of São Benedito are smallholders, and they and some of their ancestors appear to have resided on the island since the late nineteenth century, at least. Most of our research was done with six families which were part of the largest kindred group of the community. Our eldest informant in the São Benedito, Seu Didi, a 80-year-old man, said he was born in Ituqui as was his father who was the major founder of what is today known as the community of São Benedito. Close contacts were maintained with other communities, especially the neighbors São José and São Raimundo. In fact, there was no major physical division between these communities, making them appear as one continuing community. During the dry season, the exposed levee links these communities making visits easier and constant events. However, there was attention to boundaries of the community in regard to self-identity and participation in community events and rituals.

Most economic and social activities in São Benedito revolve around the domestic universe of the households. Usually, households had a more permanent population represented by the family head and the youngest offsprings. However, it was not uncommon to find multiple family households with elders and young couples living together. The main household in our study was Marta (53 year-old) and Sávio's (50 year-old). In 1995, they lived in a house with six of their nine children. These were the twins, Peto and Monica (11 year-old), Bela (18 year-old), Pipa (15 year-old), Beto (21 year-old), and Elison (22 year-old). Sávio and Marta's eldest daughter married a year before and moved to a *colônia* (new settlement) in the nearby upland area. Augusto (23 year-old), the couple's eldest son, also married and was living in the community with his wife, Lana (29 year-old), and their first baby. Another daughter was living in Santarém with some relatives.

Mobility between the rural and urban areas of the Municipality of Santarém is very intense, especially for youths and single people. Time spent way from the parents' house can be of few days, months, sometimes for years. Job opportunities and education are still the main attractions and can become a reason of permanent settling away from home. Young women appear to be the ones more attracted to the city and more likely to seek for a better formal education. The opportunity to work as a maid due to the great demands in middle class households for extra labor for domestic tasks appears to be the main advantage of women (as well as girls, since many of them move still in the late childhood) over men. Additionally, many floodplain dwellers in better financial situations keep modest houses in Santarém (Murrieta, 2001; WinklerPrins 2002). Such a pattern of mobility is a common feature all over the Amazon and, in many places, makes difficult to establish a precise separation between country and city (Nugent, 1993; WinklerPrins, 1999).

The second household we worked with systematically consisted primarily of Savio's parents. When we first met Seu Didi and Dona Joana (73 year-old) in

1994, they lived in an old and small wooden house which was located in the most central place of the community, near the school, the chapel and the *barracão* (a kind of community ballroom). Their youngest son, Cassio 30, still single, was living with them. Two granddaughters were also living with the old couple and being raised by them. Seu Didi was the leader of the community for many years as well as the major representative of official authorities (police) in the community and the island.

Women in other households in the community became close informants as well, such as Sávio's sister, Antonieta, who was married to Marta's brother, Gabriel. Antonieta and her love of gardening became an important study case for us. Dona Lu, and her husband Seu Joca, Seu Didi's first cousin lived a few doors from Antonieta. Upriver from Lu and Joca lived Dona Isaura, Seu Dinho, and their six children. Further upriver lived an old couple, Seu Osmarino and Dona Luba. They lived by themselves in the old house in which they had raised ten children. Only one of their sons, Paca, still lived in the community, in the house next door to his parents, with his wife and three children.

Methodological Considerations

The main units of analysis of our research were individuals and households. We believe that these two categories are the main social elements of production and reproduction of structures (economic, symbolic, political) as well as of motivations in everyday practices (Hart, 1992, Netting, 1993). Throughout this paper, we treat individuals as the primary units by which to explain behaviors and choices, 'since these are the obvious units of social life' (Marcus & Fisher, 1986). We connect these individuals to other 'structured' social units, especially households that have strong organizational power in labor, production and meaning. We also weave in other emic and etic levels of analysis such as the community and the Island in order to illustrate and understand connections and extensions of households and relationships between individuals (Murrieta, 2000, 2001; Murrieta et al., 1999).

The ethnographic data presented in this article were mainly collected in the community of São Benedito throughout the months of August, September, October and November of 1995. During this period we lived with two families of the community (Marta/Savio and Seu Didi/Dona Joana). Complementary observation of the annual cycle of gardening activities was done during the flood seasons of 1996 (April/May) and 1997 (January). As mentioned before, we concentrated our investigations on six women already mentioned above. Complementary data from other communities were used. Although we found women in São Benedito to exhibit more expressive care and attention upon their homegardens than most women we met in other communities, nurturing household gardens is a common feature of riverine and urban households in the region (Futemma, 1995, 2000; Siqueira, 1997; WinklerPrins 2002). In the floodplain, such care is more outstanding due to the characteristics and limitations of the environment to agricultural and related practices. Participant-observation and informal and semi-structured interviews were the main methods utilized (see Bernard, 1995). Interviews were recorded on cassette tapes and/or notebooks with informants' permission. Informants' names have been changed to protect their privacy. In addition to our interviews, each of us kept a daily diary in which we narrated events and expressed feelings, impressions and thoughts in a less structured way.

In order to understand the boundaries and continuities of women's representations of social spaces and economic activities as well as their roles as motivating elements of these practices, we visited six female informants and their gardens on different occasions. We tried to capture the social and emotional triggers of all these space/activity contexts for each of our informants. Our goal with these visits was not of strict description and classification, but, mainly, of contextual qualification of what was said and done.

Agriculture and Households in the Water Land

Agriculture has been the dominant feature of São Benedito economic life as well as for most of the island for the last fifty years. A great part of this importance is related to the rise and fall of jute cultivation in the Lower Amazon. Introduced by Japanese immigrants and entrepreneurs in the late 1930s and early 1940s, it became the key commodity of the Lower Amazon for half a century (Gentil, 1988). However, parallel to this market oriented agriculture, other products such as manioc, corn, beans, and squashes (cucurbits) were cultivated for domestic consumption as well as for small scale and opportunistic commercialization. After the complete collapse of the jute market in the late 1980's,[5] it was these crops that became the bulk of agricultural production in São Benedito, as well as in most of the island (McGrath et al., 1993, Murrieta, 2000, WinklerPrins, 1999).

In August 1995, when we arrived for a longer stay, the main products cultivated in São Benedito were manioc, beans, corn and squashes. Agriculture in the floodplain is highly dependable on the flood cycles. Seasonal floods renew and refresh the soil every year with a layer of fresh alluvium. Because of this, these floodplain soils do not depend so much on vegetation cover or plant biomass decomposition to keep reasonable levels of fertility. In fact, in most of Ituqui there is no interval of fallow, or 'rest' of the land, except during flooding.

Large backyards (*quintais*) and homegardens (*jardins*) surround the houses in São Benedito. Fences established the limits between the households and more

[5] Jute cultivation came to a halt due to the combination of a shift to polypropylene sacking, bulk handling, and the end of imposition of tariffs on imported jute.

external areas. Barbed wire and stick fences were very common in São Benedito. These fences delimited most properties to restrict the movement of cattle and to facilitate control over land. The homegardens and the backyards were filled with decorative, medicinal and fruit plants.[6] These were the spaces where the women spend most of their time and concentrate the bulk of their domestic activities.

Regarding Sávio and Marta's household, the physical movement of its female inhabitants was restricted to the house, common areas of the community and the houses of other relatives. Despite a certain level of vigilance imposed on them, the girls used to spend much more time in Augusto's house, helping their sister-in-law to take care of the baby or with their nearby cousins and relatives. An interesting aspect of the way Sávio raised his daughters was that beyond the domestic tasks of cooking, cleaning, doing laundry or taking care of younger children, they were never involved with the so-called 'heavy' productive tasks, such as agriculture, even gardening. During the whole time we lived with them, we never witnessed any of the girls helping their mother in the garden. We do not think this is a common feature, since we have seen many young women, especially teenagers, working in agriculture in São Benedito and other Ituqui communities. Such a restriction appears to be related particularly to Sávio's family. In fact, Sávio verbalized once his disapproval of having unmarried young girls 'work' *(trabalhar)*. Of course, there is here a clear male bias in his demeaning of female domestic tasks. Cleaning, doing laundry, cooking, et cetera was not considered as real 'work'.

Homegardens, Aesthetic and Agriculture

The process of following the daily routine of some households in São Benedito soon revealed subtly defined social and spatial domains. In this part of the Amazon the area immediately surrounding the house, within the fence that encircles it, is planted with fruit trees, decorative plants and medicinal herbs. Two major functional spaces divided this area, the *quintal* (backyard) and the *jardim* (homegarden). Nevertheless these categories were fluid and defined slightly differently by individual gardeners. Within this backyard and/or homegarden there were sometimes small fenced-in spaces with dense and carefully arranged, suspended flowerbeds known as *jiraus* or *canteiros*. The jardins seem to perform multiple functions for the household. They played an important role as means of presentation for their housekeepers, as feminine aesthetic pleasure and a safe ground for constant and practical experimentation. Here is the unquestionable domain of women and the embodiment of an overlapping

[6] The most common plants cultivated in *jiraus* (suspended flowerbeds) are kale (*Brassica oleracea*), green onion (*Allium cepa*), green peper (*Capsicum annum*), pepper (*Capsicum frutescens*), basil (*Ocimum basilicum*), tomatos (*Lycopersicon esculentum*).

logic of functions and motivations, which have widespread effects for the survival and social dynamics of the household. The meaning of these spaces appeared to go far beyond the economic and nutritional values of the carefully cultivated spices, greens, and fruits. Additionally, part of the jardim was situated in front of the house, almost as a passage to the outside world (Nunes, 1994). According to our female informants, especially Antonieta, a garden says much about its keeper and has a special impact on people's first impressions, mainly females. 'Uma casa sem jardim me deixa triste' [A house without a garden makes me sad].

Such caring attention to home gardens struck us as intriguing since the gardens have to be redone every year because of the seasonal floods. Antonieta said many times that she had considered completely abandoning the garden because of the hassle of redoing it every year. However, every summer, she would find herself replanting the seedlings and sprouts that she took from each one of her flowers, spices, medicinal herbs and fruit trees before the flood, for which she carefully cared until the lowering of the water. The end of the flood season did not mark the end of the struggle to maintain a nice homegarden. The decrease of the water level and the long and dry summer would bring other difficulties to the women's work. We do not remember how many times we heard the complaints of young girls and boys about the daily 'drag' of helping their mothers water every single plant, especially those in the jirau. In the particularly hot summer of 1995, one of Dona Joana's granddaughters said to be losing weight (which she apparently was) due to the plant watering effort. By this time, the river level was around four meters below the edge of the shore. We could see her point.

Some of the plants have stories about their origins, utility and role from the housewife's viewpoint. They were often connected to some kind of emotional reference or occasion, medicinal or economic function and aesthetic appeal. They also play an intense role in the female reciprocity network between households in and out of São Benedito. Gifts during visits were usually plants or fruits. Often, they were the payback of a visit received previously, a promised token of friendship or a beginning of a new personal connection.

Economic motivations were also a reason for the women's insistence of redoing their gardens every year. In the constant fluctuation of seasonal natural resources and the instability of regional market and national economy, the homegardens and backyards were always potential sources of income and food. The jirau was the source of most of the greens and spices consumed in the household such as kale, tomatoes, green peppers and green onions. The size of the suspended beds varied quite widely and ranged from two or three simple rectangular wooden boxes (ranging in size from about 2 by 1 meters long to 2 by 3 meters long) to large old canoes exquisitely arranged, giving an especially aesthetic look to the garden. This variation could be clearly seen when visiting Antonieta's lush and beautifully done jirau and Marta's simple and more

Fig. 11.2 Jirau made from the remains of a canoe in Ituqui, Pará (photo by A. WinklerPrins, 1996)

modest one. The visit by one of us to Marta's homegarden (jardim) and *quintal* (backyard) illustrates keenly some of the roles present in the above description:[7]

> S. Benedito, September 1995
> ... About 8:00 a.m. we began our stroll around the backyard. Marta had just filled the last gourd with water and we went straight to the *jirau*. Marta's *jirau* spot is a small area measuring about two by three meters and is surrounded by a fence made of irregular stakes and posts. A bit further away, I see some herbaceous plants growing on a small raised platform 'that one there is a medicinal plant, *trevo raimundinho*, for urinary infections.' On the ground there is a small shrub with little red fruits, 'that one is a *malagueta* pepper tree. It sprouted by itself. The *bentevi* (or *bem-te-vi*, *Pitangus sulphuratus*, a bird) eats these plants and its droppings fell on the ground and the plant sprouted.' We continue our walk. She glances at two plants with large broad leaves and says, 'Kale, I planted it in April.' She goes over to one of the plants and pulls off two leaves. 'What are you going to do?' I ask. 'I am going to send it to Lana', she responds. Further along there is a small square *jirau* made from a wooden box, with a shrub covered in small white flowers. 'And what about that little flowering tree there?' I

[7] Some of the plants mentioned by Marta in her garden could be identified by their Portuguese common names (see Berg, 1993; Cavalcanti, 1991; Lorenzi & Matos, 2002; Mors, Rizzini, & Pereira, 2000; Smith, 1999), others are unknown at this time. Identification by common name is difficult without full botanical collection and identification. The latter was beyond the scope of our research. They are *trevo raimundinho (Justicia pectoralis)*, *pimenta malagueta* (hot pepper), couve (kale), *boldo (Plectanthrus barbatus)*, *macaxeira* (sweet manioc, *Manihot esculenta*), *uriza* (*Pogosteman heyneanus*), *marupazinho (Eleutheriine plicata)*, *cebolinha* (green onion), graviola, (soursop tree, *Annona muricata)*, *miracuriza* tree (?), *Catauarizeiro (Crataeva benthami)*, *cuieira* (calabash gourd palm *Crescentia cujete)*, *coqueiro* (coconut palm, *Cocos nucifera*), *patchulli (Vetiferia zizanioides)*, *limão galego* (lime, *Citrus medica)*, *muracuru* tree (?), *mangueira* (mango, *Mangifera indica*), *Vai-e-volta* (?), *Zina (Zinnia elegans) and Munguba (Pseudobombax munguba)*.

ask. 'My mother-in-law gave me a branch'. This one here is *boldo* she says, pointing to another plant, a small bush. 'What's that for?' I ask. 'It's for stomachache' she replies. 'That *macaxeira* (sweet manioc) is tall', I comment. 'Yes, it is, isn't it?' she agrees. I found out later that the *macaxeira* was a present from Sebastião, one of Marta's *compadres* (godfather of one of her children). 'And what about this one?' I ask. 'That one is *uriza*, it's a sweet-smelling plant and it's used for bathing.' We moved on to another plant. 'This one's called *marupazinho*', she says. 'It's for making a tea for hemorrhoids and diarrhea.' I turn to one of the planters and notice an herbaceous plant with long narrow leaves. 'Is this green onion?' I ask. 'Yes, it's going to be beautiful this summer. I'm even thinking of buying 500 grams of seed onion. It does well in the summer, but too much rain makes it rot,' she adds. The next tree is dried up. 'It's a soursop tree (*graviola*). It's dead. It died during the flood.' She knocks on the dried trunk, which now serves as a post on which to tie the clothesline. I point to two trees by the fence and ask, 'And what about those two over there?' She responds, 'They are soursops too. They already fruited this year.' Further along we passed a tree and she said, 'This is a *miracuriza* tree'. 'What's it for?' I ask. 'It gives shade and is good for firewood,' she answered. We continue walking in the yard. 'And what about this little one here?' I ask, pointing to a small tree nearby. 'It's a *cataurizeiro*. It's useful to fishermen. The fish comes to eat the seeds in the winter. Also, the fisherman uses the fruits for bait.' Farther along are two more small trees with low, spreading canopies. 'They both are gourd trees (*cuieiras*),' she says. 'And what are they for?' I ask. 'They provide the gourds from which we make our pails. I planted them,' she concludes. 'And what about that coconut tree?' I ask, looking at a tall coconut tree right in the middle of the clearing. 'Beto planted it. Eleven years ago his grandmother gave him a shoot to plant.' Later I learned that the coconut tree was planted to celebrate Peto's birth. The tree draws my attention because it is slender and its fruit oblong. I ask if it grows coconuts. 'That coconut tree doesn't give coconuts, just male coconuts,' she replies. I see another coconut tree, still just a sprout. 'And what about this little one?' I ask. She replies, 'I planted it. Dona Deodora, a lady from the other end of the community, gave it to me.' I asked a second question, 'Why did you plant this one if you already have one?' She responded, with her familiar half-smile, 'Two is better than one, and besides this one will fruit quickly. The water cures as well!' Over by the outhouse I see a shrub with long leaves, similar to those of grass; it's *patchulli*, used to scent things. Marta admitted to not using it because the roots are very tough, and difficult to pull out. Suddenly she asks, 'Do you know about *limão galego* (lime)?' 'I do,' I reply. She shows me two large green fruits. 'This is it. I got them from my mother's house. I always get them from her house to make juice (lemonade).' We continued walking and passed by the *muracuru* tree, the mango tree and more soursop trees. She points to the latter and says, 'Beto planted them. Here, it used to have a lot of them, but the floods are big, they can't sustain it.' 'And why do you plant them if the floods always destroy them?' I ask. 'Because a big flood only comes once every two or three years, and a soursop tree fruits in a year. If it weren't for the floods, I'd be making a lot of money (selling) the soursop fruits!' she answered in a disappointed voice. At this point we had already walked around the entire backyard clearing and we went over to the front of the house, where most of the ornamental plants are (*jardim*). We walked by another shrub. 'They say this one's called *vai-e-volta* (come-and-go). The next one is called *zina*; the girls brought it for me,' she says. 'Those there in the front planters' she points to a small row of plants next to the front fence, with long stems and small, slightly closed flowers with pink petals 'I don't know their name. I brought them from Maria Jose's house.' 'Did you ask for them or did she give them to you?' I asked. 'I asked for them,' she says. She enters her house and asks if I've finished. We end the session...

As demonstrated above, flowers appeared to embody, for women, a more emotional character than other plants. Such a perspective often conflicted with a more 'utilitarian' approach of some men. We recall with some level of nostalgia, the funny and playful arguments between Seu Didi and Dona Joana in the veranda of their old house about the utility of her little exquisite roses. Seu Didi was always arguing that such an effort to maintain the flowers were silly and useless since they had no 'utility' and were going to die anyway when the water rose. Even worse, the roses attracted insect pests which he detested. Dona Joana counter-pointed, trying to argue within his logical frame, by saying that on many occasions, she obtained some economic return from her roses. She recounted one occasion when she offered a seedling of a particularly beautiful rose to a visitor from the city and when the visitor returned to São Benedito, she brought a lovely and 'expensive' tray for her. The argument did not convince Seu Didi and she ended the discussion by saying that she loved her plants and would not give them up.

Another important feature of homegardens is their role as loci of experimentation. In other words, through this almost poetic blend of aesthetic and emotional motivations, important knowledge and information was exchanged among women and opportunistic experiments could take place. Talking to Dona Joana, we came to know that her many little 'tricks' about cultivation were learned through her practice of gardening and social interaction with other females. One example was particularly striking. During strolls in her garden, we were curious about the mulch she put in the box of her jirau and of some of her flowers. She explained to us that she once noticed that many plants die during hot summers and she thought it was because they were not retaining enough water. Talking about that with a lady friend (*colega*), she was advised to put small pieces of rotten wood (especially from the *Munguba* tree) with other organic material around the base of the plant and in the case of the jirau, to cover the surface completely. She said that the result could not have been better. Later on, we saw the same technique in other homegardens. We cannot claim that she was the vehicle of the innovation but somehow the idea had spread among the women.

Also, homegardens could become a social space of subtle power struggle between men and women. As we said before, there is no doubt that the authority and privileges of men are bigger than those of women. However, such an inequality does not happen without some level of contestation from women (what could be already observed in Dona Joana's household). Homegardens were an enlightening arena to understand the nature of this process.

One afternoon while visiting Lu, the wife of Joca (Seu Didi's cousin), we gained some definite insights about power struggles between males and females over agricultural meaning and knowledge. This family provided us with an interesting case study because of its odd composition and, especially, the dominant character of the male head. Joca and Lu had five daughters and only one son. The boy had been sent to the city while he was still a child to study and hopefully find a career. Therefore, all the work in the house and in the gardens was done by the part of the family that remained on the Island, who were all women, the wife and daughters. Here, the only tasks that completely exclude the women were those related to raising cattle. Despite her apparent submission to her husband, Lu showed us on many occasions that some situations did not go on without contestation. 'You know, we are not women (referring to herself and the girls) to wait around for men to work (work meaning heavy agricultural tasks).'

That afternoon, we were visiting the home and the manioc fields and she was telling us that once she had an argument with Joca about the use of mulch for retaining soil moisture by covering the area around the base of the manioc plants (the principle was very similar to what we saw at Dona Joana's). Joca disagreed strongly, saying that it would attract ants, which could kill the crop. A few weeks later, Joca received a visit from an *extensionista* (agricultural extension agent) from EMATER (regional government extension agency) who was assessing the manioc fields of São Benedito and giving technical advice. Joca was very proud of his technical knowledge. The two men were talking in the garden when Lu interrupted and asked about her idea of covering the base of the manioc plants and related Joca's counter arguments. She told me with some pride and a mocking smile how the extensionista congratulated her and referred to Joca's practice of cleaning the ground around the base of the manioc as inappropriate. [8]

Such challenges and confrontations were not rare but changed extensively in tune depending on the personalities of the spouses. Gardening brought an important element to the subtle power struggles between genders and provided, even if only a bit, an opportunity to increase the role of women in the household decision making process.

Considerations and Conclusions

In this article, we have attempted to include certain aspects of structural continuities of everyday life, as well as social, emotional and sensuous motivations. In most studies on homegardens thus far, it has been easy to enclose household dynamics and subsistence patterns within exclusively utilitarian rationales and to ignore or undermine the everyday contradictions of improvisations, repetitions and the so-called 'non-rational' choices. However, in doing so, important elements of social dynamics are missed, as well as the potential impact these can have. Such a concern follows a contemporary effort of many theorists in the social sciences to understand social phenomena as a dialogue between different levels of society, e.g. economic and political structures and the context of

[8] Mulching is not common at all on Ituqui Island. Crops are planted in tidy rows with all slash removed. The lack of interest in mulching is partly an aesthetic consideration ('clean' fields are 'prettier' – *mais bonito*) but also a fear of snakes and other creatures hiding in the mulch material (WinklerPrins, 1999). From the perspective of soil moisture conservation mulching is increasingly encouraged in this region by extension agents as the dry season can be very dry in this part of the Amazon Basin.

practice per se (Bourdieu, 1983a, 1983b; Giddens, 1984, 1993). According to these views, social practice and representations cannot be understood as unidirectional relations. Instead, they should be seen as multilevel and multi-scale interactions among social spheres of action, individual actors and material constraints (Bourdieu, 1983a, 1983b; Giddens, 1984, 1990, 1993 and others). In sum, social preferences and choices often challenge the economic and ecological rationales and invert the functional-utilitarian equation without neglecting them completely.

We have given some examples of such interactions in São Benedito. We have tried to demonstrate some basic structural and cognitive aspects of gender and class divisions between and within households, and how this might vary and be challenged according to personal circumstances, inclinations, and accumulated experience. The role of women and the sometimes not-so-obvious system of restrictions and control over female space, actions and body have a special meaning in our considerations. An important example of this is women's relationship with their homegardens on Ituqui Island. Flowers and other plants are not only the manifestation of feminine senses of aesthetics or complementary economic strategies, but, also, source of meanings and practices in which status, conflicts and aspirations are constantly negotiated and manipulated between genders. Thus, homegardens go beyond a simple embodiment of the female condition in Ituqui. They work as effective tools of intra- and inter-household negotiations in which economic survival, gender differences, social status, and emotionality play essential roles.

At a different level, all the examples above show how the contextualization of practice is triggered and how it triggers memories of experiences and social events in which important pieces of social relationships and ecological information about the landscape and technology are encoded. These processes can be clearly identified in Marta's extensive plant repertoire. The flow of information and the confluence of an affective map that enlarges social alignments and relationships play a definite role, but, again, are not exclusive motivations; rather, they point to an overlapping and mutual process of causes and effects.

Finally, we have connected these structural aspects with the basic economic activities supporting household organization and survival. Sávio and Marta's household, as well as others', show a clear historical ability to respond to contextual changes in the local and regional political, socio-economic, and ecological landscapes. As we have hopefully shown in this study, women, through their aesthetic inclination, cultivation of social networks, and economic contributions, became critical elements in household social and material reproduction. However, such a role does not flow peacefully and is constantly under pressure of male authority and utilitarian ethos. Finally, it is important to notice that this entangled process of 'multi-causalities' and 'effects' is not clearly distinguishable and has an ever-changing character. Such dynamics are easily visible in the personalized character of people's agencies that lay behind these actions, choices, decisions and everyday improvisations.

Acknowledgments We would like to thank IPAM for its financial and logistical support, and Perpétuo Socorro de Sousa, André Souza e Edmar Dantas for their friendship and help during fieldwork. Many thanks to Dr. Walter Neves and the Laboratory for Human Evolutionary Studies (Instituto de Biociências – USP). We are very grateful for the critical readings of Pennie Magee, Barbara Piperata, Darna Dufour and José Reinaldo Lopes. We would also like to thank the following institutions for their financial and logistic support: IPAM (Instituto de Pesquisas Ambientais da Amazônia), Museu Paraense Emílio Goeldi, University of Colorado, University of Wisconsin-Madison, ITC- International Institute for Aerospace Survey and Earth Sciences, Enschede, The Netherlands, and Michigan State University. We are especially grateful to CNPq (Brazilian Council for the Development of Science and Technology, processo 200337-2), FAPESP (processos 96/7487/2 e 00/04102-2) e IPAM for sponsoring Rui Murrietás fieldwork on Ituqui Island, doctoral and post-doctoral studies. Also, our gratitude to the National Science Foundation, the Association of American Geographers, and the American Association of University Women for sponsoring Antoinette WinklerPrins's fieldwork and dissertation work. Finally, our friendship and gratitude to the people of Ituqui Island. Without them it would have been impossible to finish this work.

References

Anderson, A., & Ioris, E. (1992). The logic of extraction: Resource management and resource generation by extractive producers in the estuary. In K. H. Redford & C. Padoch (Eds.), *Conservation of neotropical forests* (pp. 158–174). New York: Columbia University Press.

Alvares, M. L., & D'Incao, M. A. (Eds). (1995). *A mulher existe? Uma contribuição ao estudo da mulher e gênero na Amazônia*. Belém: Museu Paraense Emilio Goeldi.

Berg, Elizabeth van den. (1993). *Plantas medicinais na Amazônia*. Belém: Museu Paraense Emílio Goeldi/PR/MCT/CNPq

Bernard, R. (1995). Research methods in anthropology: Qualitative and quantitative approaches. Walnut Creek/London: Altamira Press.

Bourdieu, P. (1983a). *Distinctions: A social critique of the judgment of taste*. Cambridge: Harvard University Press.

Bourdieu, P. (1983b). Esboço de uma teoria da prática. In R. Ortiz (Ed.), *Bourdieu* (pp. 46–81). São Paulo: Editora Ática.

Brondízio, E., & Siqueira, A. (1997). From extractivists to forest farmers: changing concepts of caboclo agroforestry in the Amazon estuary. *Research in Economic Anthropology, 18*, 233–279.

Brondízio, E., & Neves, W. A. (1996). Populações caboclas do estuário do Amazonas: A percepção do ambiente natural. In Clodovaldo P. (Ed.), *Uma estratégia latino Americana para a Amazônia* (Vol. 1, pp. 167–182). São Paulo: Unesp/Memorial.

Castro, F. (2000). *Fishing accords: The political ecology of fishing intensification in the Amazon*. Ph.D. Dissertation, School of Public and Environmental Affairs, Indiana University.

Cavalcante, P. B. (1991). *Frutas comestíveis da Amazônia*. Belém: Museu Paraense Emilio Goeldi/MCT/Cejup.

Denevan, W., & Padoch, C. (Eds.). (1987). *Swidden-fallow agroforestry in the peruvian Amazon*. New York: New York Botanical Garden.

Frechione, J., Posey, A. D., & Silva, L. F. (1989). Perception of Ecological Zones and Natural Resources in the Brazilian Amazon: An Ethnoecology of lake Coari. *Advances in Economic Botany, 7*, 260–282.

Furtado, L. (1993). *Pescadores do Rio Amazonas: Um estudo anthropológico da pesca ribeirinha numa área amazônica*. Belém: Museu Paraense Emilio Goeldi.

Futemma, C. R. T. (1995). *Agriculture and caboclo household organization in the lower Amazon basin: Case studies*. Unpublished M.Sc. Thesis. Tulane University, New Orleans.

Futemma, C. R. T. (2000). *Collective action and assurance of property rights to natural resources: A case study from the lower Amazon region, Santarém*. Ph.D Dissertation, School of Public and Environmental Affairs, Indiana University, Bloomington.

Futemma, C. R.T., Castro, F., Silva-Forsberg, M. C., & Ostrom, E. (2002) The emergence and outcomes of collective action: An institutional and ecosystem approach. *Society and Natural Resources, 15*, 503–522.

Gentil, J. M. L. (1988). A juta na agricultura de Várzea na Área de Santarém – Médio Amazonas. *Boletim do Museu Paraense Emilio Goeldi, Antropologia, 4*(2), 118–199.

Giddens, A. (1984). *The constitution of society*. Berkeley: University of California Press.

Giddens, A. (1990). *The consequences of modernity*. Stanford: Stanford University Press.

Giddens, A. (1993). *New rules of the sociological method*. Cambridge: Polity Press.

Greenberg, L. S. Z. (1996). *You are what you eat: Ethnicity and change in Yucatec immigrant house lots, Quintana Roo, Mexico*. Unpublished Ph.D. Dissertation, Department of Geography, University of Wisconsin, Madison.

Guillaumet, J. L., Grenard, P., Bahri, S., Grenard, F., Lourd, M., dos Santos, A.A., & Gely, A. (1988). Les lardins-vergers familiaux d'Amazonie centrale: Un exemple d'Utilisation de l'Espace. *Turrialba, 40*(1), 63–81.

Hart, G. (1992). Imagined units: Constructions of 'the household' in the economic theory. In S. Ortiz & S. Lee (Eds.), *Understanding economic process* (pp. 111–131). Lanham/New York/London: University Press of America.

Hiraoka, M. (1985). Cash cropping, wage labor, and urbanward migrations: Changing floodplain subsistence in the Peruvian Amazon. In E. Parker (Ed.) *The Amazon caboclo: historical and contemporary perspectives* (vol. 32, pp. 199–242). Virginia: Studies in Third World Societies.

Hiraoka, M. (1992). Caboclo and ribereño resource management in amazonia: A review. In K.H. Redford & C. Padoch (Eds.), *Conservation of neotropical forests* (pp. 134–157). New York: Columbia University Press.

Hiraoka, M. (1999). Miriti (*Mauricia flexuosa*) palms and their uses and management among ribeirinhos of the Amazon estuary. In C. Padoch, J.M. Aires, M. Pinedo-Vasques, & A. Handerson (Eds.), *Varzea: Diversity, development and conservation of Amazonian whitewater floodplains* (pp. 169–163). New York: New York Botanical Garden Press.

Keys, E. (1999). Kaqchikel gardens: Women, children, and multiple roles of gardens among the Maya of highland Guatemala. *Conference of Latin Americanist Yearbook, 25*, 89–100.

Kumar, B. M., & Nair, P. K. R. (2004). The enigma of tropical homegardens. *Agroforestry Systems, 61*, 135–152.

Landauer, K., & Brazil, M. (Eds.). (1990). *Tropical home gardens*. Tokyo: United Nations University Press.

Lerch, N. C. (1999). *Home gardens, cultivated plant diversity, and exchange of planting material in the Pacaya-Samira national reserve area, northeastern Peruvian Amazon*. Unpublished M.A. thesis, Department of Geography, McGill University, Montreal.

Lima, R. M. B., & Saragoussi, M. (2000). Floodplain home gardens on the central Amazon floodplain. In W. J. Junk, J. J. Ohly. M. T. F. Piedade, & M. G. M. Soares (Eds.), *The central Amazon floodplain: Actual use and options for sustainable management* (pp. 243–268). Leiden: Backhuys Publishers.

Lorenzi, H., & Matos, F. J. A. (2002). *Plantas medicinais no Brasil: Nativas e exóticas*. São Paulo: Instituto Plantarium de Estudos da Flora.

Marcus, G. E., & Fisher, M. J. (1986). *Anthropology as cultural critique*. Chicago: University of Chicago.

McGrath D. G., Castro, F., Futemma, C., Amaral, B., & Calabria, J. (1993). Fisheries and the evolution of resource management on the Lower Amazon Floodplain. *Human Ecology, 21*(2), 167–195.

McGrath, D. G. (2000). Avoiding a tragedy of the commons: Recent developments in the management of Amazonian fisheries. In A. Hall (Ed.), *Amazonia at the crossroads: The challenge of sustainable development* (pp. 171–187). London: Institute for Latin American Studies.

Moran, E. F. (1981). *Developing Amazonia*. Bloomington: Indiana University Press.

Moran, E. F. (1990). *A ecologia humana das populações da Amazônia*. Rio de Janeiro: Editora Vozes.

Moran, E. F. (1993). *Through Amazonians eyes: The human ecology of amazonian populations*. Iowa City: University of Iowa Press.

Mors, W., Rizzini, C. T., Pereira, N. A., De Filipps, R. A. (2000). *Medicinal plants of Brazil*. Michigan: Algonac.

Motta-Maués, M. A. (1993). *'Trabalhadeiras' & 'Camarados': relações de gênero, aimbolismo e ritualização numa comunidade Amazônica*. Belém: Editora Universitária UFPA.

Murrieta, R. S. S. (2000). *The dilemma of the 'Chibé'-eater: Food choice, ecology and everyday life among peasant communities in the lower Amazon, Brazil*. Unpublished Ph.D. Dissertation, Department of Anthropology, University of Colorado, Boulder.

Murrieta, R. S. S. (1998). O dilema do papa-chibé: Consumo alimentar, nutrição e práticas de intervenção na Ilha de Ituqui, Baixo Amazonas, Pará. *Revista de Antropologia, 41*, 97–150.

Murrieta, R. S. S. (2001). Dialética do sabor: Alimentaçáo, ecologia e vida cotidiana em comunidades ribeirinhas da Ilha de Ituqui. *Revista de Antropologia, 44*, 39–88.

Murrieta, R. S. S., Dufour, D. L., & Siqueira, A. D. (1999). Food consumption and subsistence in three caboclo populations on Marajó Island, Amazônia, Brazil. *Human Ecology, 27*(3), 455–475.

Netting, R. M. (1993). *Smalholders, householders: Farm families and the ecology of intensive, sustainable agriculture*. Stanford: Stanford University Press.

Niñez, V. K. (1984). *Household gardens: Theoretical considerations on an old survival strategy*. Lima: International Potato Center.

Nugent, S. (1993). *Amazonian caboclo society: An essay on invisibility and peasant economy*. Providence: Berg Publications.

Nunes, B. (1994). Casa, praça, rua, jardim e quintal. *Ciência e Trópicos, 22*(2), 253–264, Recife.

Pace, R. (1997). The Amazon caboclo: What's in a name? *Luso-Brazilian Review, 34*(2), 81–89.

Padoch, C., & de Jong, W. (1991). The house gardens of Santa Rosa: Diversity and variability in an Amazonian agricultural system. *Economic Botany, 45*(2), 166–175.

Parker, E. P. (1985a). The Amazon caboclo: An introduction and overview. In E. Parker (Ed.), *The Amazon caboclo: Historical and contemporary perspectives* (pp. xvii–li). Studies in Third World Societies, V. 32, College of William and Mary, Williamsburg, VA.

Parker, E. P. (1985b) Caboclization: Transformation of the Amerindian in Amazonia, 1615–1800. In *The Amazon caboclo: Historical and contemporary perspectives* (pp. 1–49). Studies in Third World Societies, V. 32: College of William and Mary, Williamsburg, VA.

Raffles, H., & WinklerPrins, A. M. G. A. (2003). Further reflections on amazonian environmental history: Transformations of rivers and streams. *Latin American Research Review, 38*, 165–187.

Rocheleau, D., & Edmunds, D. (1997). Women, men and trees: Gender, power and property in forest and agrarian landscapes. *World Development, 25*(8), 1351–1371.

Santos Vieira, R. (1992). *Várzeas Amazônicas e a Legislação Ambiental Brasileira*. Manaus: Instituto Brasileiro do Meio Ambiente e Recursos Naturais Renováveis.

Siqueira, A. D. (1997). *The ecology of food and nutrition: Patterns of land use and nutritional status among caboclo populations on Marajó Island, Pará, Brazil*. PhD Dissertation, Department of Anthropology, Indiana University, Bloomington.

Slinger, V. A. (2000). Peri-urban agroforestry in the Brazilian Amazon. *The Geographical Review, 90*(2), 177–190.

Smith, N. J. H. (1999). *The Amazon river forest: A natural history of plants, animals and people*. Oxford: Oxford University Press.

Smith, N. J. H. (1996). Homegardens as a springboard for agroforestry development in Amazonia. *International Tree Crops Journal, 9*, 11–30.

Thomasson, D. A. (1994). Montserrat kitchen gardens: Social functions and development potential. *Caribbean Geography, 5*(1), 20–31.

WinklerPrins, A. M. G. A. (1999). *Between the floods: Soils and agriculture on the lower Amazon floodplain, Brazil.* Unpublished Ph.D. Dissertation, Department of Geography, Madison: University of Wisconsin.

WinklerPrins, A. M. G. A. (2001). Why context matters: Local soil knowlwdge and management among an indigenous peasantry on the lower Amazon floodplain, Brazil. *Etnoecologica, 5*, 6–20.

WinklerPrins, A. M. G. A. (2002). House-lot Gardens in Santarém, Pará, Brazil: Linking Rural with Urban. *Urban Ecosystems, 6*, 43–65.

WinklerPrins, A. M. G. A. & Souza, P. S. M. G. A. (2005). Surviving the city: Urban home gardens and the economy of affection in the Brazilian Amazon. *Journal of Latin American Geography, 4*, 107–126.

Part V
Diet and Health

Chapter 12
Bread of the Land: The Invisibility of Manioc in the Amazon

Cristina Adams, Rui Murrieta, Andrea Siqueira, Walter Neves and Rosely Sanches

Abstract A detailed analysis on the Amazonian riverine populations' diet is presented, based on quantitative data collected by the authors amongst these populations over the last 13 years, with special emphasis on caloric and protein intake. The five communities studied in this article are sited at the estuary (Marajó Island) and Lower Amazon (Ituqui Island) regions. Grounded on almost unique quantitative data on the Amazon region riverine populations, the authors analyse in detail the influence of several local factors in these populations' food intake variability: ecosystem, micro-environment, seasonality, history and social organisation. In a synthetic way, it can be stated that, in the case of the estuarine flood plain, the main caloric sources are represented by manioc flour and the *açaí*, whereas in the seasonal flood plain, manioc flour, fish and sugar play such role. Fish represents the greatest protein source, independently of the ecosystem considered. All five communities feature protein intake levels above the minimum recommended by FAO. Regarding calories, only one of the communities (one established in the estuary) meets the internationally recommended minimum for energy consumption. At the end of the article, the authors discuss the anthropological implications of the ubiquity and historical longevity of the association between manioc flour and fish as the mainstay of riverine subsistence.

Keywords Manioc · Cassava · Diet · Subsistence · Seasonality · Food consumption

Introduction

When the Europeans arrived in the Americas, Amerindians had already been growing manioc (*Manihot esculenta* Crantz) for 5,000 years (Hillocks, Thresh, & Bellotti, 2002) across a vast area that stretched from Central America over the

C. Adams (✉)
School of Arts, Sciences and Humanities at the University of São Paulo, Brazil
e-mail: cadams@usp.br

C. Adams et al. (eds.), *Amazon Peasant Societies in a Changing Environment*,
DOI 10.1007/978-1-4020-9283-1_13,

Antilles and down the entire Atlantic coast of South America (Ribeiro, 1987; Sauer, 1987; Schmidt, 1958; Silva, 2005). In the sixteenth century, the tuber was spread across Africa and Asia by the Portuguese merchants (FAO, 1977; Hillocks et al., 2002), becoming one of the most important staple crops in the modern world (Dufour, 1985; Jackson, 1993).[1]

Most of world manioc production comes from smallholders located in areas on the economic and ecological fringe. The technology normally used follows traditional pre-industrial methods, especially in the poorer regions of the tropics (Henry & Hershey, 2002; Hillocks et al., 2002). The success of manioc under these conditions can be attributed to the fact that the plant flourishes in the poor tropical soils, where it can remain for a long period of time, constituting a reliable source of energy for populations that would otherwise have no way of ensuring food security (Prudencio & Al-Hassan, 1994).

In Pre-Columbian Brazil, manioc flour (*farinha*) already comprised the staple food of many indigenous societies (Cooke & Piperno, 1993; Dufour, 1994; McKey & Beckerman, 1993; Silva, 2005). Since the first century of colonization, the crop began to be consumed by the new populations migrating into the territory (Zeron, 2000). The failure of the European foodstuffs brought over by the colonizers to acclimatize to the new environment led to the search for substitutes, of which manioc was the best known. The most widespread and amply consumed staple in the colony, its derivatives were considered a substitute for European bread (Adams, 2000; Barreto, 1939; Cândido, 1964; Cascudo, 1983; Ferreira Filho, 1938; Gravata, 1938; Mussolini, 1980; Zeron, 2000).[2] Ubiquitous on the tables of all social classes – whites, slaves,[3] *mestiços* (*caboclos*, *caiçaras* and *caipiras*)[4] (Adams, 2002b, 2003; Cândido, 1964; Castro, 1965; Franke, 1987; Mussolini, 1980; Silva, 2005; Zeron, 2000) – and in the most diverse forms (flour, porridge, cakes, pancakes, beverages, *maniçoba* [manioc

---

[1] Today, manioc (also known as cassava) is grown in over 90 countries, providing nourishment and sustenance to more than 500 million people in developing nations. The total area of manioc crop is somewhere in the region of 16 million hectares (50% in Africa, 30% in Asia and 20% in Latin America), with global production reaching 152 million tons (Emperaire, 2002a; Hillocks et al., 2002).

[2] Numerous authors describe manioc flour as the 'Bread of the Tropics', 'Bread of the Earth', 'Bread of the Poor' (Adams, 2000; Barreto, 1939; Cândido, 1964; Ferreira Filho, 1938; Mussolini, 1980: 224), or even as 'Brazilian bread' (Gravata, 1938).

[3] Franke (1987: 460) shows how the African slaves were reduced to a diet monotonously based on manioc during the colonial period of sugar-cane monoculture in the Brazilian northeast.

[4] In the late seventeenth century, a system of plantations was established in the Southeast that saw autochthonous crops, including corn, beans and manioc, grown along the inroads to the savannahs to ensure staple provisions for the Bandeirante explorers. For Cândido (1964), this measure, allied with the fact that the São Paulo population had absorbed the techniques of the Indians, which favour mobility and ecological balance, helped establish the staple diet of the common Paulista (people from São Paulo), a characteristic of the bushman lifestyle. According to the author, 'beans, corn and manioc, indigenous plants, formed what we could call the alimentary triumvirate of the bushman's diet, later altered with the substitution of the latter with rice' (1964: 52).

shoots mixed with seasoned meats]), manioc flour played a central role in the population's diet (Silva, 2005).

This importance seems to have prevailed until the early twentieth century, when the role of manioc went into decline, both ideologically and materially,[5] and it acquired the stigma of being a food of the poor, the stuff of the backwoods of developing countries (Grenand, 1993).[6] The publication of Josué de Castro's classic *The Geography of Hunger* (Castro, 1965) in Brazil in the 1960s cemented the image of manioc as the staple of the poorest layers of society, and especially so in the Amazon. In this work, the author characterizes the Amazonian region as a uniform dietary area with manioc flour as its emblem. According to his research, the Amazonian diet was characterized by a lack of variety, by nutritional deficiencies (of protein, mineral salts and vitamins) and by low caloric content.[7] In Castro's opinion (1965), the incipient regional agriculture was not capable of providing an adequate diet, and the blame lay with the 'demographic desert' of the Amazon.[8]

Josué de Castro's ideas were consonant with the discussions then underway in American anthropology on human occupation in the Amazon. The first explicative models on the adaptation and social evolution of Amazonian populations also stressed the importance of limiting environmental factors[9] over more historical/social approaches; blaming the non-development of dense, complex and stratified Amazonian societies (Carneiro, 1970; Gross, 1975; Lathrap, 1968; Meggers, 1954, 1987; Neves, 1996; Ross, 1978b; Steward, 1949) upon limited caloric intake on one side and protein shortfalls on the other. However, from the 1980s onwards, these approaches came under increasing attack on the grounds of their ecological determinism (Beckerman, 1979; Neves, 1989; Roosevelt, 1991), vis-a-vis the importance of historical and sociopolitical factors (Beckerman, 1991; Neves, 1998; Roosevelt, 1994). Manioc always played a central role in the debate (Adams, Murrieta, & Sanches, 2005; Carneiro, 1961, 1970; Meggers, 1987; Moran, 1989; Murrieta, 1998; Murrieta, Dufour, & Siqueira, 1999).

---

[5] Nardi Filho, for example, mentions the fact that manioc tilling was 'disdained for its lowliness, perhaps in virtue of its purely Brazilian origin [. . .] Foodstuff of the poor, standing 'provisions' for times of famine in the sunbaked northeast, bitter and sweet cassava were not worth planting on better lands' (1939: 791; apud Vaughan, 1939: 742).

[6] However, various authors have attacked this view (Barickman, 1998; Murrieta, 1998, 2000, 2001a), arguing that manioc is more than an emergency foodstock.

[7] However, the author observes that the low caloric consumption was not so critical because it was compensated for by the slower metabolism (20% lower) caused by the heat and humidity (Castro, 1965).

[8] It is interesting to note that until 2003 the National Household Census, conducted by the IBGE since 1981, covered all regions of Brazil except the rural areas of the North (Rondônia, Acre, Amazônia, Pará, Roraima, Amapá and Tocantins). In other words, the transformations in the rural Amazonian world *were rendered invisible* to the rest of the nation.

[9] Castro also considered the type of occupation in Amazonia to have been limited by forces of nature and by the 'oppressive obstacles of the environment' (1965: 42).

More recently, there has been growing scientific interest in aspects of nutrition and food consumption among indigenous (Coimbra Jr., Flowers, Salzano, & Santos, 2004; Dufour, 1992, 1994; Holmes, 1983; Holmes & Clark, 1992; Santos & Coimbra Jr., 1994, 1998) and peasant populations in the Amazon (Adams, 2002a; Adams et al., 2005; Giugliano, Shrimpton, Arkcoll, Giugliano, & Petrere Jr., 1978; Giugliano, Giugliano, & Shrimpton 1981; Giugliano, Shrimpton, Marinho, & Giugliano 1984; Migliano, 2000; Murrieta, 1994, 2000; Murrieta & Dufour, 2004; Murrieta, Brondizio, Siqueira, & Moran, 1989, 1992; Murrieta, Neves, & Dufour 1998; Murrieta et al., 1999; Silva, 2001; Silva, James, & Crews 2006; Siqueira, 1997). The as yet relative sparsity of data urges caution when it comes to making generalizations on nutrition and food consumption among Amazonian populations. Even so, some regularities have been identified: moderate levels of malnutrition (higher rates of below-average height per age); dependence on local products for staple foods, such as fish and manioc flour; high rates of intestinal parasitical infection; relatively lower calorie consumption compared with protein intake; greater variations in food consumption among households than among communities; and signs of increaseing consumption of industrialized foods (Adams, 2002a; Adams et al., 2005; Giugliano et al., 1978, 1981, 1984; Holmes, 1983; Murrieta, 1998, 2000; Murrieta & Dufour, 2004; Murrieta et al., 1989, 1999; Murrieta, Adams, Bakri, Strumpf, & Sousa, 2006; Silva, 2001; Silva et al., 2006).

In this chapter we will present a comparative analysis of food consumption in five caboclo communities from the state of Pará; three from the estuarine floodplains (Marajó Island) and two from the seasonal floodplains (Ituqui Island). We will discuss the consumption of macronutrients (calories and proteins), with emphasis on the nutritional importance of manioc in the caboclo diet. The comparison between populations settled in different ecosystems will allow us to highlight some of the abovementioned regularities, especially the central role of manioc in caboclo diet[10] and its combination with varied protein sources.

[10] This comparative study (Adams, 2002a) was only possible thanks to a database developed by various researchers involved in two distinct projects. The first of these, by Walter Neves, Rui Murrieta, Andréa Siqueira, Eduardo Brondizio and Renate Viertler in Marajó Island, Pará (Murrieta et al., 1999; Neves, 1992), was carried out between January 1991 and July 1992. Some of the researchers involved kept the project going in subsequent years (Brondizio, 1996; Siqueira, 1997). The second project, coordinated by Rui Murrieta and Cristina Adams, was conducted in Ituqui Island between 1995 and 1997, based on the methodological design of the first study (Adams, 2002a; Adams et al., 2005; Murrieta, 1998, 2000, 2001a; Murrieta & Dufour, 2004). The results for calorie and protein consumption obtained in 1997 in Aracampina (ARA) and São Benedito (SB), Ituqui Island (Adams, 2002a), are discussed here in comparison with those obtained for these same communities by Murrieta (2000) and Murrieta & Dufour (2004) in 1995/96 and by Murrieta et al. (1999) and Siqueira (1997) in Paricatuba (PA), Praia Grande (PG) and Marajó-Açu (MA), Marajó Island, in 1991.

Evaluation of Food Consumption

The methodology used to evaluate food consumption in the study communities was '24-hour food recall' (Adams 2002a; Murrieta, 1994, 2000; Murrieta & Dufour, 2004; Murrieta et al., 1999). This technique involves semi-structured interviews with the adult responsible for preparing meals in the household. Throughout the interview, this individual is encouraged to recall all the foods consumed by the household over the last 24 hours. The interviews were conducted over seven consecutive days during both the Amazonian seasons (summer/dry and winter/rainy). Preferably, the interviews took place in the kitchen so as to help in the recollection of the foodstuffs consumed the previous day and for convenience in weighing and measuring the dishes used to prepare the meal. These weights were obtained in grams using portable scales. Also collected was information on the origins of the foods consumed (purchase, cultivation, extraction, gift, exchange, sale) and on the number of participants present at those meals (Adams 2002a; Murrieta, 2000; Siqueira, 1997).

The food recall data was gathered on Marajó Island in 1991 (winter and summer) at the communities of Paricatuba, Praia Grande and Marajó-Açú (municipality of Ponta de Pedras – PA), and have already featured in other publications (Brondizio & Siqueira, 1997; Murrieta et al., 1999). Two communities were examined on Ituqui Island – Aracampina and São Benedito (municipality of Santarém, Pará) – in two stages (1995/96 and 1997). The first data gathering was made in October 1995 (summer) and in April 1996 (winter) and some of this data has already been published (Murrieta, 1998, 2001a, 2001b; Murrieta & Dufour, 2004; Murrieta et al., 1998). The second survey was taken in early May (winter) and November (summer) of 1997 (Adams 2002a), and a portion of these findings were also recently published (Adams et al., 2005).

The households were selected on the basis of their representativeness in terms of family composition, economic and subsistence activities, socioeconomic status and willingness to participate in the survey (Murrieta, 1994, 2000; Murrieta et al., 1999; Siqueira, 1997). From the total of 86 households (HDs) in Ilha de Marajó, 16 were selected for the survey (Table 12.1): 6 in both Praia Grande (PG) and Marajó-Açú (MA), and 4 in Paricatuba (PA), together totaling 112 individuals (Murrieta, 1994; Murrieta et al., 1999; Siqueira, 1997). For the 1995/96 survey at Ilha de Ituqui, 8 HDs were selected in Aracampina (from a total of 73) and 4 from São Benedito (total of 35), comprising a combined sample of 80 individuals (Murrieta, 2000; Murrieta & Dufour, 2004). In 1997, the survey re-visited the same households as covered by the previous collection (Adams 2002a), totaling 95 individuals in either season. However, the specially high floods that year made it difficult to reach all of the families originally planned, as many abandoned their homes and moved to Santarém or to the uplands, a common strategy in years of severe flooding (Futemma, 1995; WinklerPrins, 1999). As a result, the 1997 survey included 2 HDs during winter (15 individuals) and 7 in the summer

Table 12.1 Number of households, individuals and days investigated in each community and at each stage (Adams 2002a; Murrieta, 2000; Siqueira, 1997)

Ecosystem (Floodplain)	Year	Season	Community	N° HDs	Total days surveyed
Estuarine	1991	Summer	PG	6	42
			MA	6	42
			PA	4	28
		Winter	PG	6	42
			MA	6	42
			PA	4	28
Seasonal	1995	Summer	SB	4	28
			ARA	8	56
	1996	Winter	SB	4	28
			ARA	7	49
	1997	Summer	SB	4	24
			ARA	7	44
		Winter	SB	2	14
			ARA	2	14

(44 individuals) in Aracampina, and 2 HDs in winter (15 individuals) and 4 in summer (21 individuals) in São Benedito.

The quantitative data on food intake obtained through the interviews was converted into calorie and protein values in accordance with the Brazilian food composition tables (Franco, 1987; IBGE, 1999; INPA, 1998; Philppi, 2002; Unicamp, 2004). The weekly food intake for each household was totaled and compared against the nutritional requirements proposed by the FAO/WHO/UNU (1985) (Adams 2002a; Murrieta, 2000; Murrieta et al., 1999; Siqueira, 1997), according to the gender, age, body weight and level of physical activity of all the permanent members of the household present during the survey. Energy expenditure was estimated as per values fixed by the FAO/WHO/UNU (1985) for moderate physical exertion, though values for pregnant or breastfeeding women were compensated to account for their heightened energy requirements.

The units of analysis adopted were the population, the household (HD) (Murrieta, 1994, 2000; Siqueira, 1997)[11] and the ecosystem (Adams 2002a). Population was defined as a group of people living in a given geographic area, delimited in both etic and emic terms (Murrieta et al., 1999: 457). Household (HD) was defined as the spatial unit in which the food is consumed, though the flux of foodstuffs inside and outside the HD and consumption in other locales were not ignored. In this study, the populations examined were grouped

[11] The individual was also considered, though only the results for household consumption are discussed here. The individual food consumption research served to demonstrate the differences in consumption between men and women and to present a more accurate picture of the adequateness of energy and protein intake (Murrieta, 2000, Murrieta & Dufour, 2004).

under a broader unit – the ecosystem (estuarine and seasonal) – so as to highlight the relations and interdependence among the environmental components (Odum, 2001).

Estuarine Floodplains – Marajó Island

The populations studied on the estuarine floodplains are located in the southeast of Marajó Island, in the municipality of Ponta de Pedras, Pará. The local ecosystem could be described as an ecotone between the two macro-environments of Marajó – the natural fields and the forests. The main vegetal formations in the region are floodplain forests (*mata de várzea* or *mata de maré*), upland forest (rainforest or *terra firme*), grasslands and mangroves. The estuarine floodplain is an ecosystem adapted to seasonal and daily floods. The most important vegetal species in this environment, given its contribution to domestic consumption and its economic value on the regional market, is açaí (*Euterpe oleracea* Mart.) (Brondizio & Siqueira, 1997; Murrieta et al., 1992, 1999). During the study period, there were patches of upland/terra firme forest in two of the focus communities: Praia Grande and Paricatuba. These forests contain more biomass and biodiversity than floodplain forest. Grassland is encountered all across the region and some of these tracts have been used by the locals as pastureland and for gathering. The mangroves on the riverbanks and in Marajó bay play a crucial role as nurseries for fish and crustaceans, the most important protein sources in the region. In addition to the natural landscapes, there are also various anthropogenic formations, such as brushwood, stands of coconut trees and pasture (Brondizio, Moran, Mausel, & Wu, 1996; Murrieta et al., 1992. 1999).

The annual mean temperature for the municipality of Ponta de Pedras is approximately 27° C, with precipitation of some 3,000 mm per year. As such, like most of the Amazon, the region has two main seasons: a dry summer (May to November) and wet winter (December to early May). At the beginning of the 1990s, the population of Ponta de Pedras was 16,506, with a population density of 5.72 inhabitants per km^2 (IBGE, 1991). The populations selected for study were Marajó-Açú, Paricatuba and Praia Grande. The results from the research conducted in Ponta de Pedras have already been the subject of various publications and theses (Brondizio, 1999; Brondizio & Neves, 1996; Brondizio & Siqueira, 1997; Brondizio, Moran, Mausel, & Wu, 1994; Brondizio et al., 1996; Murrieta, 1994; Murrieta et al., 1989, 1992, 1999; Pucciarelli, Neves, Melcher, & Murrieta, 2005; Silva, 2001; Silva & Eckhardt, 1994; Silva et al., 1995; Siqueira, 1997; Siqueira et al., 1993) that offer detailed descriptions of the communities and the study area. In the present article, therefore, we shall merely give a brief overview of the region and its communities with these works as a reference.

The population of Marajó-Açú lives along the banks of a river of the same name and, at the time of the study, was composed by 371 individuals in 46 households (1991). At the time, the main subsistence activities were the cultivation of açaí,[12] the demand for which was growing on the regional urban market, prawn fishing and river trading with other areas of Marajó Island. Agriculture was restricted to a handful of manioc crops. Crop-sharing was the prevailing agrarian solution and cause of occasional conflicts with the landowners (Brondizio & Siqueira, 1997; Brondizio et al., 1994; Murrieta, 1994; Murrieta et al., 1989, 1999).

The population of Paricatuba was distributed along the banks of the Paricatuba River and on the adjacent uplands. At the time the study was carried out, the population was of 144 individuals in 19 households. The key productive activities were subsistence agriculture (mostly manioc, rice, beans and corn), fishing, hunting of small game and the cultivation of açaí and other fruits. Most of the inhabitants were owners of their lands. Though the general pattern was similar to that at Marajó-Açú (acaí, prawn fishing and river trading), the households in Paricatuba seemed more independent of the local market and more integrated amongst themselves. This integration was characterized by a social network for exchanging foodstuffs, manpower rally-rounds and access to staple resources based on kinship bonds (Murrieta, 1994; Murrieta et al., 1999; Siqueira et al., 1993).

The population of Praia Grande was located on an upland stretch along the shores of Marajó Bay. In 1991, there were 111 individuals in 21 households. Historically, their subsistence system was based on slash-and-burn agriculture, fishing, hunting and the cultivation and extraction of market-bound produce (mallow *Urena sp* and rubber *Hevea spp*). However, in the 1960's, an agricultural cooperative set up by the Catholic Church began to change the pattern of local subsistence, with manioc largely abandoned in favour of new, mechanized cultivars (Bahia coconut, corn, rice and beans) and ranching. Nevertheless, despite the best efforts of the Church, the residents and the external input of resources from interventionist projects, every attempt to implement a self-sustainable system failed. Indeed, one of the side-effects of these attempts was environmental degradation most likely due to the use of inadequate technologies and the expansion of land for pasture (Brondizio & Neves 1996; Brondizio et al., 1994; Murrieta, 1994; Murrieta et al., 1992, 1999; Siqueira, 1997).

Seasonal Floodplains – Ituqui Island

Ituqui is a 30,000-hectare island on the main channel of the Lower Amazon River, located 30 km from the town of Santarém (PA). Unlike the region of Marajó, which draws heavy rains, the Santarém region lies in the Amazon's 'dry

[12] Brondizio & Siqueira (1997) argue that the selection, management and planting practices for açaí in the region of Ponta de Pedras (PA) should be categorised as forest cultivation rather than as a simple system of extraction (also see Brondizio, 1996; and Brondizio, in this volume).

belt' (WinklerPrins, 1999), where the seasons are more distinct and annual rainfall is less than 2,000 mm.[13] Average annual precipitation in the Santarém region is 1,973 mm; the dry season lasts from July to December and the rainy season (winter) from January to June. On Ituqui, approximately 80% (or 1,578 mm) of annual rains fall during the wet season and only 20% (395 mm) during the summer, when planting takes place. There are also micro-regional differences and Ituqui Island seems to be drier than the surrounding uplands, thus creating difficult conditions for agriculture (WinklerPrins, 1999, 2001, 2002; WinklerPrins & McGrath, 2000).

Ituqui Island is covered by a mosaic of forest, savannah and pasture highly adapted to seasonal flooding. During the rainy season, the waters rise and inundate most of the island, radically altering its environment (Murrieta & Dufour, 2004; WinklerPrins, 2002). The highest water levels recorded during the study period were 7.3 m (1995), 7.9 m (1996) and 9.0 m during the great floods of 1997 (Murrieta 2000; WinklerPrins, 1999). Over these periods, the only visible vegetation was floating gramineous and other macrophytes and the boughs of partially submerged trees (Murrieta 2000; Murrieta & WinklerPrins, 2003; Murrieta & Dufour, 2004; WinklerPrins, 1999). The savannahs and forests remain water-free throughout the summer, allowing for agricultural and ranching activities. As an environment that experiences constant modifications through the depositing of fresh sediments in some areas and soil erosion in others, this is an environment characterized by rich biological diversity (Castro, 1999; Murrieta, 2000; Murrieta & Dufour, 2004; WinklerPrins, 1999, 2002).

The population of Ituqui Island is spread across eight communities: Aracampina, São Benedito, Santana,[14] São Raimundo, São Benedito, Fé em Deus, Conceição and Nova Vista. Total population at the time of the study was roughly 2,000 inhabitants (6.7 inh./km^2), though 70% of the area was occupied by 12 large cattle ranches (Câmara & McGrath, 1995). The majority of the population of Ituqui Island was composed of smallholders whose families had probably lived there since the late nineteenth/early twentieth centuries (Murrieta, 2000, 2001a). In the early stage of occupation, the settlers planted cacao (*Cocoa sp*), raised cattle and seasonally tapped rubber (*Heavea brasiliensis*). The definitive collapse of the rubber trade in the Amazon after the Second World War and the difficulties in planting cacao drove the inhabitants to plant jute (*Corchorus capsularis*), which dominated the economic life of Ituqui for nearly 50 years (Gentil, 1988; Murrieta 2000; Murrieta; WinklerPrins, 1999). When jute production collapsed in the 1980s, fishing and cattle raising became the main economic activities on the island (Futemma, 1995; Murrieta, 2000; Murrieta & Dufour, 2004). Fishing in Ituqui occurs in two key ecozones: the river and the island lakes (Murrieta, 2000).

[13] This differentiation, however, has received relatively little attention in the literature, leading to an erroneous generalization for the entire Amazon region (WinklerPrins, 1999).

[14] Though Santana is not situated on the island properly speaking, most of its population uses its environments for fishing and agriculture.

The two communities chosen for the survey, Aracampina and São Benedito, represent the extreme environmental poles of Ituqui Island (Murrieta, 2000; Murrieta & Dufour, 2004). Aracampina is situated on the banks of the main flow of the Amazon River, on the lowest part of the island, where it is exposed to the seasonal floods and the river's currents (Castro, 1999). In the latter half of the 1990s, this was the most populous community (550) and the one that occupied most space (Castro, 1999; Futemma, 1995). Due to its location, the population of Aracampina has reaped the benefits of the silting process of the river, which has deposited a small island in front of the community, where they can raise cattle, fish and grow beans. In addition, ready access to the main body of the Amazon River gave the residents of Aracampina the option of seasonal fishing of large catfish (in summer), which could supplement family income (Castro, 1999).

São Benedito is located on the opposite side of the island, on the floodplain of the Paraná do Ituqui, a branch of the Amazon River. In 1997, there were 164 inhabitants in 36 households. This community possesses one of the largest areas of arable and ranch land in the region and is situated close to two lakes. In addition, it is surrounded by a mosaic of micro-environments including secondary flooded forest, swampland, grasslands and pasture (Murrieta, 2000; Murrieta & WinklerPrins, 2003; Murrieta & Dufour, 2004). Given their localization, the residents of São Benedito were less engaged in river fishing than their counterparts at Aracampina. Though both areas have similar soil, the ground in São Benedito tends to be more humid (WinklerPrins, 1999, 2001; WinklerPrins & McGrath, 2000).

Planting in Ituqui Island took place in the summer, when the *restinga* was exposed. The main cultivars were manioc (*Manihot esculenta*), bean (*Phaseolus vulgaris*), pumpkin (*Cucurbita spp.*) and watermelon (*Citrullus lanatus*) (Murrieta, 2000; WinklerPrins, 1999). Manioc was planted on the higher ground as soon as the waters drained, as it took six months to mature. Harvesting was at the onset of the rainy season, just before the floods, normally in January, and processing was done immediately. The unpredictability of the floods made planting manioc a high-risk activity (Murrieta, 2000). Once transformed into flour, the manioc was stocked in sacks. Most of the production went for household consumption, while the beans and the cucurbitaceans (pumpkin and watermelon) were destined for sale in Santarém (WinklerPrins, 1999).

Fishing was one of the most important activities in Ituqui, being most productive in the inland lakes during the dry season and dwindling during the rains as some species left the lakes. Deforestation and over-fishing were affecting the island's fish-stocks and attempts were being made to control access to them (McGrath et al., 1993; WinklerPrins, 1999). Cattle were raised as a sort of savings fund by the people of Ituqui, and in 1994, despite the difficulties in keeping livestock, especially during the floods, 70% of the households in both communities owned heads (McGrath, Clancy, & Murrieta, 1995, cited in WinklerPrins, 1999). The two communities differ in distribution of incomes per economic activity. Fishing was dominant in Aracampina (53.1% of income), while agriculture prevailed in São Benedito (27%). This difference can be partially explained by the lesser availability of arable land in Aracampina and by the identity embraced by its people. Moreover, the soil in Aracampina was slightly inferior in agricultural quality and the community was not regularly served by boat routes, which made it harder to sell the produce (WinklerPrins, 1999, 2001, 2002; WinklerPrins & McGrath, 2000). A large part of the incomes in São Benedito came from pensions (21.6% versus 7.8% in Aracampina) and salaries (10.8% in São Benedito against 4.7% in Aracampina)[15] (McGrath et al., 1995, cited in WinklerPrins, 1999).

Caloric/Proteinic Contribution of Manioc and Other Foodstuffs

Calories

The food recall results for Ituqui and Marajó islands testify to the nutritional importance of manioc in the diets of the studied populations. Manioc proved the main source of calories in both ecosystems (Table 12.2), being consumed largely as manioc flour (farinha)[16] (Adams, 2002a; Murrieta, 1994, 2000; Murrieta et al., 1999; Siqueira, 1997). Though it is true that the importance of manioc (and derivatives) to the diet of these populations is evident, it must be stressed that its role far exceeds that of a 'secure source of calories' or 'emergency foodstuff', but is actually an acquired taste, a *habitus*, to use the term of Pierre Bourdieu (1983a, b),[17] as discussed by Murrieta (2001a). Manioc flour is highly appreciated and its quality is easily recognizable in its texture and taste, whether taken as an accompaniment to meat or eaten constantly throughout the day as a snack (Murrieta, 2001a), a habit inherited from the Indians and the European colonizers (Silva, 2005; Zeron, 2000: 35). Manioc flour was also taken in sweetened coffee, or as porridge with assorted flavors from earliest infancy (Murrieta, 2001a).

It was noted that manioc consumption was slightly lower on the seasonal than on the estuarine floodplain, probably a reflex of the annual Amazon floods, which shorten the agricultural period on Ituqui to roughly six months. On Marajó Island, only Praia Grande (33.2%) presented values similar to those of Aracampina (34.0 and 35.2%) and São Benedito (25.2 and 31.1%). It is

[15] These pensions were either rural pensions or those from fishing colonies, while the salaries were largely from teaching or civil service posts.

[16] Manioc flour is consumed in the form of various types of porridge, such as *carimã* (finely-sieved manioc flour and dough mixed with sugar and salt), *chibé* (lightly sieved manioc flour soaked in lukewarm water and cooked with salt in boiling water) or *crueira* (thick flour made from the residues of water flour production) (Murrieta, 2001a).

[17] In Bourdieu's definition (1983a), a *habitus* is 'a system of predispositions maintained and reproduced by social agents; at once a set of transposable dispositions and a template of past experience converted into perceptions, appreciations and practices' (cited in Murrieta, 2001a).

Table 12.2 Main calorie sources (% of total consumption) per community (both seasons combined) on the estuarine (Marajó Island) and seasonal floodplains (Ituqui Island) (Adams, 2002a, Murrieta, 2000; Siqueira, 1997)

	Estuarine floodplain			Seasonal floodplain			
	PA	MA	PG	ARA (95/6)	SB (95/6)	ARA (97)	SB (97)
Cassava	45.8	41.5	33.2	34.0	25.2	35.2	31.1
Fish	–	–	–	17.5	26.0	11.5	24.6
Açaí-palm	14.0	30.0	18.7	–	–	–	–
Sugar	–	–	–	11.5	11.5	11.2	10.0
Cereals	–	–	–	8.2	10.5	10.4	5.7
Dairy	–	–	–	7.3	2.7	4.7	1.3
Vegetal oil	–	–	–	4.2	2.6	3.6	3.1

interesting to recall that Praia Grande was the community that had abandoned the traditional patterns of subsistence. On the other hand, the community of Marajó-Açú also experienced a production shortfall for self-sufficiency, but nevertheless still presented reasonably high values of caloric contribution for manioc (41.5%).

In addition to the variations between the two ecosystems, there were also oscillations within these ecosystems. On Marajó Island, manioc accounted for 45.8% of calorie intake in Paricatuba, where manioc was still one of the main subsistence activities (Siqueira, 1997), while in Praia Grande, where the traditional plantations had long since been abandoned in favour of mechanized farming, the figure dropped to a mere 33.2% (Murrieta et al.,1992). On the seasonal floodplains in 1996, the difference between Aracampina and São Benedito measured nine points in favour of the former. Moreover, on Ituqui Island, striking annual variations could be noted for the community of São Benedito: while in 1996 manioc had accounted for 25.2% of calorie intake, this had risen to 31.3% by 1997, despite the fact that the floods were more severe that year, with water levels rising so quickly that many families lost part of their crop because they did not have enough time to process the flour. These figures point to a complex amalgam of factors involved in nutritional choices, the importance of micro-ecological variations and the need for longitudinal studies that factor-in the medium and long-term environmental and socio-economical variations for these populations.

The second highest calorie source for these populations varied depending on the ecosystem. In the estuary, it was açaí (*Euterpe oleraceae* Mart., a species that does not naturally occur on the Ituqui floodplains), while on the seasonal plains it was fish.[18] Unlike with manioc, açaí yields in the three estuarine communities were sufficient for domestic consumption (and sale in Marajó-Açú and

[18] Though 'fish' is taken here as a single category, it actually encompasses a range of species, and this diversity would seem to play an important role in both consumption and in individual preferences (Murrieta & Dufour, 2004).

Paricatuba) during the study period (Murrieta et al., 1999; Siqueira, 1997), reflecting the so-called process of *açaízation* (for a more detailed analysis of the process in these communities, see Brondizio in this volume).

The importance of fish as a calorie source in Ituqui warrants further attention. Fish contribution was higher in São Benedito than in Aracampina, especially if we consider the period 1995/96, when fish provided almost as much calorie intake as manioc (26.0 and 25.2%, respectively). This difference would appear to be associated with the productivity of the island lakes (Murrieta, 2000; Murrieta & Dufour, 2004). Preliminary investigations suggest that the lakes at São Benedito are more productive than those at Ituqui (Murrieta, 2000). Other factors, such as domestic variations in the importance of fishing as an activity and in personal aptitudes may also have contributed (Murrieta & Dufour, 2004). Indeed, calculations based on the São Benedito data for 1997 (Adams, 2002a) showed that the HDs in which fish was the biggest calorie contributor were precisely those whose owners had a reputation for being 'good fishermen' (for more on nutritional choices and fish, see Murrieta, 1998, 2000, 2001a, 2001b).

Seasonal analysis (Adams, 2002a; Murrieta, 2000; Murrieta & Dufour, 2004) shows a reduction in the importance of fish as a calorie source during winter in Aracampina, which was compensated for by extra sugar. In the summertime, the importance of manioc as a calorie source remained practically constant in both communities, but there was a tendency toward more diversification in Aracampina in terms of secondary calories and more dependence on industrialized foods (cereals and sugar) (Murrieta, 2000; Murrieta & Dufour, 2004).

The dependency observed in Ituqui is reflected in the position sugar occupies in the calorie-intake of both communities (Table 12.2), varying between 10.0 and 11.5% over the two years of the study. Sugar would seem to be a reliable calorie source during the winter in Ituqui and to fuel those long hours of toil under often harsh environmental conditions (Murrieta, 2000a; Murrieta & Dufour, 2004). Though there is no systematized data available for the estuarine floodplain, we believe that sugar contribution would also have been high there, particularly as Siqueira's 1997 study showed that the people of Marajó used sugar in similar fashion to those of Ituqui to sweeten coffee.

Proteins

In general, protein sources were more varied than those for calories in both ecosystems (Table 12.3) (Adams, 2002a, Murrieta, 2000; Siqueira, 1997). Regardless of the ecosystem studied, fish was the principal protein source in all communities, but especially on the seasonal floodplains[19] (Ituqui is located

[19] The Amazonian floodplains have one of the highest per capita fish consumptions in the world. The estimated consumption for 17 riverine communities in the region is 379 g per person per day (Cerdeira et al., 1997, cited in Castro, 1999).

Table 12.3 Main protein sources (% of total consumption) per community (both seasons combined) on the estuarine (Marajó Island) and seasonal floodplains (Ituqui Island) (Adams, 2002a, Murrieta, 2000; Siqueira, 1997)

	Estuarine floodplain			Seasonal floodplain			
	PA	MA	PG	ARA (95/6)	SB (95/6)	ARA (97)	SB (97)
Fish	13.3	34.3	23.1	54.3	64.1	43.5	67.3
Prawn	12.9	8.9	13.3	–	–	–	–
Pork	16.4	4.4	6.0	–	–	–	–
Beef	5.5	18.8	16.4	–	–	3.5	0.6
Game	17.2	2.6	0.5	–	–	13.6	1.0
Chicken	–	1.7	6.5	–	–	4.8	8.9
Cereals	–	–	–	5.4	5.4	4.2	2.1
Milk	–	–	–	11.6	3.1	4.0	1.1
Eggs	1.6	1.4	0.3	–	–	0.4	1.5

in a region highly productive in terms of fisheries) (Murrieta, 2001b). The protein contribution, allied with its caloric importance, makes fish the cornerstone of the caboclo diet.

The most striking difference between the ecosystems concerns secondary protein sources. Prawn, a typical estuarine species, is the second largest protein supplier on Marajó Island, followed by pork and beef (Siqueira, 1997, pp. 251–2). In Ituqui, however, the chief secondary sources are meats, especially game and chicken (Table 12.3),[20] cereals and milk, depending on the community/year considered. The use of coop chickens as a protein source seems to be particularly important during winter, when fishing is harder (Murrieta & Dufour, 2004).

As with calorie-intake, there were intra-ecosystem variations in the relative contributions of key protein sources. In the estuary, fish consumption was much higher in Marajó-Açú (34.3%) than in Paricatuba (13.3%) and Praia Grande (23.1%), with the inverse holding for prawns: approximately 13% in Praia Grande and Paricatuba and 8.9% in Marajó-Açú (Siqueira, 1997). Likewise, fish consumption was higher in São Benedito than in Aracampina both in 1995/6 and in 1997. The data for the seasonal floodplains (Adams, 2002a, Murrieta, 2000) shows that there can also be considerable annual and seasonal variations in protein sources within the same community.

[20] In the case of São Benedito, this value includes domestic duck, which was consumed in large quantities at two of the HDs during the week of the survey in virtue of Mother's Day lunch (winter). One of the household heads killed two ducks from her domestic stock the day before the festivities, kept one and gave the other to her sister, the head of another HD. High game consumption in Aracampina was also occasional. The summer survey coincided with an unusually large supply of *tracajá* turtle (*Podocnemis unifilis*) at one HD. As the meat of this aquatic turtle is a favourite of the family, it was consumed in large quantities.

Greens, Fruits and Vegetables

There was low consumption of greens and vegetables in both ecosystems, despite the relative abundance in Praia Grande and Aracampina due to the community vegetable gardens implanted by the Catholic Church and a local NGO.[21] In fact, in Ituqui, greens and vegetables are relegated to the category of 'non-foods' and are used as seasonings for the main elements of meals in order to break the monotony and diversify flavor (Murrieta, 2001a). The vegetables and greens most commonly used by the estuarine communities were pumpkin, squash and redroot (Siqueira, 1997). In Ituqui (1997), they were pumpkin, French basil, chicory, kale, squash and cabbage. The most often used seasonings in Marajó were onion, pepper, annatto, common basil and chicory (Siqueira, 1997) and in Ituqui (Adams, 2002a) onion, scallion (spring onion), coriander, chili pepper, green pepper, tomato, paprika, cumin and pepper (the latter three being industrialized products purchased in Santarém).

In Ituqui, the consumption of fruits was seasonal, given the periodicity of native fruit-bearing trees and the fragility of exotic species in the face of the flood cycle. In the estuary, the most commonly consumed fruits were banana, pineapple, papaya, avocado, orange, coconut, lemon and maracujá (Siqueira, 1997). On the seasonal floodplains (1997) the favorite fruits were banana, lemon, orange, mango, melon and, especially, watermelon (Adams, 2002a).

Calorie and Protein Requirements: the Importance of Manioc in the Local Diet

A comparison of average calorie and protein consumption of the households with the international recommendations of the FAO/WHO/UNU (1985) and Franco (1987) turns up a similar pattern for both floodplains (Fig. 12.1). Despite protein intake being above recommended levels, calorie consumption was revealed to be too low in almost all cases (Adams, 2002a; Murrieta, 2000; Murrieta & Dufour, 2004; Siqueira, 1997).[22] Of all the communities surveyed, Marajó-Açú was the only one to reach the recommended levels (Siqueira, 1997). In general terms, the estuarine floodplain communities presented a level of caloric consumption closer to the desired. However, the reverse was true

[21] At the time of the 1997 phase, the intervention programmes of a local NGO were in full swing, including the community vegetable gardens in Aracampina, with potential impacts on direct nutrient consumption in the HDs surveyed (Murrieta, 2000).

[22] Recommended consumption was calculated based on the permanent members of the household, considering gender, age, reproductive status and level of activity. However, it is known that the number of participants at each meal is variable. In this sense, the results presented should be considered as relative averages for food availability at the households in relation to the number of fixed residents. The results should not therefore be used as a sole measure for individual consumption and nutritional status (Murrieta et al., 1999).

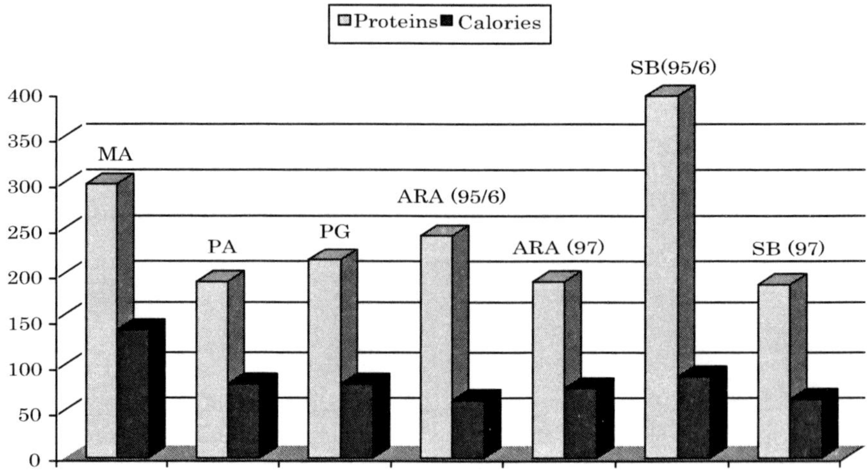

Fig. 12.1 Average calorie and protein consumption (%) per household in relation to the minimum international requirements (FAO/WHO/UNU, 1985, and Franco, 1987) (Adams, 2002a, Murrieta, 2000; Siqueira, 1997)

when it came to protein intake, where the seasonal floodplains presented a slightly higher value.

What most calls attention on the seasonal floodplains is the annual variation in calorie/protein consumption amongst the communities. In 1995/96 (Murrieta, 2000; Murrieta & Dufour, 2004), calorie and especially protein intake in São Benedito was more satisfactory than values for the following year. In Aracampina, protein consumption fell against figures for 1995/96, while energy consumption improved. Even so, calorie consumption for this community was still below recommended levels, while protein intake was on a par with Paricatuba (Fig. 12.1), the community with the lowest values in the estuarine region (Siqueira, 1997). With the exception of energy consumption in Aracampina, all of the values for Ituqui Island decreased from 1995/96 to 1997, a fact that could perhaps be attributed to the harsher conditions of the winter that year. Indeed, the floods of 1996 were relatively low (Murrieta, 2000) and the residents of São Benedito were able to harvest and process the manioc crop in time.

Considering that the combination of manioc and fish or prawns is the staple food of these communities, the seasonal, annual and inter-ecosystem variations observed in caloric consumption most likely reflect the productive instability of the manioc crop due to important environmental components, chiefly on the seasonal floodplain (WinklerPrins, 1999). The seasonal flooding that characterizes this ecosystem affects both farming and other productive activities (fishing and ranching), causing direct impact on food consumption. The dwindling of ecological zones suitable for use that occurs with the floods also narrows diet, both directly (less resource availability) and indirectly (lower incomes)

(Murrieta & Dufour, 2004). This instability affects household manioc production, which forces many families to buy supplementary flour for large parts of the year, especially in summer, when the domestic stock has already been consumed (Murrieta, 2000).

On the other hand, the inexistence of this seasonal limitation in the estuary gives the populations of Ponta de Pedras access to a greater diversity of ecological zones throughout the year. The diet based on manioc, açaí and fish/prawns indicates that the strategy adopted by the estuarine communities consists in drawing from a broader selection of ecological possibilities (Murrieta et al., 1999). Açaí is collected from the flooded forest, manioc is cultivated in the uplands, fish are caught in the bay and rivers, and the prawns are fished in the mangroves that run along the riverbanks and the shores of Marajó bay (Murrieta et al., 1999). The only estuarine community that presents caloric intake from manioc products similar to those for Itaqui was Praia Grande (Table 12.2), which shifted from traditional subsistence and discontinued the cultivation of the tuber (Murrieta, 1994; Siqueira, 1997), thus increasing food insecurity (Emperaire, 2002b).

However, the distinct environmental characteristics in themselves are not enough to explain the differences in pattern of nutrient consumption among the communities tracked. As observed in earlier studies (Murrieta, 1994; Murrieta et al., 1999; Siqueira, 1997), despite the geographical proximity of the communities on Marajó Island, a great many socio-economic and cultural differences exist amongst them. One example would be the importance of food exchanges and reciprocity networks in the communities of Paricatuba and Praia Grande in comparison with Marajó-Açú (Murrieta et al., 1999; Siqueira, 1997).

On Ituqui Island, Aracampina and São Benedito also display countless differences in terms of local environmental (micro-ecological) characteristics, subsistence strategy, history of community occupation and the political and institutional relations each maintains with Santarém, all of which are reflected in food consumption patterns (Murrieta, 2000). In Aracampina, the shift from manioc cultivation to other activities, such as fishing, can be explained not only in virtue of the slightly poorer soils and greater vulnerability to flooding (environmental factors), but also by the fact that there are no regular boats to and fro (Murrieta, 2000; Murrieta & Dufour, 2004; WinklerPrins, 1999), as well as by the general regional economic context (Gentil, 1988).

Conclusions – the Invisibility of Manioc

The results presented in this comparative analysis bring us to the discussion that raged among Meggers (1954, 1987), Lathrap (1968), Carneiro (1970), Gross (1975), Ross (1978b) and Beckerman (1979) on the limiting factors to human occupation in the Amazon: carbohydrates or proteins? Everything would indicate that fish resources meant that proteins were no impediment to human occupation, at least on the white water rivers floodplains.[23]

Our research indicates that calorie sources (mainly manioc) truly are the most critical factors for subsistence on the Amazonian floodplains, as Meggers (1958, 1987) believed. Even so, the caloric consumption on the diets studied varied from 60 to 155% of the minimum international requirements (FAO/WHO/UNU, 1985), with averages of 100.8% for the estuary and 72.7% for the seasonal floodplains. Considering that the method used underestimates caloric intake, as it does not include foods consumed outside the household, it is likely that these values are higher, and that even the populations on the seasonal floodplains manage to obtain the calorie levels recommended by the international health agencies from the surrounding landscape, configuring a more satisfactory nutritional scenario than that observed in our surveys.

The results of our research lead us to propose that, despite the socio-economic invisibility of the manioc grown and consumed by traditional populations in the Amazon, the combination of this tuber and its derivates with fish and/or prawns forms the nucleus of the riverine adaptive strategy, at least partially explaining the impressive historical resilience of these communities on the Amazon floodplains and, perhaps, other landscapes of the region too (Murrieta, 2000, 2001a; Murrieta & Dufour, 2004; Murrieta et al., 1999). It is beyond question that manioc and fish constitute the staple diet of various indigenous communities along the major watercourses (rivers and shores) of the Amazon (Beckerman, 1993). The use of sophisticated manioc detoxification techniques, on one hand, and the consumption of fish as a protein source, on the other, are further factors that contribute to the way the human organism metabolizes the tuber's cyanide content. This compound is toxic in large quantities, though in small doses can be readily metabolized by methionine and cystine, amino acids amply found in fish (Spath, 1981). In other words, far from being a retrograde diet, manioc must have always been the core cultivar in the caboclo subsistence system. The abidance of this adaptive nucleus over the course of two centuries becomes even more significant when we consider the powerful dynamic provoked by the numerous transformations that accompanied the national and transnational societies present during that same period (Beckerman, 1993; Moran, 1993; Nugent, 1993; Pinton & Emperaire, 2000; Ross, 1978a; Santos, 1980; Weinstein, 1985).

Bearing in mind that the caboclo populations are part of that swathe of the Brazilian population historically discriminated against, one might expect that the pillar of its agricultural system, manioc, would have also been the butt of prejudiced observations and wrongheaded interpretations (the same applies to açaí, as Brondizio shows in this volume). Despite the dramatic historical and social changes Brazil has undergone over the last 500 years, many of which were

[23] It is worth noting that, in some cases, such as in São Benedito in the summer of 1995, fish can surpass even manioc as the most important calorie source. As such, the dual role of fish in the caboclo diet as source of both protein and calories (Murrieta, 2000; Murrieta & Dufour, 2004) deserves systematic investigation.

felt in the Amazon as economic cycles imposed from the outside (Beckerman, 1993; Moran, 1993; Nugent, 1993; Pinton & Emperaire, 2000; Ross, 1978a; Santos, 1980; Weinstein, 1985), we can raise three hypotheses to explain such a long-lasting presence of manioc in the caboclo diet:

1. Genetic limitation: specific environmental characteristics related to the tropics would have reduced the potential availability of carbohydrate-rich cultivars for domestication. In fact, the subsistence of most human societies living in tropical environments relies on a limited number of vegetal species high on carbohydrate content (rice, corn, manioc, taro, yam or banana) that serve as staples (Cattle & Schwerin, 1985). In other words, despite the near incalculable biodiversity that surrounds them, tropical forest farmers find it hard to domesticate or semi-domesticate cultivars that could help vary their caloric base (De Garine Hugh-Jones, & Prinz, 1993) in virtue of characteristics inherent to the biology of the available species. Given the low-tech agricultural system in place in the Amazon, it is possible that manioc is the only viable solution from a genetic perspective.
2. Astuteness: caboclo flexibility and invisibility (Nugent, 1993) represent the core of the adaptive strategy these societies employ when it comes to absorbing external impacts and attempts at economic manipulation by the regional, national and transnational elites, and the ubiquitous association between manioc and fish may be just one facet of that strategy. By maintaining an historical identity of opposition (Harris, 1999), the caboclo has ensured himself control over his platform for social and biological reproduction, preventing his subsistence from falling completely into the hands of external politico-economic forces (Emperaire, 2002a).
3. Acquired taste: the consumption of manioc and its countless derivatives became an essential element of the *habitus* (see Bourdieu, 1983a, b) of caboclo society (Murrieta, 2001a). According to this logic, Murrieta (2001a: 54) argues that the role played by manioc and its derivatives cannot be viewed solely from the perspective of natural ontology. Its role goes far beyond that of 'emergency foodstock', or a 'secure source of calories', having actually become a social construction over time, an acquired taste 'intimately tied up with a certain sense of place'. As such, even if the fish/manioc binomial was originally fixed in caboclo society for practical reasons associated with basic sustenance in a complex and unstable political context, its abidance for over two hundred years also rested upon symbolic values, values that can only be attained through a constructivist epistemology.

It is important to note that the abovementioned hypotheses are by no means mutually exclusive. As Grundmann & Stehr (2000) so aptly contend, it is absolutely possible to bring nature and society into approximation without lapsing into naturalist fallacy. While classical anthropology excluded the physical environment, that exclusion is now obsolete and urgently needs to be abandoned. In this sense, the major challenge facing all those who share this view is to breach the barriers imposed by the social sciences of the classical tradition without risking a relapse into environmental and biological determinisms.

Acknowledgements This work would have been impossible without the concession of data collated on Marajó Island by Renate Brigite Viertler, Hilton Pereira da Silva and Eduardo Sonnewend Brondizio and the dedication and support of Perpétuo Socorro de Sousa in the gathering of field data at Ituqui. In England, the discussions with Roy F. Ellen (UKC) and the suggestions of Simon Strickland (UCL) helped correct errors that had occurred along the way and enriched the theoretical discussion. In São Paulo, the theoretical and methodological debate with Solange Lima was a constant wellspring of ideas and suggestions. The doctoral theses of Fabio de Castro, Celia Futemma and Antoinette WinklerPrins brought considerable support to this work. Funding from Fapesp and Capes made much of this project possible. To all, our sincerest thanks.

References

Adams, C. (2000). *Caiçaras na mata Atlântica: Pesquisa científica versus planejamento e gestão ambiental*. São Paulo: Annablume.

Adams, C. (2002a). *Estratégias adaptativas de duas populações caboclas (Pará) aos ecossistemas de várzea estuarina e estacional: Uma análise comparativa*. Ph.D. Dissertation, University of São Paulo

Adams, C. (2002b). Identidade caiçara: Exclusão histórica e sócio-ambiental. In U. P. Albuquerque (Ed.), *Atualidades em etnobiologia e etnoecologia*. Palestras Convidadas do IV Simpósio Brasileiro de Etnobiologia e Etnoecologia (pp. 27–43). Recife: Sociedade Brasileira de Etnobiologia e Etnoecologia.

Adams, C. (2003). The pitfalls of synchronicity: A case study of the caiçaras from the Atlantic Rainforest of South-eastern Brazil. In E. Berglund & D. G. Anderson (Eds.), *Ethnographies of environmentalists: Understanding conservation strategies and the distribution of privilege* (pp. 19–31). London: Berghahn.

Adams, C., Murrieta, R. S. S., & Sanches, R. A. (2005). Agricultura e alimentação em populações ribeirinhas das várzeas do Amazonas: Novas perspectivas. *Ambiente & Sociedade*, *8*, 1–22.

Barickman, B. J. (1998). *A Bahian counterpoint: Sugar, tobacco, cassava, and slavery in the Recôncavo, 1780–1860*. Stanford: Stanford University Press.

Barreto, B. M. (1939). O valor da mandioca: 'pão dos trópicos'. *Chácaras e Quintais*, *60*, 354.

Beckerman, S. (1979). The abundance of protein in Amazonia: A reply to Gross. *American Anthropologist*, *81*, 533–60.

Beckerman, S. (1991). A Amazônia estava repleta de gente em 1492? In W. A. Neves (Ed.), *Origens, adaptações e diversidade biológica do homem nativo da Amazônia* (pp. 143–59). Belém: Museu Paraense Emílio Goeldi/CNPq.

Beckerman, S. (1993). Major patterns in indigenous Amazonian subsistence. In C. M. Hladick, A. Hladick, O. F. Linares, H. Pagezy, A.Semple, & M. Hadley (Eds.), *Tropical forests, people and food: Biocultural interactions and applications to development* (pp. 411–24). Paris: Unesco/Parthenon.

Bourdieu, P. (1983a). *Esboço de uma teoria da prática*. São Paulo: Ática.

Bourdieu, P. (1983b). *Distinctions: A social critique of the judgment of taste*. Cambridge: Harvard University Press.

Brondizio, E. S. (1996). *Forest farmers: Humans and landscape ecology of caboclo populations in the Amazon Estuary*. Ph.D. Dissertation, Indiana University.

Brondizio, E. S. (1999). Agroforestry intensification in the Amazon Estuary. In T. Granfelt (Ed.), *Managing the globalized environment: Local strategies to secure livelihoods* (pp. 88–113). London: Intermediate Technology Publ.

Brondizio, E. S., & Neves, W. A. (1996). Populações caboclas do Estuário do Amazonas: A percepção do ambiente natural. In: C. Pavan (Ed.), *Uma estratégia latino-americana para a Amazônia* (pp. 167–82). São Paulo: Mmarhal/Memorial/Unesp.

Brondizio, E. S., & Siqueira, A. D. (1997). From extractivists to farmers: Changing concepts of caboclo agroforestry in the Amazon estuary. *Research in Economic Anthropology*, *18*, 233–79.

Brondizio, E. S., Moran, E. F., Mausel, P., & Wu, Y. (1994). Land use change in the Amazon Estuary: Patterns of caboclo settlement and landscape management. *Human Ecology*, *22*, 249–78.

Brondizio, E. S., Moran, E. F., Mausel, P., & Wu, Y. (1996). Land cover in the Amazon Estuary: Linking of the thematic mapper with botanical and historical data. *Photogrammetric Engineering & Remote Sensing*, *62*, 921–9.

Câmara, E. P. L., & McGrath, D. G. (1995). A viabilidade da reserva de lago como unidade de manejo sustentável dos recursos da várzea amazônica. *Boletim do Museu Paraense Emílio Goeldi Nova Série Antropologia*, *11*, 87–132.

Cândido, A. (1964). *Os parceiros do Rio Bonito. Estudos sobre o caipira paulista e a transformação dos seus meios de vida.* São Paulo: Duas Cidades.

Cascudo, L. C. (1983). *História da Alimentação no Brasil.* Belo Horizonte: Itatiaia, São Paulo: Edusp. Vol 1.

Carneiro, R. L. (1970). A theory of the origin of the state. *Science*, 169: 733–8.

Carneiro, R. L. (1961). Slash-and-burn cultivation among the Kuikuru and its implications for cultural development in the Amazon Basin. *The evolution of horticultural systems in native South America, causes and consequences, a symposium*, Caracas.

Castro, F. (1999). *Fishing accords: The political ecology of fishing intensification in the Amazon.* Ph.D. Dissertation, Indiana University.

Castro, J. (1965). *Geografia da fome. O dilema brasileiro: pão ou aço.* São Paulo: Brasiliense.

Cattle, D. J., & Schwerin, K. H. (Eds.). (1985). *Food energy in tropical ecosystems.* New York: Gordon & Breach.

Cerdeira, R. G. P., Ruffino, M. L., & Isaac, V. J. (1997). Consumo de pescado e outros alimentos pela população ribeirinha do Lago Grande de Monte Alegre, PA – Brasil. *Acta Amazonica*, *27*, 213–27.

Coimbra Jr., C. E. A., Flowers, N. M., Salzano, F. M., & Santos, R. V. (Eds.). (2004). *The Xavante in transition: Health, ecology, and bioanthropology in Central Brazil.* Ann Arbor: The University of Michigan Press.

Cooke, R., & Piperno, D. (1993). Native American adaptations to the tropical forests of Central and South America before European colonization. In C. M. Hladik, A. Hladik, O. F. Linares, H. Pagezy, A. Semple, & M. Hadley (Eds.), *Tropical forests, people and food: Biocultural interactions and applications to development* (pp. 25–36). Paris: Unesco/Partenon.

De Garine, I., Hugh-Jones, S., & Prinz, A. (1993). Cultural factors in food choices: Background. In C. M. Hladik, A. Hladik, O. F. Linares, H. Pagezy, A. Semple, M. Hadley (Eds.), *Tropical forests, people and food: Biocultural interactions and applications to development* (pp. 525–32). Paris: Unesco/Partenon.

Dufour, D. L. (1985). Manioc as a dietary staple: Implications for the budgeting of time and energy in Northwestern Amazon. In: D. J. Cattle & K. H. Schwerin (Eds.), *Food energy in tropical ecosystems* (pp. 1–20). New York: Gordon and Breach.

Dufour, D. L. (1992). Nutritional ecology in the tropical rain forests of Amazonia. *American Journal of Human Biology*, *4*, 197–207.

Dufour, D. L. (1994). Diet and nutritional status of Amazonian peoples. In A. Roosevelt (Ed.), *Amazonian indians from prehistory to present: Anthropological perspectives* (pp. 151–75). Tucson/London: University of Arizona Press.

Emperaire, L. (2002a). O manejo da agrobiodiversidade: O exemplo da mandioca na Amazônia. In N. Bensusan (Ed.), *Seria melhor mandar ladrilhar? Biodiversidade como, para que, por quê* (pp. 189–201). Brasília: Editora UNB.

Emperaire, L. (2002b). A agrobiodiversidade em risco: O exemplo das mandiocas na Amazônia. *Ciência Hoje, 32*, 28–33.

FAO. (1977). Cassava processing. *FAO Plant Production and Processing Series*. Roma: FAO.

FAO/WHO/UNU. (1985). *Energy and protein requirements*. Geneva: World Health Organization.

Ferreira Filho, J. C. (1938). Manual da mandioca: A mais brasileira das plantas uteis. Parte 1: Culturas, inimigos, indústria. *Chácaras e Quintais, 57*, 129–202.

Franco, G. (1987). *Tabela de composição química dos alimentos*. Rio de Janeiro: Livraria Atheneu.

Franke, R. W. (1987). The effects of colonialism and neocolonialism on the gastronomic patters of the Third World. In M. Harris & E. B. Ross (Eds.), *Food and evolution: Toward a theory of human food habits* (pp. 455–79). Philadelphia: Temple University Press.

Futemma, C. R. T. (1995). *Agriculture and caboclo household organization in the Lower Amazon Basin: Case studies*. M.Sc Dissertation, Department of Anthropology, University of Tulane.

Gentil, J. (1988). A juta na agricultura de várzea da área de Santarém, Médio Amazonas. *Boletim do Museu Paraense Emílio Goeldi Nova Série Antropologia, 4*, 118–205.

Giugliano, R., Giugliano, L. G., & Shrimpton, R. (1981). Estudos nutricionais das populações rurais da Amazônia. 1 – Várzea do rio Solimões. *Acta Amazonica, 11*, 773–88.

Giugliano, R., Shrimpton, R., Arkcoll, D. B., Giugliano, L. G., & Petrere Jr., M. (1978). Diagnóstico da realidade alimentar e nutricional do estado do Amazonas, 1978. *Acta Amazônica, 8*, 1–54.

Giugliano, R., Shrimpton, R., Marinho, H. A., & Giugliano, L. G. (1984). Estudos nutricionais das populações rurais da Amazônia. II. Rio Negro. *Acta Amazonica, 14*, 427–49.

Gravata, A. G. (1938). Manual da mandioca: A mais brasileira das plantas uteis. Quarta Parte: As industrias da mandioca. *Chácaras e Quintais, 57*, 352–60, 499–517, 631–48, 727–41, 892–902.

Grenand, F. (1993). Bitter manioc in the lowlands of tropical America: From myth to commercialization. In: C. M. Hladick, A. Hladick, O. F. Linares, H. Pagezy, A. Semple, & M. Hadley (Eds.), *Tropical forests, people and food: Biocultural interactions and applications to development* (pp. 447–62). Paris: Unesco/Parthenon.

Gross, D. (1975). Protein capture and cultural development in the Amazon Basin. *American Anthropologist, 77*, 526–49.

Grundmann, R., & Stehr, N. (2000). Social science and the absence of nature: Uncertainty and the reality of extremes. *Social Science Information, 39*, 155–79.

Harris, M. (1999). The Brazilian floodplains: Where cholera does not kill caboclos. In S. Day, E. Papataxiarches, & M. Stewart (Eds.), *Lilies of the field. Marginal People who live for the moment*. Oxford: Westview Press.

Henry, G., & Hershey, C. (2002). Cassava in South America and the Caribbean. In: R. J. Hillocks, J. M. Tresh, & A. C. Bellotti (Eds.), *Cassava: Biology, production and utilization* (pp. 17–40). Wallingford: CAB International.

Hillocks, R. J., Thresh, J. M., & Bellotti, A. C. (Eds.). (2002). *Cassava: Biology, production and utilization*. Wallingford: CAB International Publishing.

Holmes, R. (1983). Nutritional status and cultural change in Venezuela's Amazon territory. In J. Hemming (Ed.), *The frontier after a decade of colonisation* (pp. 237–87). Manchester: Manchester University Press.

Holmes, R., & Clark, K. (1992). Diet, acculturation and nutritional status in Venezuela's Amazon territory. *Ecology of Food and Nutrition, 27*, 163–87.

IBGE. (1991) *Censo demográfico*. Instituto Brasileiro de Geografia e Estatística. (1991). *Tabelas de composição de alimentos*. Rio de Janeiro: Instituto Brasileiro de Geografia e Estatística.

IBGE. (1999) *Censo demográfico*. Instituto Brasileiro de Geografia e Estatística. (1999). *Tabelas de composição de alimentos*. Rio de Janeiro: Instituto Brasileiro de Geografia e Estatística.

INPA. (1998). *Composição de alimentos da Amazônia*. Manaus: Instituto de Pesquisas da Amazônia/Coordenação de Pesquisas em Ciências da Saúde (CPCS).
Jackson, F. L. C. (1993). The influence of dietary cyanogenic glycosides from cassava on human metabolic biology and microevolution. In C. M. Hladik, A. Hladik, O. F. Linares, H. Pagezy, A. Semple, & M. Hadley (Eds.), *Tropical forests, people and food: Biocultural interactions and applications to development* (pp. 321–38). Paris: Unesco/Partenon.
Lathrap, D. W. (1968). The 'hunting' economies of the tropical forest zone of South America: An attempt at an historical perspective. In R. B. Lee & J. Devore (Eds.), *Man, the hunter* (pp. 23–9). Chicago: Aldine.
McGrath, D., Castro, F., Futemma, C. R., Amaral, B. D., & Calabria, J. A. (1993). Fisheries and the evolution of resource management on the Lower Amazon Varzea. *Human Ecology, 21*, 167–95.
McGrath, D. G., Clancy, C., & Murrieta, R. S. S. (1995). *Demographic and economic census of Ituqui Island, Santarém, Pará, Brazil*. Santarém: Projeto Várzea.
McKey, D., & Beckerman, S. (1993). Chemical ecology, plant evolution and traditional manioc cultivation systems. In: C. M. Hladik, A. Hladik, O. F. Linares, H. Pagezy, A. Semple, & M. Hadley (Eds.), *Tropical forests, people and food: Biocultural interactions and applications to development* (pp. 83–112). Paris: Unesco/Partenon.
Meggers, B. (1954). Environmental limitation on the development of culture. *American Anthropologist, 56*, 801–24.
Meggers, B. (1958). Ambiente y cultura en la Cuenca del Amazonas: Revision de la teoria del determinismo ambiental. In: *Estudios sobre Ecologia Humana*. Conferencias en la Sociedad de Antropologia de Washington.
Meggers, B. (1987). *Amazônia, a ilusão de um paraíso perdido*. Belo Horizonte: Itatiaia/Edusp.
Migliano, A. B. (2000). *Amazônia: Interações ecológicas e estratégias adaptativas dos caboclos do Médio Solimões*. MA dissertation, University of São Paulo.
Moran, E. F. (1989). Models of native and folk adaptation in the Amazon. *Advances in Economic Botany, 7*, 22–9.
Moran, E. F. (1993). *Through Amazonian eyes*. Iowa City: University of Iowa Press.
Murrieta, R. S. S. (1994). *Diet and subsistence: Changes in three caboclo populations on Marajó Island, Amazonia, Brazil*. M.A Dissertation, University of Colorado.
Murrieta, R. S. S. (1998). O dilema do Papa-Chibé: Consumo alimentar, nutrição e práticas de intervenção na Ilha de Ituqui, Baixo Amazonas, Pará. *Revista de Antropologia, 41*, 97–150.
Murrieta, R. S. S. (2000). *The dilemma of the 'Chibé'-eater: Food choices, ecology and everyday life among peasant communities in the Lower Amazon, Brazil*. Ph.D. Dissertation, University of Colorado.
Murrieta, R. S. S. (2001a). Dialética do sabor: Alimentação, ecologia e vida cotidiana em comunidades ribeirinhas da Ilha de Ituqui, Baixo Amazonas, Pará. *Revista de Antropologia, 44*, 39–88.
Murrieta, R. S. S. (2001b). A mística do Pirarucu: Pesca, ethos e paisagem em comunidades rurais do Baixo Amazonas. *Horizontes Antropológicos, 7*, 113–30.
Murrieta, R. S. S., & WinklerPrins, A. M. G. A. (2003). Flowers of water: Homegardens and gender roles in a riverine caboclo community in the Lower Amazon, Brazil. *Culture & Agriculture, 25*, 35–47.
Murrieta, R. S. S., & Dufour, D. (2004). Fish and farinha: Protein and energy consumption in Amazonian rural communities on Ituqui Island, Brazil. *Ecology of Food and Nutrition, 43*, 231–55.
Murrieta, R. S. S., Neves, W. A., & Dufour, D. (1998). Status nutricional infantil em três comunidades ribeirinhas da Ilha de Ituqui, Amazônia, Brasil. *Bol. Mus. Para. Emílio Goeldi, série Antropologia, 14*, 3–19.
Murrieta, R. S. S., Dufour, D. L., & Siqueira, A. D. (1999). Food consumption and subsistence in three caboclo populations on Marajó Island, Amazonia, Brazil. *Human Ecology, 27*, 455–75.

Murrieta, R. S. S., Brondizio, E. S., Siqueira, A. D., & Moran, E. F. (1989). Estratégias de subsistência de uma população ribeirinha do Marajó-Açú, Ilha de Marajó, Brasil. *Boletim do Museu Paraense Emílio Goeldi Nova Série Antropologia 8*, 147–63.

Murrieta, R. S. S., Brondizio, E. S. S., Siqueira, A. D., & Moran, E. F. (1992). Estratégias de subsistência da comunidade de Praia Grande, Ilha de Marajó, Pará, Brasil. *Boletim do Museu Paraense Emílio Goeldi Nova Série Antropologia, 8*, 185–201.

Murrieta, R. S. S., Adams, C., Bakri, M., Strumpf, R., & Sousa, P. S. (2006). Caboclo household food intake in Two Amazonian ecosystems: A comparative analysis. *American Journal of Physical Anthropology, 129*, 135–136.

Mussolini, G. (1980). *Ensaios de antropologia indígena e caiçara.* Rio de Janeiro: Paz e Terra.

Nardi Filho, F. (1939). Mandioca e mentira. *Chácaras & Quintais, 59.*

Neves, E. G. (1998). Twenty years of Amazonian archaeology in Brazil (1977–1997). *Antiquity, 72*, 625–32.

Neves, W. A. (1989). Teorias de determinismo ecológico na Amazônia: um caso de marginalidade da comunidade científica nacional. In W. A. Neves (Ed.), *Biologia e ecologia humana na Amazônia: Avaliação e perspectiva* (pp. 59–76). Belém: SCT/PR CNPq/ Museu Paraense Emílio Goeldi.

Neves, W. A. (1992). *Antropologia ecológica de populações ribeirinhas do Estuário do Amazonas: Subsistência e adaptação.* Relatório Científico. São Paulo: Museu Paraense Emílio Goeldi/Universidade de São Paulo.

Neves, W. A. (1996). *Antropologia ecológica. um olhar materialista sobre as sociedades humanas.* São Paulo: Cortez.

Nugent, S. (1993). *Amazonian caboclo society: An essay on invisibility and peasant economy.* Oxford: Berg.

Odum, E. P. (2001) *Fundamentos de ecologia.* Lisboa: Fundação Calouste Gulbenkian.

Philppi, S. T. (2002). *Tabela de composição de alimentos: Suporte para decisão nutricional.* São Paulo: Universidade de São Paulo.

Pinton, F., & Emperaire, L. (2000). A farinha de mandioca, um elo dos sistemas extrativistas. In L. Emperaire (Ed.), *O extrativismo na Amazônia Central* (pp. 57–67). São Paulo: Unesp/ Imprensa Oficial do Estado.

Prudencio, Y. C., & Al-Hassan, R. (1994). The food security stabilization roles of cassava in Africa. *Food Policy, 19*, 57–64.

Pucciarelli, H. M., Neves, W. A., Melcher, S. S., & Murrieta, R. S. S. (2005). Nutritional status and sexual dimorphism in three Amazonian caboclo communities. *Antropologie, República Tcheca, XLIII*, 25–37.

Ribeiro, B. G. (1987). *O índio na cultura brasileira: Pequena enciclopédia da cultura brasileira.* Rio de Janeiro: Unibrade/Unesco.

Roosevelt, A. C. (1991) Determinismo ecológico na interpretação do desenvolvimento indígena da Amazônia. In W. A. Neves (Ed.), *Origens, adaptações e diversidade biológica do homem nativo da Amazônia* (pp. 103–41). Belém: Museu Paraense Emílio Goeldi/CNPq.

Roosevelt, A. C. (1994). *Amazonian Indians from prehistory to present: Anthropological perspectives.* Tucson/London: University of Arizona Press.

Ross, E. (1978a). The evolution of the Amazon peasantry. *Journal of Latin American Studies, 10*, 30–62.

Ross, E. (1978b). Food taboos, diet, and hunting strategy: The adaptation to animals in Amazon cultural ecology. *Current Anthropology, 19*, 1–36.

Santos, R. (1980). *História econômica da Amazônia.* Belém: Sejup.

Santos, R. V., & Coimbra Jr., C. E. A. (1994). Contato, mudanças socioeconomicas e a bioantropologia dos Tupi-Mondé da Amazônia brasileira. In R. V. Santos & C. E. A. Coimbra Jr. (Eds.), *Saúde e povos indígenas* (pp. 189–213). Rio de Janeiro: Fiocruz.

Santos, R. V., & Coimbra Jr., C. E. A. (1998). On the (un)natural history of the Tupi-Mondé Indians: Bioanthropology and change in the Brazilian Amazon. In A. H.Goodman & T. L.

Leatherman (Eds.), *Building a new biocultural synthesis* (pp. 269–93). Ann Arbor: The University of Michigan Press.

Sauer, C. O. (1987). As plantas cultivadas da América do Sul. In D. Ribeiro (Ed.), *Suma etnológica brasileira* (pp. 59–93). Petrópolis: Finep.

Schmidt, C. B. (1958). *A lavoura caiçara*. Rio de Janeiro: Ministério da Agricultura, Serviço de Informação Agrária.

Silva, H. P. (2001). *Growth, development, nutrition and health in caboclo populations from the Brazilian Amazon*. Ph.D. Dissertation, Ohio State University.

Silva, H. P., & Eckhardt, R. B. (1994). Westernization and blood pressure variation in four Amazonian populations. *Collegium Anthropologicum, 18*, 81–7.

Silva, H. P., Crews, D. E., & Neves, W. A. (1995). Subsistence patterns and blood pressure variation in two rural caboclo communities of Marajó Island, Pará, Brazil. *American Journal of Human Biology, 7*, 535–42.

Silva, H. P., James, G. D., & Crews, D. E. (2006). Blood pressure, seasonal body fat, heart rate, and ecological differences in caboclo populations of the Brazilian Amazon. *American Journal of Human Biology, 18*, 10–22.

Silva, P. P. (2005). *Farinha, feijão e carne-seca: um tripé culinário no Brasil colonial*. São Paulo: Editora Senac.

Siqueira, A. D. (1997). *The ecology of food and nutrition: Patterns of land use and nutritional status among caboclo populations on Marajó Island, Pará, Brazil*. Ph.D. Dissertation, Indiana University.

Siqueira, A. D., Brondizio, E. S., Murrieta, R. S. S., Silva, H. P., Neves, W. A., & Viertler, R. B. (1993). Estratégias de subsistência da população ribeirinha do Igarapé do Paricatuba, Ilha do Marajó, Brasil. *Bol. Mus. Para. Emílio Goeldi, série Antropologia, 9*, 153–70.

Spath, C. D. (1981). Getting to the meat of the problem: Some comments on protein as a limiting factor in Amazonia. *American Anthropologist, 83*, 377–9.

Steward, J. H. (1949). South American cultures: An interpretative summary. In: J. H. Steward (Ed.), *Handbook of South American Indians*. Washington: Smithsonian Institution.

UNICAMP. (2004). *Tabela brasileira de composição de alimentos*. Campinas: Unicamp/NEPA.

Vaughan, B. (1939). Rendimentos culturais da mandioca. *Chácaras & Quintais, 60*, 742–3.

Weinstein, B. (1985). Persistence of caboclo culture in the Amazon: The impact of the rubber trade, 1850–1920. In E. P. Parker (Ed.), *The Amazon caboclo: Historical and contemporary perspectives* (pp. 89–113). Williamsburg: College of William and Mary.

WinklerPrins, A. M. G. A. (1999). *Between the floods: Soils and agriculture on the Lower Amazon Floodplain, Brazil*. Ph.D. Dissertation, University of Wisconsin.

WinklerPrins, A. M. G. A. (2001). Why context matters: Local soil knowledge and management among an indigenous peasantry on the Lower Amazon floodplain, Brazil. *Etnoecológica, 5*, 6–20.

WinklerPrins, A. M. G. A. (2002). Recent seasonal flooplain-upland migration along the Lower Amazon river. *The Geographical Review, 92*, 415–31.

WinklerPrins, A. M. G. A., & McGrath, D. G. (2000). Smallholder agriculture along the Lower Amazon floodplain, Brazil. *PLEC News and Viwes, 16*, 34–42.

Zeron, A. C. (Ed.). (2000). *Equipamentos, usos e costumes da casa brasileira: Alimentação*. São Paulo: Museu da Casa Brasileira.

Chapter 13
Socio-Ecology of Health and Disease: The Effects of Invisibility on the Caboclo Populations of The Amazon

Hilton Silva

Abstract Following the issues of the material supporting bases of *caboclo* societies and their biosocial consequences, this chapter presents a judicious biomedical work on three riverine communities, associating the nutritional and parasitological status analyses for these communities' child and youth fractions. The Caxiuanã community is sited in the National Forest of the same name, in a black water river system, 400 km Southwest of the capital city Belém. This population's subsistence is based on slash-and-burn agriculture, fishing, *açaí* (*Euterpe oleracea*) extraction and wild game hunting. On Ituqui Island, sited 900 km from Belém in the Lower Amazon, a region also studied by several contributions featured in this volume, the two communities evaluated by the author were Aracampina and Santana. The Praia Grande community is located in a transitional area between floodplain and upland, on Marajó Island, and has its subsistence based on mechanized agriculture. Silva shows with his biomedical research the high level of malnutrition in the three communities, both chronic and acute. Even more alarming are the high indices obtained by the author regarding the infection of these communities' individuals by intestinal parasites. Silva's expectation is that this information may come to ground the development of public health policies that are integrated and participative, as well as of the widening of sanitary interventions geared towards the improvement of the studied population's and other Amazonian riverine populations' quality of life.

Keywords Diet · Nutrition · Food · Disease · Biomedicine · Parasites

The term caboclo has been the subject of considerable discussion as can be seen from the chapters in this volume (for example: Guzmán, Murrieta and WinklerPrins, Nugent), and in numerous publications in Brazil and abroad

---

H. Silva (✉)
Professor of Biological Anthropology, Federal University of Rio de Janeiro, Rio de Janeiro, Brazil
e-mail: hdasilva@acd.ufrj.br

C. Adams et al. (eds.), *Amazon Peasant Societies in a Changing Environment*,
DOI 10.1007/978-1-4020-9283-1_14,

(Oliveira Filho, 1979; Parker, 1985; Chibnik, 1991; Nugent, 1993; Lima-Ayres, 1992; Murrieta, Dufour, & Siqueira, 1999; Silva, 2001). In this chapter, caboclo is used to refer to genetically admixed populations that traditionally live in rural areas of the Brazilian Amazon (Silva & Eckhardt, 1994; Silva 2001; Silva, Crews, & Neves, 1995; Silva, James, & Crews, 2006). They inhabit all of the Amazonian ecosystems and originated from the genetic convergence of Indigenous, African and European groups in the North of Brazil over the last two centuries (Lima-Ayres, 1992; Nugent, 1993; Murrieta, 1998). Historically, the main subsistence activities of these groups have been fishing, agriculture, the extraction of forest resources, hunting, livestock raising, and trading (Parker, 1985; Nugent, 1993; Siqueira, 1997).

Many caboclo populations live in hard to reach rural areas with little or no public infrastructure or medical assistance, education or environmental sanitation services. As a result, the data available on their standards of health, physical growth, nutritional conditions and quality of life are limited (Silva, 2002b; Silva et al., 2006). This chapter raises questions about the relations between the socio-ecological conditions and aspects of health among four caboclo populations studied during the periods 1992–94 and 1996–97 in different environments in the state of Pará, Brazil, in a bid to trace the bioanthropological profiles of these groups and discuss how they reflect their socio-political invisibility.

The huge changes wrought upon the environment and ecology of the Brazilian Amazon over the course of the twentieth century (Clusener-Godt & Sachs, 1995; Sponsel, 1995; Silva, 2001) make this a particularly special region for the study of human populations in transition from a more traditional lifestyle, that is, predominantly based on subsistence activities, to a more 'westernized' way of life, based on the market economy. These studies can potentially deepen our understanding of the mechanisms of biological and cultural adaptation to the different ecosystems (Baker & Little, 1976; Silva & Crews, 1995) whilst also providing information that could prove extremely useful to health promotion programmes (McElroy & Townsend, 1996; Hahn, 1999).

Through the study of four caboclo populations (Aracampina, Caxiuanã, Praia Grande and Santana) this chapter will broaden the spectrum of information available in the area of rural public health in Brazil, thus contributing to the understanding of how diverse environmental and socio-economic conditions affect populations with similar genetic characteristics living in natural environments. The results presented herein may help in the development of rural health programmes with a strong emphasis on local participation and cultural sensibility, envisaging improved conditions of life and health for the caboclo populations.

The Amazon, Health and Caboclo Populations

In the last 50 years, the Brazilian Amazon has been under searing pressure from the federal government to 'integrate' economically with the rest of the nation. The construction of the Transamazon highway in the 1970s, along with the policy of building huge hydroelectric plants like Tucuruí and Balbina, and other so-called 'megaprojects', represent the considerable effort made to bring people into the region and develop its economy (Moran, 1981, 1983; Nugent, 1993). The hope was that these projects would help establish industries, ranches, logging companies and other enterprises in a bid to improve the economy and strangle the chances of interference from any bordering countries that may wish to control the natives or the region's wealth of natural resources (Sponsel, 1986, 1995; Nugent, 1993; Pichón, 1996). However, the real results of these and other federal government drives to colonize and 'develop' the Amazon have been environmental degradation, the extinction of flora and fauna, rural exodus, increased population density in peri-urban areas, the slumification of families driven from the land and deep-set alterations in the traditional socio-economic relations between the local populations (Schmink & Wood, 1984; Silva, 1997; Murrieta, 1998, 2000).

In the case of the caboclos, despite being all but ignored by the authorities, populations have been growing in numbers and economic importance in the Amazon since the late nineteenth century hey-day of the rubber trade (Wagley, 1963, 1974; Chibnik, 1991; Murrieta, 1994, 1998, 2000) and now comprise the largest rural group in the north of Brazil (IBGE, 1996; Murrieta, 1998). Nevertheless, in spite of all this, they have received little attention from researchers and politicians due to their genetically and culturally admixed characteristics and their loose politico-social organization, which have conspired to turn this segment of the Brazilian population into what some authors describe as an 'invisible' people (Chibnik, 1991; Nugent, 1980, 1993).

Particularly in the area of health, despite some progress made over the twentieth century, the situation of the northern populations compares raggedly with other regions of the country, with the rural north faring even worse than its urban counterpart, proof of the paltry investments directed toward the most vulnerable groups of the Amazon, including the caboclos (Silva, 2004a, 2004b). For these peoples, access to health services and good sanitary conditions is still extremely limited, and this is reflected in the high rates of illness among children and adults (Monteiro, Benício, & Gouveia, 1992; Jatene, Britto, Moura, Sá, & Diniz, 1993a; Jatene, Britto, Alves, Braglia, & Nogueira 1993b; Monteiro, 1997; Silva, 2001, 2002b).

In relation to the nutritional situation, while there is research attempting to describe the general nutritional situation and the growth characteristics of the Brazilian population (INAN, 1990a, 1990b; Coitinho, Leão, Recine, & Sichieri, 1991; Monteiro et al., 1992; Monteiro, 1997), these have largely been concentrated in areas of greater population density and better infrastructure, such as the cities of São Paulo, Rio de Janeiro, Belo Horizonte, Brasília and the urban areas of the North and Northeast. As a result, there is no extensive and comprehensive research on the current situation of health, growth and nutrition in rural populations of the Amazon Basin (Jatene et al., 1993a; Silva, 2001), and only a handful of studies on these traits in caboclo groups (Neves, 1992; Siqueira, 1997; Silva, 1995, 1999, 2001, 2002a, 2002b; Murrieta, Neves, & Dufor, 1998, Murrieta et al.,

1999; Adams, 2002; Murrieta & Dufour, 2004; Silva et al., 2006; Silva & Padez, 2006), including the research described in this chapter.

Studying Caboclo Groups

Given the extensive and scattered nature of human occupation in the Amazon, and the logistical difficulties in conducting research in the region, only rarely studies like this, involving various populations from different environments, are developed in the area. The data were collected by the author in Caxiuanã, Aracampina and Santana, and by Poema – Pobreza e Meio Ambiente na Amazonia (Project Poverty and the Environment in the Amazon), run by the Federal University of Pará/UFPA, in Praia Grande. These two data sets were combined in this chapter, as together they present a general panorama of the recent research in health/illness among caboclo groups.

Methodological Considerations

All of the populations dealt with here are located in the state of Pará, the second largest in Brazil, with an area of 1,253,164.5 km^2. The four areas involved in this analysis have a combined population of approximately 1,300 people, of which roughly 50% were children and adolescents at the time of the investigation (under 18) (IBGE, 1991, 1996; Silva 2001).

The data on health and growth was gathered in accordance with internationally adopted standard procedures (Weiner & Lourie, 1969; 1981; Cuff, 1995; Frisancho, 1999). Anthropometry and clinical exams have been the main instruments used by bioanthropologists and health professionals to evaluate health and nutritional welfare (Eveleth & Tanner, 1976; WHO, 1995; Frisancho, 1999; Hahn, 1999) and the anthropometric measurements employed here have been widely used by the international scientific community (Gómez et al., 1956; Weiner & Lourie, 1969; Friedlaender, 1975; Giles & Friedlaender, 1976; Baker & Little, 1976; Bogin, 1993; WHO, 1995; Frisancho, 1999). The bioanthropological data sets were supplemented with demographic, ethnographic and social information to help provide a more accurate characterization of the study populations. All the data gathering procedures followed the national and international regulations to protect human rights.

The anthropometric variables used in the analysis were height, weight and tricipital, subscapular and suprailiac skinfolds. Standing height is the measurement most widely applied for estimating linear growth (WHO, 1995; Bogin, 1999; Frisancho, 1999) and this was ascertained using a GNUEPEL® anthropometer. In Caxiuanã, Aracampina and Santana, weight was measured using a HEALTH-O-METER® portable scale, while in Praia Grande an Inmetro-calibrated FILIZOLA® scale was used. The margin of error for both was

+/– 100 grams. Tricipital skinfolds were measured using a LANGE® plicometer on the right arm. All measurements were taken three times, in millimeters, and the figures recorded were the means of the three sets (Silva, 2001). Information on weight-per-age, where low values indicate acute malnutrition, and height-per-age, where low values suggest chronic malnutrition, allows us to compare the data for the children studied against the reference values used by the World Health Organization (NCHS, 1997; Baumgartner, Roche, & Himes, 1986, modified by Kuczmarski et al., 2000), and against the findings of other studies conducted nationally. The data was analyzed per age group using the US CDC Epi-Info software package, following Goulart (1997). In order to simplify the analyses, comparison of weight-per-age and height-per-age data against international references was made separately for the datasets from Caxiuanã and Praia Grande, but jointly for those from Aracampina and Santana (under the heading Ituqui). All other data referred to here was inputted and analyzed using the SPSS/PC® statistical package (Green, Salkind, & Akey, 1997).

The clinical tests involved physical assessment and also a coproparasitological investigation of the study populations. This method was chosen on the premise that enteroparasitological infections are a good health indicator, as levels of infection can be associated with socio-sanitary conditions and access to essential medical services (Fitton, 1999; Silva, 2001). The fecal samples were analyzed using standard procedures (Direct, Faust, Baerman methods, etc.). The samples were collected in appropriate sterile plastic containers and sent for analysis at nationally certified laboratories in Belém. A total of 622 samples were tested: 326 from Santana, 123 from Aracampina, 116 from Caxiuanã and 57 from Praia Grande. The material was analyzed for the presence of nematodes, cestodes and protozoa.

The natural environments considered by this study are representative of the largest ecosystems inhabited by the caboclos. These include areas 100% floodplain, like Aracampina; transitional areas between floodplain and uplands, like Santana and Praia Grande; and entirely upland/terra firme areas, like Caxiuanã. In general, floodplains are areas that undergo seasonal flooding and spend various months per year under water. The uplands, on the other hand, tend to be areas covered by forest and are never inundated (Moran, 1981, 1983). The study also considered the main water systems in the Amazon. Santana, Aracampina and Praia Grande belong to white water ecosystems, while Caxiuanã is located in a black water river system. Black water rivers are nutrient-poor, while white water rivers are traditionally richer in nutrients and biomass (Moran, 1974, 1991).

It is recognized that populations like the caboclos, which are going through intense processes of *in situ* socio-economic modification, are difficult to classify in discrete categories, as they avail of myriad subsistence strategies (Moran, 1974; Brondízio, Morán, Mausel, & Wu, 1994; Brondízio, 1996; Siqueira, 1997). However, level of contact with the market economy (local and regional), as well as the degree of dependence upon external products, whether foods or consumer goods, can be indicators of relative levels of 'traditionality' and 'transitionality' for certain riverine populations. In this study, the groups were organized according to the classification proposed by Pollard, Brush, & Harrison (1991) and adapted by Silva (2001). Groups that predominantly practice subsistence activities are classified as 'Traditional', whilst those more engaged with the market economy, but who do not live in urban or peri-urban areas are considered 'Transitional'. Those intensely involved with the market economy and living in urban areas are classified as 'Modernized'. However, this classification is not without its limitations, as it classifies populations as per degree of involvement with the market economy mainly, but it is nonetheless methodologically sufficient for the purposes of this chapter. According to this classification, Caxiuanã can be considered the most traditional population, while Aracampina and Santana are the most transitional. Both of these, however, like the also transitional Praia Grande, have their own distinct socio-ecological characteristics.

Populations, the Environment and Subsistence

Caxiuanã

The Caxiuanã National Forest (NF) is a 330,000-hectare protected area mainly covered by tropical forest (85%), flooded forest (12%) and secondary vegetation and unforested areas (3%) (MPEG, 1993; Silva et al., 1995a). The NF is located in the municipalities of Melgaço and Portel, approximately 400 km southwest of Belém. The area is situated in the Caxiuanã Bay, a black water river system with a relatively acidic pH (MPEG, 1993, 1994; Lisboa, 2002). Daily tidal variations have little influence on water levels in Caxiuanã (MPEG, 1994).

The Ferreira Penna Scientific Station (FPSS), which occupies 10% of the Caxiuanã National Forest, has a medical post with a local nurse's aid, and is the only source of allopathic medical care in the NF. Increased interaction between the residents of the NF and the FPSS, as well as the research and ecotourism activities run there, will probably have some impact on various aspects of the area and its inhabitants in the future. However, as the Station only became fully operational during the data collection period, in 1996, it was too early to be able to discern its impact on the lifestyles and welfare of the local population.

Of the four populations covered in this chapter, the people of Caxiuanã have been the least studied. During the field work, there were 212 people living in 29 households inside the NF and all of these were included in the research. The population of Caxiuanã is made up of 103 women (49%) and 109 men (51%).

Subsistence in Caxiuanã is based on slash-and-burn agriculture, fishing, gathering of forest products and hunting. The main products extracted are açaí (*Enterpe oleracea*) and Brazilnut (*Bertholetia excelsia*), while the most prevalent agricultural products are manioc (*Manihot esculenta*), corn (*Zea mays*), tobacco (*Nicotina tabacum*), sugar cane (*Saccharum officinarum*), banana (*Musa sp.*) and other edible fruits (Silveira, Quaresma, Guapindaia, & Machado, 1997; Lisboa, 2002).

The diet in Caxiuanã consists mainly of manioc flour, fish and game, and is enriched during the dry season by açaí. Depending on their access to the closest towns, some families supplement their daily meals with industrialized products like coffee, refined sugar, biscuits, canned foods, jerked beef, rice and chicken, duck or pork (Murrieta, Batistoni, Pedrosa-Jr., 2004). The 14 families who live closest to the FPSS work there on an occasional or temporary basis and sell some of their produce to the Station. Some of the men also work for the logging industry outside the National Forest. Salaries from the loggers, retirement pensions and, more recently, services rendered to the FPSS, are the prime sources of cash income for the population (Silva, 2001).

Given its subsistence orientation and limited insertion in the regional markets, the population of Caxiuanã is considered the most traditional of the four groups studied.

Ituqui Island

Two of the populations studied are located within the area of influence of Ituqui Island, in the municipality of Santarém, some 900 kilometers from Belém. Ituqui belongs to a white water ecosystem. The island has an area of some 20,000 hectares and is located 40 km away from Santarém, the equivalent of three to four hours by boat on the Amazon River (McGrath, 1994). The entire island is flooded on a seasonal basis, and remains practically underwater for four to six months of the year. During the dry season, the fertile floodplain soil is used for tillage and cattle-raising. During the rainy season, most of the forests and savannas become one large swamp (Murrieta & Dufour 2004). The annual floods reach 9 to 10 meters, or more. The river currents and the seasonal floods have enormous impact on the environment and the lives of the local people. The two populations included in the research were chosen for their contrasting geographical locations and socio-economic characteristics within the area.

Aracampina

Aracampina is located on the north of the island and is entirely exposed to the influence of the *várzea* (floodplain). Geographically, it is the largest settlement on Ituqui, with houses spread out in almost linear fashion along a 2 km stretch of beach. During the collection of data (1996 and 1997), approximately 380 people were living in the 74 households of the village. Of this total, 284 people (74.7%), in 67 dwellings, participated in the study, amounting to 147 women (52.5%) and 133 men (47.5%).

Of the four groups studied, Aracampina is the one most affected by seasonality. The entire area is flooded during the rainy season, which usually begins in December and ends in June or July. During this period, the Amazon River rises various meters and submerges almost the entire island, connecting the inland lakes (McGrath, 1994; McGrath, Castro, Futema, Amaral, & Calabria 1993; Murrieta, 2000, 2001). As in Caxiuanã, there is no running water or sewage system in Aracampina. Water for drinking, bathing and washing domestic utensils comes directly from the river (Silva, 2001).

Like most of the island, subsistence in Aracampina is based on fishing, but agriculture and trade also play a relevant role. Many families (74.3%) fish for domestic consumption and for sale year round. The majority also have lots on the uplands or floodplain, where they grow manioc, beans, corn and other crops. Some families raise cattle on the floodplain and almost all keep small farm animals (chickens, pigs, ducks and goats) for domestic consumption and, occasionally, for sale.

Food consumption patterns in Ituqui vary depending on the locale and time of year. However, this variety is limited and some products are ubiquitous to all households (Murrieta, 2001; Murrieta & Dufour, 2004). The Aracampina diet is based on fish and manioc flour and is occasionally complemented with rice, beans, chicken or, more rarely, pork or beef. According to Murrieta and Dufour (2004), coffee and sugar are also important foodstuffs for the populations of Ituqui. In the rainy season, the diet varies a little, with the inclusion of more industrialized products, such as biscuits, powdered milk, chocolate and pasta. From time to time the diet also includes meats from hunted game, especially turtle (*Podocnemis expansa*), yellow-spotted turtle (*Podocnemis unifilis*), alligator (*Melanosuchus niger*) and, less frequently, manatee (*Trichechus inunguis*) (Silva, 2001). The children that attend the local school receive the daily school meal provided by the municipal government, though supply is often interrupted. Given its subsistence orientation and insertion on the regional market, the population of Aracampina can be considered transitional.

Santana

Santana represents a more advanced set of transitional (westernized) characteristics, given its higher population density, better infrastructure, urbanization and intense commercial contact with the city of Santarém. Unlike Caxiuanã and Aracampina, Santana is accessible by land through a two-and-a half-hour journey on a 100 km of unpaved road, or it can be reached by boat, taking roughly 40 minutes in a straight line from Santarém. At the time of data gathering, the village consisted of 77 houses with a total population of approximately 477 people, of which 398 (83.4%) participated in the research: 202 women (50.7%) and 196 men (49.3%).

The area has an urban organization, with defined roads and houses grouped into three distinct neighborhoods. Like Aracampina, Santana is also located in a white water ecosystem, but the houses there are built on the highest banks of the river, where they are free from flooding all year round.

There is no sewage system in Santana and all the water used by the houses goes to the fields or into the river. Water for domestic use is drawn from four manually-operated wells. As there is no running water, stocks for daily use have to be pumped by hand and carried back to the house as needed. Many of those who live along the river use it for bathing and other daily needs instead of going to the wells. Others use the wells for drinking water and the river for all other necessities (Silva, 2001).

Santana is the only population in Ituqui that has its own medical post, which keeps a stock of emergency medications, such as antidotes for snake bites, treatments for minor general accidents such as scratches and dog bites, antibiotics and some specific medicines for child diarrhea. These medicines are distributed free of charge by the village health agents. The post also attends people from the nearby communities.

Subsistence activities in Santana include fishing for domestic consumption and sale, the sale of groceries, cultivation of manioc and other agricultural produce for domestic consumption and sale, cattle-raising, administrative and teaching posts, and fruit picking (mango, orange and papaya) for domestic consumption and sale. The men often work as contractors for logging companies at other localities or for large ranches in the vicinity. Retirement pensions are another important source of income for various families. All of the households keep chickens, ducks, sheep or pigs for domestic consumption or for sale when they need an immediate influx of cash. Children and teens enrolled at the local school also receive a daily school meal as Aracampina.

The diet in Santana is based on fish, manioc flour and rice, with beef, chicken, pasta, beans, tinned foods and meats from hunted game occasionally consumed. The same general pattern is maintained in both seasons. However, like in Aracampina, there is more nutritional variety in the rainy season, with the addition of other industrialized products, such as powdered milk and chocolate, biscuits, crackers, as well as seasonal fruits like mango, orange and papaya, and greens, rarely mentioned during the dry season. As in Aracampina, sugar and coffee are taken daily by both the adults and children of Santana.

Praia Grande

Praia Grande is located in a transitional area between floodplain and upland on Marajó Island, approximately 10 km from the town of Ponta de Pedras, which is easily accessible by dirt road. At the time of the research, between 1992 and 1994, there were 19 houses spread out along a stretch of roadside, as opposed to along the riverbank as in the other three communities. When the data was gathered, there were approximately 120 people living in Praia Grande (Poema,

1994). However, for logistical reasons, only children under the age of five were examined – 25 girls (43.8%) and 32 boys (56.2%), together representing some 56% of the population. This area is also classified as transitional, though it stands out from the other three insofar as its agricultural system was substantially modified by the introduction of mechanization in the early 70s (Neves, 1992; Murrieta, Brondízio, Siqueira, & Moran, 1992; Murrieta, 1994).

As with the other three populations, at the time of the study there was no running water or sewage disposal system in Praia Grande and most of the population (70%) drew its water from open wells dug near their houses. The rest collected water from the nearby streams (Poema, 1994). Most of the households had latrines with a dug-out septic tank. There was no electricity in the village, except for some houses that had generators running on gasoline or car batteries so they could watch television. As there was no medical post in the village, the sick had to be transferred to a hospital in Ponta de Pedras.

Subsistence in Praia Grande is based on mechanized agriculture under a cooperative system, producing coconut, corn, rice and beans for commerce or home consumption. In addition to agriculture, many families bring in extra provisions planting manioc, collecting açaí, keeping a few heads of cattle, fishing and doing odd jobs on the nearby cattle ranches (Neves, 1992; Murrieta et al., 1992; Murrieta, 1994).

Though the diet in Praia Grande is also composed mostly of manioc flour with fish, beef, pork or chicken, they probably have the most varied diet of the four communities studied. Other important items in their diet are rice, beans, corn, canola or soya vegetable oil. Additional food items include açaí, sundry tropical fruits (banana, avocado, cashew, orange, lime, pineapple, etc.), guava jelly, tinned meats and sardines, prawns, coconut, coconut water, biscuits and crackers, sweetened coffee, powdered chocolate, milk, vegetables and wheat-bread. Children enrolled in school receive a daily school meal as in the other two areas (Murrieta, 1994; Murrieta et al., 1999).

In general, the samples of the four populations can be considered representative of each group during the study period, as the percentage of participants in each case was significant (>50%).

Nutritional Status, Growth and Health

Child nutrition and health indicators are among the most sensitive barometers of a nation's health (Califano, 1979; WHO, 1995). Patterns of physical growth are especially crucial public health indicators, as they are a measure of the resources invested in the social welfare of the population (Malina, Bouchard, & Beunen, 1988; Monteiro, 1991, 1999; World Bank, 1993). Various studies have established that health and illness are intimately connected with (in)adequate nutrition, as access to food is an important determinant of survival rates, especially among children (Gonzales, Valera, Rodriguez, Vega, & Guerra-Garcia, 1984; Haas,

1990; Probart, 1993). The consequences of nutritional deficits during childhood reverberate throughout the individual's lifetime, even long after the initial causes of the shortfall have changed (Frisancho, 1993; Cuff, 1995; Moore et al., 1999; Bogin, 2001). Therefore, understanding the prevalence of nutritional deficiencies, considering their impacts on the morbimortality rates of the population under study, is a fundamental factor in investigating the health of a group or nation (INAN, 1990a; Monteiro, 1997, 1999; Silva, 2001).

The most characteristic and often used measure of infantile malnutrition is delayed/stunted growth (Falkner & Tanner, 1986; Tanner, 1990; Frisancho, 1999). In a classic article from 1956, Gómez and collaborators established that low weight-for-age is one of the best predictors of infant mortality and recommend that this measure be used to assess the prevalence of undernourishment in study groups. Some years later, Waterlow (1973) indicated that weight-per-age could not distinguish sufficiently well between acute and chronic malnutrition and suggested height-for-age as an alternative gauge for chronic malnutrition (stunting) and weight-for-age as an indicator of acute malnutrition (wasting). Other authors also use weight-for-height as a pointer for nutritional status, with low values signaling a combination of acute and chronic nutritional deficiency (Anjos, Meirelles, Knackfuss, Cardoso, & Costa, 1989; Blanco, Valera, Torun, & Fajardo, 1995; Espinosa, 1996). The use of combinations of these and other variables to determine nutritional status has received considerable attention from the scientific community (Fidanza, 1991; WHO, 1995; Frisancho, 1999; Kuczmarski et al., 2000).

There are currently various studies on physical growth and health among Brazilian Amerindians (Oliveira, 1952; Salzano & Callegari-Jacques, 1988; Santos, 1993; Santos & Coimbra Jr., 1994; Coimbra Jr., Santos, & Escobar, 2003; Leite, 2004) and a host of others on Amazonian ecosystems and the economic interactions between the inhabitants of the region and their environment (Wagley, 1974; Moran, 1981, 1983; Gentil, 1988; Lima-Ayres, 1992; Nugent, 1993). However, there are still few studies on growth, nutrition and health among caboclos. Among the most recent are Siqueira (1997), Murrieta et al. (1998), Alencar, Yuyama, Nagahama, and Parente (1999), Silva (1995, 1997, 2001, 2002a, 2002b), Pucciarelli, Neves, Melcher, & Murrieta (2005), Silva et al. (2006), Adams (2002), Piperata (2007, 2008) and Piperata & Dufour (2007). All of these use anthropometric measurements as markers of nutritional status and health. In this chapter, the anthropometric measurements of children will be used as one of the indicators of the general nutrition and health situation among the study populations. The other indicator will be the level of infection by intestinal parasites.

Results

The Situation of Growth and Nutrition

When the anthropometric data of children from the four populations is compared by sex and age-group, there are no statistically relevant differences between the values for weight, height and skinfolds between boys and girls (0–11 years). As a result, in the analyses that follow, the data sets were combined and presented together.

Without considering the effects of seasonal variations, for which there is no available data for analysis of all four groups, the children of Caxiuanã are, in general, shorter than those from Aracampina, Santana and Praia Grande at all age groups. The situation is more complex in relation to weight. In the 0–2 age group, the children of Caxiuanã are heavier than those from Santana, and in the 3–5 and 6–8 age groups, they are heavier than those of Aracampina. Weight differences at all age groups between the children of Aracampina, Santana and Praia Grande were negligible.

In comparison with the data from the four caboclo populations, Engstron and Anjos (1999) examined the effects of maternal nutrition on nutritional health in children based on a general sample of the Brazilian population and found that 15.9% of boys and 12.9% of girls suffer from chronic growth delay. They also identified a strong association between the nutritional status of the mother and the delayed growth of her children. In a case study involving 264 first-year primary school children in the city of Osasco, São Paulo, Lei, Freitas, Chaves, Lerner, and Stefanini (1997) identified close links between sanitary conditions, family incomes and delayed growth. Anjos et al. (1989), studying some aspects of growth and nutrition in children from different socio-economic backgrounds in urban Rio de Janeiro, found a clear correlation between nutritional status and socio-economic position. The prevalence of chronic undernourishment in the lower income brackets was 6.2% and acute malnutrition 3.5%. On the other hand, they detected only 0.5% chronic malnutrition and no cases of acute malnutrition among the higher income range. The percentage of caboclo children suffering from malnutrition (acute and chronic) is much higher than the rates encountered by the abovementioned studies, with the situation proving particularly worrying in Caxiuanã, where the children presented the worst values among the caboclo in general (Table 13.1).

Data for girls from low-income families living in the peri-urban region of Santarém (Silva, 1995) reveals a nutritional pattern comparable to those found at Aracampina, Santana and Praia Grande. However, other studies have shown significant variability in the prevalence of undernourishment in the Amazon, with wide seasonal variations in malnutrition rates.

According to Jatene et al. (1993a), the Northern region has the highest national rates of malnutrition for children in the 0–5 age bracket (an average of 42.3% in urban areas). However, Monteiro et al. (1993), based on national data from 1989, point toward a prevalence of only 10.6% of undernourished children in the region, a value closer to that obtained by Giugliano, Albuquerque, and Shrimpton (1978) for the city of Manaus. Unfortunately, in most of the studies carried out, the samples for the Amazon region are taken almost exclusively from urban and peri-urban areas, with only rare exceptions. For example, Alencar et al. (1999) examined 283 children (0–5 years old) in rural and urban areas in Amazonas state and found that 41.7% suffered from

Table 13.1 Comparison of some Amazonian and Brazilian samples with the caboclo populations studied in terms of rates of malnutrition (acute and chronic)

Population	Age bracket (years)	Source	Malnutrition %*
Caxiuanã	0–11	This study	57.2
Ituqui	0–11	This study	49.3
Praia Grande	0–5	This study	59.1
Brazil, North	0–10	Monteiro et al., 1993	10.6
Brazil, North, Urban	0–5	Jatene et al.,1993a	42.3
Ituqui	0–9.9	Murrieta et al., 1998	19.8
Ituqui	2–9.9	Adams, 2002	24.5
Marajó	0–9.9	Adams, 2002	41.5
Marajó	0–10	Pucciarelli et al., 2005	25.8
Amazonas, Urban	0–5	Alencar et al., 1999	41.7
Amazonas, Rural	0–5	Alencar et al., 1999	25.6
Novo Airão, Amazonas (AM)	0–9	Alencar et al., 1999	49.0
S. Gabriel da Cachoeira, AM	0–9	Alencar et al., 1999	35.6
Rio Solimões, AM	7–11	Giugliano et al., 1981	54.7
Rio Negro, AM	7–11	Giugliano et al., 1984	63.3
Manaus, AM	7–11	Giugliano et al., 1978	13.0

* *Malnutrition = −2 SD on average (NCHS, 1977).*

Table 13.1 refers to some of the main studies on growth and nutrition patterns involving non-indigenous Amazonian populations. However, as can be seen from the table and the text, the age groups studied by the various authors, the tools used and the form in which the results were presented vary widely. In virtue of this, in order to facilitate comparison between the different publications, I opted to present only combined (average) data for acute and chronic malnutrition, as they are presented in this way in many of the cited articles. I understand that, while this form of presentation has its limitations from an epidemiological perspective, it ought not compromise the interpretation of the results in this case, seen as the objective of the comparison is merely to demonstrate the historical precariousness of the general situation of health/nutrition among the study populations, which results, largely, from their socio-political 'invisibility'. Presenting what little data there is on acute and chronic malnutrition separately would not substantially change the general overview provided by this analysis.

malnutrition (acute and chronic) in the urban areas against 25.6% in the rural. They also demonstrated that children were particularly susceptible to malnutrition during the first year after birth, probably in association with weaning, a common finding in many populations (Ferraroni, Filho, & Ferraroni, 1979; Espinosa, 1996; Murrieta et al., 1998).

In another study, Murrieta et al. (1998), analyzing anthropometric data for children from two rural populations on Ituqui Island, found that only 2.1% of the 232 sample presented low weight-for-height values (acute and chronic undernourishment), a percentage lower than those found for the caboclo populations in this chapter. For those authors, only 19.8% of the children presented low height-per-age measurements (chronic malnutrition), again, a number far lower than that described here. However, like this study, Murrieta et al. (1998) could find no discernible differences in the nutritional status of boys and girls from the same age group.

Comparing children from the Islands of Ituqui and Marajó, Adams (2002) found chronic malnutrition rates of approximately 28% in Aracampina and São Benedito (Ituqui), against 46% in Marajó-Açú, Paricatuba and Praia Grande (Marajó). The same study identified rates of 21% of acute malnourishment in Ituqui and 37% in Marajó. On the other hand, Pucciarelli et al. (2005) arrived at percentages as low as 6% for children with acute malnutrition (0–10) in Praia Grande, and 28% with chronic malnutrition, though results for other areas in Marajó indicated much higher percentages, somewhere in between those recorded by Adams and those presented in this chapter.

The highest levels of malnutrition were described in the state of Amazonas. In a study of children in rural areas of the Solimões River, Giugliano, Giugliano, and Shrimpton (1981) found that 54.7% of the children suffered from malnutrition (acute and chronic), while on the Negro River, the rates were as high as 63.3% (Giugliano, Shrimpton, Marinho, & Giugliano, 1984). The levels of child malnutrition for the Solimões River are lower than those observed in Caxiuanã and Praia Grande and higher than those for Ituqui. Nevertheless, the Negro River presents higher rates than those attributed to the four populations of this study.

There are as yet no definitive national standards for growth patterns and nutritional status for the entire Brazilian population, but only some studies comparing the nutritional situations of children from different parts of the country (INAN, 1990a, 1990b; Monteiro et al., 1993; Ferreira et al., 1997; Alencar et al., 1999; Monteiro, 1999) and, as shown above, there are still very few publications on physical growth in rural Amazonian populations. Table 13.1 summarizes the results of some of the works available on child nutrition of non-Indigenous populations in the Amazon, in Brazil and those presented in this chapter.

As Table 13.1 shows, the caboclo populations described in this chapter present higher levels of malnutrition (acute and chronic) than most of the other groups described in the literature, except those of Giugliano et al. (1981, 1984) for the Solimões and Negro Rivers. According to the analyses made by Silva (2001), if we consider only the international WHO/NCHS norms for growth, 79.6% of the children of Caxiuanã could be classified as suffering from some degree of chronic nutritional deficiency and 34.8% as suffering from acute malnutrition (average of 57.2%). In Ituqui, 72.1% have some level of chronic nutritional deficiency while 26.6% suffer from acute deficiencies (average of 49.3%). In Praia Grande, some 59.1% of children under 5 present some degree of malnutrition (acute or chronic).

Nevertheless, although the values for acute and chronic malnutrition among the four populations described are high, even when compared with other rural Amazonian populations, it is possible that they have been somewhat overestimated, as the samples were relatively small and the reference values used were international standards (WHO/NCHS), which means they might not be

Table 13.2 ANOVA of the sum of tricipital, subscapular and suprailiac skinfolds (Sumskfd) in the children of Ituqui and Caxiuanã, with both sexes combined

Age group	Variable	F	p
0–2 years	Sumskfd	4.29	0.01
3–5 years	Sumskfd	1.83	0.16
6–8 years	Sumskfd	4.73	0.01
9–11 years	Sumskfd	4.59	0.01

wholly adequate to the Brazilian population (Silva, 2001). On the other hand, the considerable variation in the nutritional situation of children and adults in different seasons of the year (Silva, 2001; Adams, 2002; Piperata, 2007), associated with the possibility of inter-observational error in the anthropometric measurements and with the different periods, locations and samples on which the above-cited works were based, means that we cannot conclude that the data presented here is incorrect. Only continued anthropometric studies in the region could definitively confirm or refute these results.

In relation to body mass, skinfold data is only available for Caxiuanã, Aracampina and Santana (Silva, 2001). The sum of skinfolds in the children of Caxiuanã is generally lower than in children of Aracampina and Santana. One-way Anova shows that the 0–2, 6–8 and 9–11 age-groups present significant statistical differences between the populations, but the 3–5 age-group does not differ among the three populations. Of the age-brackets which did show differences, the children of Aracampina obtained the highest values while those of Caxiuanã presented the lowest (Table 13.2). The skinfold data corroborates the growth data, indicating that the children of the groups that practice more traditional subsistence activities have lower adipose reserves, a fact that would also agree with the findings of Pucciarelli et al. (2005). The values for the 3–5 age-bracket need to be investigated further, as they may be the result of the sample size or of other factors as yet not included in the analysis.

The Parasitological Situation

Intestinal parasites are a serious health problem in developing countries (Filho, Ferraroni, & Montoril, 1978; Ferraroni et al., 1979; Santos, Rosa, Jesus, & Loureiro, 1992). The various illnesses related to parasitological infection of the intestines cause anemia, electrolytic loss through diarrhea, vomiting and digestive problems, while hampering the immune system, reducing normal food absorption and weakening the organism as a whole, causing severe damage for both adults and children that can even lead to death, especially among the latter (Fitton, 1999).

Complementing the information gleaned from the anthropometric data, the health situation in the four caboclo communities was also evaluated through coproparasitological tests. Representative samples were collected from children and adults in Ituqui and Caxiuanã. In Praia Grande, for logistical reasons, samples were only taken for children under the age of 5. Analysis of the fecal samples showed that the study population tested negative for only two parasites (*S. mansoni* and *T. saginata*). Ten species are present in Caxiuanã, nine in Santana, eight in Aracampina and six in Praia Grande. In Caxiuanã, the three most common types of parasite are *A. lumbricoides, E. nana* and *E. histolytica*. In Aracampina, they are *G. lamblia, E. histolytica* and *B. hominis*. In Santana, *E. histolytica, G. lambia* and *E. nana* are the most prevalent, while in Praia Grande the most common are *G. lamblia, A. lumbricoides* and *T. trichiuris* (Table 13.3).

Note that the three most common types of parasite in Aracampina and Santana are protozoa, while in Caxiuanã the most prevalent pest is a helminth, followed by a protozoan. In Praia Grande, protozoa are most common, followed by the helminthes. Also observable is that *G. lambia*, while one of the five most widespread parasites among the floodplain populations and in Praia Grande, is rare in Caxiuanã. Men and women are infected with the same frequency in Caxiuanã, Santana and Aracampina, and multiple parasites are widespread in all four communities. In Praia Grande, 47.2% of those examined have more than one species of parasite, followed by 43% in Caxiuanã and 37% and 20.8% in Aracampina and Santanam respectively (Table 13.3).

Among the caboclos, the frequency with which such signs and symptoms as mucocutaneous anemia, abdominal pain, diarrhea and nausea appear among the population is an indicator of the heavy burden these intestinal parasites represent for these groups, regardless of age (Silva, 2001). Even considering that

Table 13.3 Percent of population infected with intestinal parasites in the four study groups.

Parasite	Caxiuanã	Aracampina	Santana	Praia Grande
	% of population	% of population	% of population	% of Population*
Ascaris lumbricoides	57.8	13.8	6.1	35.1
Strongyloides stercoralis	5.2	0	0.6	1.7
Trichuris trichiura	18.1	0	0	28.0
Ancylostoma duodenale	19.8	6.5	5.2	26.3
Enterobius vermicularis	0	0.8	0.2	0
Escherichia coli	16.4	1.1	0	28.0
Entamoeba histolytica	31.0	30.1	48.7	10.5
Giardia lamblia	5.2	39.8	25.4	47.3
Endolimax nana	37.1	0	13.1	3.5
Blastocisty hominis	20.4	21.1	6.7	1.7
Multiple parasites	43.0	37.0	20.8	47.2

* *The percentages in this column refer only to children under age 5. For logistical reasons, on the original project, only the 0–5 age group was tested in Praia Grande.*

the data for Praia Grande is partial, as it covers only children age five years or younger, it is clear that intestinal parasites are frequent in all groups, with little quantitative difference between infections in men and in women. This was to be expected, as the determinants of contamination by the most common parasites concern environmental exposure and daily activities rather than the age or gender of the host (Smith, 1970; Fitton, 1999).

Studies on intestinal parasite infections among other rural Amazonian populations allow for some comparison with the populations described in this chapter. In the town of Tucuruí, in Pará, Vasconelos (1981) recorded that 46% of the 110 children tested were infected by parasites, 25.4% with the helminthes and 15.4% with protozoans. Filho et al. (1978), Dias, Filho, Paes, Farias, and Aguiar (1982) and Bóia et al. (1999) also encountered higher rates of helminthes than protozoans in their studies of rural and urban populations in the state of Amazonas. All of the studies cited demonstrate that intestinal parasites generally affect roughly 50% of the populations studied, a value very close to those observed in Aracampina, Santana, Praia Grande and Caxiuanã.

South American Indian populations also frequently suffer from intestinal parasite infection. Results show that the prevalence of intestinal parasites among the Cofán, Ticuna, Kashinawa and Yanomani, for example, is generally close to the rates for helminth infection in caboclos, though the occurrence among Indians is higher for the most 'westernized' groups (Lawrence et al., 1983; Fitton, 1999), the reverse of the trend identified among the caboclos, which might be explained by wider access to health services among the last.

Some specific trends emerge when considering the different environments the caboclos inhabit. The main parasites in Aracampina and Santana, which share an environment similar to that of the Cofán Indians in Ecuador, are protozoa. On the other hand, in Caxiuanã, the main parasite is *A. lumbricoides*, just as observed in other non-floodplain areas. In Praia Grande, however, despite the prevalence of *G. lamblia*, infections are evenly distributed between helminthes and protozoa. Levels of protozoan infection among caboclos are very similar to those for Indigenous groups. If we consider that *E. nana* and *B. hominis* have little effect on human health, *E. histolytica* and *G. lamblia* become the most frequent and widespread parasites among caboclo populations. This same trend has also been observed in other rural Amazonian groups (Dias et al., 1982; Lawrence et al., 1983; Bóia et al., 1999).

Socio-ecological particularities also account for differences in levels of parasitosis among the four groups. In general, Caxiuanã and Praia Grande have the highest rates of infection, by the largest number of species, and the most frequent occurrence of multiple parasite infection. Santana, on the other hand, has the lowest percentage of multiple parasite infection, the smallest number of species considered pathogenic and, with the exception of amoebiasis, the lowest figures for all of the parasites identified. The situation in Santana is likely due to a combination of factors, including the health awareness campaigns run by the NGO Projeto Saúde e Alegria (PSA), better public health infrastructure, higher per capita income, resulting in better access to commercial anti-parasitic medicines, and a geographical location less influenced by the floods. The intermediary position of Aracampina reflects its situation as the most transitional group from a socio-economic perspective. Its geographical location, lack of sanitary infrastructure and the periodical flooding of the houses and environs, including the toilets, contribute to the grave levels of parasitic contamination observed there. On the other hand, traditional remedies produced by the São Sebastião de Aracampina Mothers' Club, the environmental and health education programs run at the local school, and some access to commercial medicines probably help keep the infection levels somewhere between those of Caxiuanã and Santana. During the research period, Caxiuanã and Praia Grande received no health interventions whatsoever, so the situations encountered there reflect their absolute abandonment by the public health service. In the case of Praia Grande, the data was collected before a water quality intervention by the NGO Poema. The interventions that followed the study obtained extremely positive results, demonstrating that, even with relatively few resources, it is possible to improve the health of rural Amazonian populations (Poema, 1994).

The high rates of intestinal parasitosis in caboclo populations are connected with the various epidemiological and socio-ecological characteristics of the majority of rural groups in Brazil and in other developing countries (Vasconcelos, 1981; Santos et al., 1992; Fitton, 1999). The lack of environmental sanitation allied with the want of institutionalized and wide-reaching environmental and health education programmes, the absence of health care and certain lingering traditional habits that facilitate the transmission of diseases are all factors that contribute directly to the precarious health situation observed among the rural groups of the Amazon (Silva, 2004b).

Conclusions

The study of populations in transition, whether migrant or groups experiencing change *in situ*, is an unrivalled opportunity to investigate the plasticity and the mechanisms of human adaptation to different environments, and this has been one of the main objectives of biological anthropology (Friedlaender, 1975; Baker, 1978, 1988; Little & Garruto, 2000). At the same time, these studies offer information that can make an effective contribution to promoting health and improving the quality of life of the studied groups. Research into populations that have inhabited the same environments for thousands of years has shown that human beings can adapt both biologically and culturally to a vast array of ecosystems (Baker, 1969, 1978; Dressler, 1990; Silva, 1993). These studies have also demonstrated just how flexible human biology and culture actually are. However, short-term adaptation to socio-ecological transitions frequently comes at a considerable cost. Malnutrition, increased arterial pressure and adiposity, elevated stress levels (measured by the secretion of catecholamines), alcoholism and heart disease are among the damaging effects of the transition from a traditional lifestyle to a more modernized/westernized way of life (Pearson, James, & Brown, 1993; Smith, Earland, Bhatia, Heywood, & Singleton, 1993; Ulijaszek & Strickland, 1993; Dressler, 1996; Silva, 1999, 2001; Adams, 2002; Silva et al., 2006; Piperata, 2007).

Many questions on the time it takes for human organisms to adapt to new environmental challenges, to what extent such short-term adaptations could deregulate the allostasis (dynamic adaptation in the face of stressful situations and stimuli) in the future, and what possible implications these adaptive processes could have on micro-evolutionary human mechanisms remain to be answered. They can, however, be more deeply investigated through the study of caboclo populations, as they are in the throes of changing from an environment in which they have lived for generations to a 'new', 'modern' environment in which their biocultural mechanisms of adaptation are being put to the test.

The data on growth, nutritional status and health presented here shows that, contrary to the findings for Amerindian populations, the caboclos fare worse the more traditional are the subsistence activities they practice – a fact frequently observed among these peoples (Siqueira, 1997; Silva, 2001; Pucciarelli et al., 2005). It would appear that the caboclos respond to situations of socio-ecological stress differently to the Indians. This may be due to a combination of environmental, economic, genetic and cultural factors, as, given their ancestral origins and patterns of contact with the rest of Brazilian society, the caboclos present characteristics sometimes more akin to those of the Indians and sometimes closer to non-Indigenous populations. Living in such a biosocial borderland has brought significant consequences for the health and political status of these groups.

The invisibility of the rural Amazonian populations to the public authorities is clearly reflected in the sanitary and health conditions in which they live, particularly those groups dwelling in more remote areas and with less insertion in the regional markets. The difficulties of access to markets, goods and services caused by the vast Amazonian distances, the lack of effective forms of politico-social organization and participation of these groups, and the cultural differences between them and the urban dwellers end up relegating the caboclos and other vulnerable groups to the role of mere extras in regional decision-making processes (Nugent, 1993; Silva, 2004b). Their anxieties, needs and infrastructural problems are freely used as bargaining chips during electoral periods, though their votes rarely convert into real benefits or heightened citizenship once the campaigning is done.

In this chapter the effects of the different dwelling locations and patterns of subsistence on the growth, nutritional status and health of the four groups is evaluated, demonstrating the complexity of the health situation among caboclos. In addition, the data obtained from these bioculturally similar populations living in distinct ecosystems allowed for comparison with other national and regional datasets in order to determine how the caboclos fare in relation to these other groups. The results make it perfectly clear that: 1) given their biological and socio-cultural characteristics, the caboclos studied respond differently of the Amerindians in the face of the westernization process; 2) the health situation of the four populations is precarious, regardless of geographical location; and 3) there are enormous deficiencies in the public health and environmental sanitation policies for the rural populations of the Amazon.

Throughout the twentieth century, the Amazon suffered intense political, economic and social processes whose effects on its environment and human populations have yet to be fully understood. The results presented in this chapter show that interaction between human groups and their environments are extremely complex and that the caboclo populations have been rendered invisible to the Brazilian public authorities. The effects of this 'invisibility' are clearly seen when their state of health is examined.

Future Perspectives

Though this study has helped expand the scope of investigations into the growth, nutrition and health of non-Indigenous Amazonian populations, there is still much to be done. We have only scratched the surface when it comes to understanding the biosocial mechanisms involved in human adaptation to the various ecosystems around the globe (Little & Garruto, 2000; Bogin, 1993, 1999, 2001), and especially the tropical ecosystems (Silva, 2001; Piperata, 2005). In virtue of their long and pervasive presence in the Amazon, the caboclo populations are a source of precious scientific information on how human groups have adapted to the floodplains and hot, humid forests of the Amazon. As indicated by Nugent (1993), Siqueira (1997) and Silva (1999, 2001), among others, there is still a lot to learn about all the aspects of caboclo life (and other admixed populations in the Amazon and throughout the world). In this case, of particular relevance is the lack of longitudinal studies and investigations in the domestic sphere. Further studies are needed to ascertain, for example, how domestic relations influence economic activities, the distribution of resources, access to health services, goods and foodstuffs, and how these reflect upon the conditions of growth and health of individuals. Longitudinal and transversal studies on a populational and household level among caboclos could also offer highly relevant information on the biology of human plasticity.

In addition to aggregating knowledge on rural Amazonian populations, this study also aimed to help improve their health conditions. Toward this end, all of the information obtained was made available to the families, to the local governments and to non-governmental organizations working with health and public policy in Pará state. This was done through reports submitted to all of the agencies involved in the research and through direct discussions with the populations studied. I hope that this information can contribute to the development of public health policies that are integrative and participative and to a wider range of sanitary interventions geared toward improving the quality of life of the local groups and other Amazonian populations.

Acknowledgments This work would not have been possible without the collaboration of many people and institutions. Firstly, I would like to thank the families of Caxiuanã, Santana, Aracampina and Praia Grande for their encouragement and voluntary participation. I also thank the Fullbright Foundation, Laspau Programme and the CNPq for their financial support. Thanks to the Museu Paraense Emílio Goeldi and the staff at the Ferreira Penna Science Station, especially Dr. Pedro Lisboa, to Ibama in Breves, particularly Iranildo A. de Oliveira, Germásio Ferreira, Joel G. de Mello and Cândido L. Ribeiro, for their logistical and field support. My gratitude also goes to the Várzea Project/ Federal University of Pará/IPAM and to Drs David McGrath and Rui Murrieta; to the Projeto Saúde e Alegria in Santarém; Poema/UFPA, Belém; and to Profa. Rosângela Pereira of the Department of Nutrition at the Federal University of Rio de Janeiro. Special thanks to the Anthropology Department at Ohio State University and to Drs. Douglas E. Crews and Frank Poirer for their support during my stay in the USA. Finally, thanks to my field team for their dedication in even the most adverse conditions.

References

Adams, C. (2002). *Estratégias adaptativas de duas populações caboclas (Pará) aos ecossistemas de várzea estuarina e estacional: Uma análise comparativa*. Unpublished PhD Thesis, Department of Ecology, Universidade de São Paulo, São Paulo.

Alencar, F. H., Yuyama, L. K., Nagahama, D, & Parente, R. C. P. (1999). Estudo antropométrico de pré-escolares da calha do rio Negro, Amazonas, Brasil. II Barcelos. *Acta Amazônia, 29*, 293–302.

Anjos, L. A. S., Meirelles, E., Knackfuss, I., Cardoso, C., & Costa, S. G. (1989). Indicadores de gordura corporal em crianças de 7 a 11 anos vivendo em condições sócio-ambientais diferentes no Rio de Janeiro. *Ciência e Cultura, 41*, 1179–88.

Baker, P. T. (1969). Human adaptation to high altitude. *Science, 163*, 1149–56, 1969.

Baker, P. T. (Ed.). (1978). *The biology of high altitude peoples*. Cambridge: Cambridge University Press.

Baker, P. T. (1988). Human population biology: A developing paradigm for biological anthropology. In: *Trends in anthropology: Future perspectives, culture and biology* (pp. 255–63). Basil Blackwell/Unesco.

Baker, P. T., & Little, M. A. (1976). *Man in the Andes*. Stroudsburg: Dowden, Hutchison, and Ross.

Baumgartner, R. N., Roche, A. F., & Himes, J. H. (1986). Incremental growth tables: Supplementary to previously published charts. *American Journal of Clinical Nutrition, 43*, 711–22.

Blanco, M. L., Valera, Y. H., Torun, B., & Fajardo, L. (1995). *Taller sobre la evaluation nutricional antropometrica en America Latina*. Caracas: Ediciones Cavendes.

Bogin, B. (1993). *Patterns of human growth*. Cambridge: Cambridge Univ. Press.

Bogin, B. (1999). *Patterns of human growth* (2nd Ed). Cambridge: Cambridge Univ. Press.

Bogin, B. (2001). *The growth of humanity*. New York: Wiley-Liss.

Bóia, M. N., Motta, L. P., Salazar, M. S. P., Mutis, M. P. S., Coutinho, R. B. A., & Coura Jr. (1999). Estudo de parasitoses intestinais e da infecção chagásica no município de Novo Airão, Estado do Amazonas, Brasil. *Cadernos de Saúde Pública, 15*, 497–504.

Brondízio, E. S. (1996). *Forest farmers: Human and landscape ecology of caboclo populations in the Amazon Estuary*. Ph.D Dissertation, School of Public and Environmental Affairs, Indiana University, Bloomington.

Brondízio, E. S., Moran, E. F., Mausel, P., & Wu, Y. (1994). Land use change in the Amazon estuary: Patterns of caboclo settlement and landscape management. *Human Ecology, 22*, 249–78.

Califano, J. A. (1979). *Healthy people: The surgeon general's report on health promotion and disease prevention*. Washington: U.S. Department of Health, Education, and Welfare.

Chibnik, M. (1991). Quase-ethnic groups in Amazonia. *Ethnology, 30*, 167–82.
Clusener-Godt, M., & Sachs, I. (Eds.). (1995). *Brazilian perspectives on sustainable development*. Paris: Partenon.
Coimbra Jr., C. E. A., Santos, R. V., & Escobar, A. L. (2003). *Epidemiologia e saúde dos povos indígenas no Brasil*. Rio de Janeiro: Ed. Fiocruz/Abrasco.
Coitinho, C. D., Leão, M. M., Recine, E., & Sichieri, R. (1991). *Condições nutricionais da população brasileira: Adultos e idosos*. Brasília: Ministério da Saúde/Instituto Nacional de Alimentação e Nutrição.
Cuff, T. (1995). Introduction: Historical anthropometrics: Theory, method, and the state of the field. In: Komlos, J. (Ed.), *The biological standard of living on three continents: Further explorations in anthropometric history*. Boulder: Westview Press.
Dias, L. C. S., Filho, J. D., Paes, M. G., Farias, N. A., & Aguiar, J. C. S. (1982). Prevalência de parasitas intestinais em habitantes do rio Negro, estado do Amazonas, Brasil. *Acta Amazonica, 12*, 65–70.
Dressler, W. W. (1990). Social class, skin color, and arterial blood pressure in two societies. *Ethnicity and Disease, 1*, 60–77.
Dressler, W. W. (1996). Social identity and arterial blood pressure in the Africa-American community. *Ethnicity and Disease, 6*, 176–89.
Engstron, E. M., & Anjos, L. A. (1999). Deficit estatural nas crianças brasileiras: Relação com condições sócio-ambientais e estado nutricional materno. *Cadernos de Saúde Pública , 15*, 559–67.
Espinosa, M. V. P. (1996). *La crianza de los niños menores de seis años en Latinoamerica*. Santiago: Organización de los Estados Americanos.
Eveleth, P., & Tanner, J. M. (1976). *Worldwide variation in human growth*. Cambridge: Cambridge Univ. Press.
Falkner, F., & Tanner, J. M. (Eds.). (1986). *Human growth: A comprehensive treatise*. New York: Plenum Press.
Ferraroni, M. J. R., Filho, M. M., & Ferraroni, J. J. (1979). Parasitas intestinais numa população humana da cidade de Nova Olinda do Norte, Amazonas. *Acta Amazonica, 9*, 657–9.
Ferreira, H., Albuquerque, M., Ataide, T., Morais, M., Mendes, M., Siqueira, T., et al. (1997). Estado nutricional de crianças menores de dez anos residentes em invasão do 'Movimento Sem-Terra', Porto Calvo, Alagoas. *Cadernos de Saúde Pública, 13*, 137–9, 1997.
Fidanza, F. (1991). *Nutritional status assessment: A manual for population studies*. London: Chapman and Hall.
Filho, M. M., Ferraroni, J. J., & Montoril, D. A. A. (1978). Diagnóstico sócio-parasito-sanitário da cidade de Nova Olinda do Norte – Amazonas. *Acta Amazonica, 8*, 91–8.
Fitton, L. J. (1999). *Is acculturation healthy? Biological, cultural, and environmental change among the Cofán of Ecuador*. Unpublished Ph.D. Dissertation, Department of Anthropology, The Ohio State University, Columbus.
Friedlaender, J. S. (1975). *Patterns of human variation: The demography, genetics, and phenetics of Bougainville Islanders*. Cambridge: Harvard University Press.
Frisancho, R. A. (1993). *Human adaptation and accommodation*. Ann Arbor: Univ. Michigan Press.
Frisancho, R. A. (1999). *Anthropometric standards for the assessment of growth and nutritional status*. Ann Arbor: Univ. Michigan Press.
Gentil, J. M. L. (1988). A juta na agricultura de várzea na área de Santarém, Médio Amazonas. *Boletim do Museu Paraense Emílio Goeldi Série Antropologia, 4*, 118–99.
Giles, E., & Friedlaender, J. (Eds.). (1976). *The measures of man: Methodologies in human biology*. Cambridge: Schenkman Publishing Co.
Giugliano, R., Albuquerque, H., & Shrimpton, R. (1978). Estudo antropométrico, clínico e de padrões alimentares em um grupo de escolares de Manaus, 1976. *Acta Amazonica, 8*, 75–82.

Giugliano, R., Giugliano, L., & Shrimpton, R. (1981). Estudos nutricionais das populações rurais da Amazônia. I. Várzea do rio Solimões. *Acta Amazonica, 11*, 773–88.
Giugliano, R., Shrimpton, R., Marinho, H., & Giugliano, L. (1984). Estudos nutricionais das populações rurais da Amazônia. II. Rio Negro. *Acta Amazonica, 14*, 427–50.
Gómez, F., Galvan, R. R., Frenk, S., Munoz, J. C., Chavez, R., & Vázquez, J. (1956). Mortality in second and third degree malnutrition. *Journal of Tropical Pediatrics, 2*, 77–83.
Gonzales, G., Valera, J., Rodriguez, L., Vega, A., & Guerra-Garcia, R. (1984). Secular change in growth of native children and adolescents at high altitude Huancayo, Peru (3,200 meters). *American Journal of Physical Anthropology, 64*, 47–51.
Goulart, M. A. E. (1997). A avaliação nutricional infantil no software EPI INFO (versão 6.0) considerando-se a abordagem coletiva e individual, o grau e o tipo da desnutrição. *Journal of Pediatrics, 73*, 225–30.
Green, S. B., Salkind, N. J., & Akey, T. M. (1997). *Using SPSS for Windows, analyzing and understanding data.* New Jersey: Prentice Hall.
Haas, J. D. (1990). Mortality and morbidity consequences of variation in early child growth. In: A. C. Swedlund & G. J. Armelagos (Eds.), *Disease in populations in transition: Anthropological and epidemiological perspectives.* New York: Bergin and Garvey.
Hahn, R. A. (Ed.). (1999). *Anthropology in public health: Bridging differences in culture and society.* New York: Oxford University Press.
IBGE (Instituto Brasileiro de Geografia e Estatística). (1991). *Anuário Estatístico do Brasil.* Brasília: Ministério da Economia, Fazenda e Planejamento.
IBGE (Instituto Brasileiro de Geografia e Estatística). (1996). *Anuário Estatístico do Brasil.* Brasília: Ministério da Economia, Fazenda e Planejamento.
INAN. (1990a). *Pesquisa Nacional Sobre Saúde e Nutrição (resultados preliminares)* (2nd ed.). Brasília: Instituto Nacional de Alimentação e Nutrição do Ministério da Saúde.
INAN. (1990b). *Perfil de crescimento da população brasileira de 0 a 25 anos.* Brasília: Instituto Nacional de Alimentação e Nutrição do Ministério da Saúde.
Jatene, S. R., Britto, R. C., Moura, E. A. F., Sá, E. V., & Diniz, A. (1993a). *A meia vida da criança na Amazônia.* Belém: Unamaz/UFPA.
Jatene, S. R., Britto, R. C., Alves, Z. J., Braglia, M. A., & Nogueira, P. E. (1993b). *Crianças no Pará: A explosão da pobreza.* Belém: Unamaz/UFPA.
Kuczmarski, R. J., Ogden, C. L., Grummer-Strawn, L. M., Flegal, K. M., Guo, S. S., Wei, R., et al. (2000). CDC Growth Charts: United States. *Advanced Data from Vital and Health Statistics, 8*, 1–27.
Lawrence, D. N., Neel, J. V., Abadie, S. H., More, L. L., Adams, L. J., Healy, G. R., et al. (1983). Estudos epidemiológicos entre populações ameríndias da Amazônia. III. Parasitoses intestinais em povoações recentemente contatadas e em aculturação. *Acta Amazonica, 13*, 393–407.
Lei, D. L. M., Freitas, I. C., Chaves, S. P., Lerner, B. R., & Stefanini, M. L. R. (1997). Retardo do crescimento e condições sociais em escolares de Osasco, São Paulo, Brasil. *Cadernos de Saúde Pública, 13*, 277–83.
Leite, M. S. (2004). *Iri' Karawa, Iri' Wari': Um estudo sobre práticas alimentares e nutrição entre os Wari' (Pakaanova) do Sudoeste Amazônico.* Unpublished PhD Dissertation, – Fundação Oswaldo Cruz, Escola Nacional de Saúde Pública, Rio de Janeiro.
Lima-Ayres, D. M. (1992). *The social category caboclo: History, social organization, identity, and outsider's social classification of the rural population of an Amazonian Region (The Middle Solimões).* Ph.D. Dissertation, University of Cambridge, Cambridge.
Lisboa, P. L. B. (Ed.). (2002). *Caxiuanã: Populações tradicionais, meio físico e diversidade biológica.* Belém: Museu Paraense Emílio Goeldi/MCT.
Little, M. A., & Garruto, R. M. (2000). Human adaptability research into the beginning of the third millennium. *Human Biology, 72*, 179–99.
Malina, R. M., Bouchard, C., & Beunen, G. (1988). Human growth: Selected aspects of current research on well-nourished children. *Annual Review of Anthropology, 17*, 187–219.

McGrath, D. (1994). *Projeto Várzea: Lake reserve and community based management of várzea resources, Ituqui, Brazil.* ODA/WWF Proposal (WWFBR0604).

McGrath, D., Castro, F., Futema, C., Amaral, B. D., & Calabria, J. A. (1993). Manejo comunitário da pesca nos lagos de várzea do Baixo Amazonas. In L. G. Furtado, W. Leitão, & A. F. Mello (Eds.), *Povos das águas: Realidade e perspectivas na Amazônia.* (pp. 213–229) Belém: Museu Paraense Emílio Goeldi/MCT.

McElroy, A., & Townsed, P. K. (1996). *Medical anthropology in ecological perspective.* Boulder: Westview Press.

Monteiro, C. A. (1991). Counting the stunted children in a population: A criticism of old and new approaches and a conciliatory proposal. *Bulletin World Health Organization, 69,* 761–6.

Monteiro, C.A. (1997). O panorama da nutrição infantil nos anos 90. *Cad. de Pol. Soc. Série Documentos para Discussão,* n. 1. Brasília: Unicef.

Monteiro, C.A. (1999). Evolução do perfil nutricional da população brasileira. *Saúde em Foco, 8,* 4–8.

Monteiro, C. A., Benício, M. H. D., & Gouveia, N. C. (1992). Saúde e nutrição das crianças brasileiras no final da década de 80. In C. A. Monteiro, M. H. D. Benício, & N. C. Gouveia (Eds.), *Aspectos da saúde e nutrição no Brasil* (pp. 19–42). Brasília: Ministério da Saúde.

Monteiro, C. A., Benício, M. H. D., Iunes, R., Gouveia, N. C., Taddei, J. A. A. C., & Cardoso, M. A. A. (1993). ENDEF e PNSN: Para onde caminha o crescimento físico da criança brasileira? *Cadernos de Saúde Pública, 9,* 85–95.

Moore, S. E., Cole, T. J., Collison, A. C., Poskitt, E. M. E., McGregor, I. A., & Prentice, A. M. (1999). Prenatal or early postnatal events predict infectious deaths in young adulthood in rural Africa. *International Journal of Epidemiology, 28,* 1088–95.

Moran, E. F. (1974). The adaptive system of the Amazonian caboclo. In: C. Wagley (Ed.), *Man in the Amazon* (pp. 136–159). Gainesville: University Press of Florida.

Moran, E. F. (Ed.). (1981). *Developing the Amazon.* Bloomington: Indiana Univ. Press.

Moran, E. F. (Ed.). (1983). *The dilemma of Amazonian development.* Boulder: Westview.

Moran, E. F. (1991). Human adaptive strategies in Amazonian blackwater ecosystems. *American Anthropologist, 93,* 361–82.

MPEG (Museu Paraense Emílio Goeldi). (1993). *Plano de Manejo e Implementação: Estação Científica 'Ferreira Penna' (ECFP), Caxiuanã, Mun. Melgaço (PA).* Belém: Ministério da Ciência e Tecnologia (MCT)/Conselho Nacional de Desenvolvimento Científico e Tecnológico (CNPq).

MPEG (Museu Paraense Emílio Goeldi). (1994). *Relatório da Estação Científica 'Ferreira Penna'/ECFPn: Da concepção ao momento atual (1989/1994).* Belém: Ministério da Ciência e Tecnologia (MCT)/Conselho Nacional de Desenvolvimento Científico e Tecnológico (CNPq).

Murrieta, R. S. S. (1994). *Diet and subsistence: Changes in three caboclo populations on Marajo Island, Amazonia, Brazil.* Unpublished M.A. Dissertation, Department of Anthropology, University of Colorado Unpublished, Boulder.

Murrieta, R. S. S. (1998). O dilema do papa-chibé: Consumo alimentar, nutrição e práticas de intervenção na Ilha de Ituqui, Baixo Amazonas, Pará. *Revista de Antropologia, 41,* 97–150.

Murrieta, R. S. S. (2000). *The dilemma of the 'Chibé'-eater: Food choices, ecology and everyday life among peasant communities in the Lower Amazon, Brazil.* Unpublished Ph.D. Dissertation, Department of Anthropology, University of Colorado, Boulder.

Murrieta, R. S. S. (2001). Dialética do sabor: Alimentação, ecologia e vida cotidiana em comunidades ribeirinhas da Ilha de Ituqui, Baixo Amazonas, Pará. *Revista de Antropologia, 44,* 39–88.

Murrieta, R. S. S., & Dufour, D. (2004). Fish and farinha: Protein and energy consuption in Amazonian rural communities on Ituqui Island, Brazil. *Ecology of Food and Nutrition, 43,* 231–55.

Murrieta, R. S. S., Brondízio, E., Siqueira, A., & Morán, E. F. (1992). Estratégias de subsistência da comunidade de Praia Grande, Ilha de Marajó, Pará, Brasil. *Boletim do Museu Paraense Emílio Goeldi Série Antropologia*, 8, 185–201.

Murrieta, R., Neves, W. A., & Dufour, D. (1998). Status nutricional infantil em três comunidades ribeirinhas da Ilha de Ituqui, Amazônia, Brasil. *Boletim do Museu Paraense Emílio Goeldi Série Antropologia, 14*, 3–19.

Murrieta, R. S. S., Dufour, D., & Siqueira, A. D. (1999). Food consumption and subsistence in three caboclo populations on Marajó Island, Amazonia, Brazil. *Human Ecology, 27*, 455–75.

Murrieta, R. S. S., Batistoni, M., & Pedrosa-Jr., N. (2004). Consumo alimentar e ecologia em populações ribeirinhas na região da Floresta Nacional de Caxiuanã. *Boletim Rede Amazônia, 31*, 85–94.

National Center for Health Statistics (NCHS). (1977). *NCHS Growth Curves for Children – Birth – 18 years*. Washington: U.S. Department of Health, Education and Welfare/Public Health Service.

Neves, W. A. (1992). *Antropologia ecológica de populações ribeirinhas do Estuário do Amazonas: Subsistência e adaptação*. Brasília: Relatório Científico apresentado ao CNPq.

Nugent, S. (1980). Amazonia: Ecosystem and social system. *Man, 16*, 62–74.

Nugent, S. (1993). *Amazonian caboclo society: An essay on invisibility and peasant economy*. Oxford: Berg.

Oliveira, H. C. (1952). O estado de saúde dos índios Karajá em 1950. *Revista do Museu Paulista, 6*, 489–508.

Oliveira Filho, J. P. (1979). O caboclo e o brabo: Notas sobre duas modalidades de força de trabalho na expansão da fronteira Amazônica no século XIX. *Encontros com a Civilização Brasileira, 11*, 101–140.

Parker, E. (1985). The Amazon caboclo: An introduction and overview. In E. Parker (Ed.), *The Amazon caboclo: Historical and contemporary perspectives* (pp. xvii–li). Studies in Third World Societies, V. 32, Williamsburg: College of William and Mary.

Pearson, J., James, G. D., & Brown, D. (1993). Stress and changing lifestyles in the Pacific: Physiological stress responses of Samoans in rural and urban settings. *American Journal of Human Biology, 5*, 49–60.

Pichón, F. J. (1996). Land-use strategies in the Amazon frontier: Farm-level evidence from Ecuador. *Human Organization, 55*, 416–24.

Piperata, B. A. (2005). The energetics of lactation among tropical horticulturists in the Brazilian Amazon. Unpublished PhD Dissertation, Department of Anthropology, University of Colorado, Boulder.

Piperata, B. A. (2008). Forty days and forty nights: A biocultural perspective on postpartum practices in the Amazon. *Social Science & Medicine, 67*, 1094–1103.

Piperata, B. A. (2007). Nutritional status of Ribeirinhos in Brazil and the nutrition transition. *American Journal of Physical Anthropology, 133*, 868–878.

Piperata, B. A. & Dufour, D. L. (2007). Diet, Energy Expenditure, and Body Composition of Lactating Ribeirinha Women in the Brazilian Amazon. *American Journal of Human Biology, 19*, 722–734.

Poema. (1994). *Projeto Pobreza e Meio Ambiente da UFPA*. Belém: Universidade Federal do Pará.

Pollard, T. M., Brush, G., & Harrison, G. A. (1991). Geographic distributions of within-population variability in blood pressure. *Human Biology, 65*, 643–61.

Probart, C. (1993). Research issues in nutrition. *Collegium Anthropologicum, 17*, 363–70.

Pucciarelli, H. M., Neves, W. A., Melcher, S. S., & Murrieta, R. S. S. (2005). Nutritional status and sexual dimorphism in three Amazonian caboclo communities. *Anthropologie, XLIII*, 25–37.

Salzano, F. M., & Callegari-Jacques, S. M. (1988). *South American indians: A case study in evolution*. New York: Oxford University Press.

Santos, R. V. (1993). Crescimento físico e estado nutricional de populações indígenas brasileiras. *Cadernos de Saúde Pública, 9*, 46–57.

Santos, R. V., & Coimbra Jr., C. E. (Eds.). (1994). *Saúde e povos indígenas*. Rio de Janeiro: Fiocruz.

Santos, E. O., Rosa, J. F. T., Jesus, I. M., & Loureiro, E. C. B. (1992). A saúde das populações da Amazônia brasileira. In L. Yarzábal, C. Espinal, & L. R. Aragón (Eds.), *Enfoque integral de la salud humana en la Amazonia* (pp. 95–96). Caracas: Imprenta UCV.

Schmink, M., & Wood, C. (1984). *Frontier expansion in Amazonia*. Gainesville: Univ. Florida Press.

Silva, H. P. (1993). *Nutritional stress and body dimensions in high altitude Quechua speaking populations of Southern Peru*. Unpublished M.A. Dissertation, State College, The Pennsylvania State University.

Silva, H. P. (1995). Physical growth and health aspects of girls from a day care center in Santarém: Comparison with other samples from Brazil and Ecuador. *American Journal of Human Biology, 8*, 129.

Silva, H. P. (1997). Health and environmental change in caboclo populations of the Brazilian Amazon. In *XIV Congress of the International Union of Anthropological and Ethnological Sciences* (p. 335). Williamsburgh.

Silva, H. P. (1999). Bioanthropological aspects of ecologically diverse Amazonian peasant populations. *American Journal Physical Anthropology, 28*, 251–2.

Silva, H. P. (2001). *Growth, development, nutrition and health in caboclo populations from the Brazilian Amazon*. Unpublished Ph.D. Dissertation, Department of Anthropology, The Ohio State University, Columbus.

Silva, H.P. (2002a). Socio-ecological stress and blood pressure variation in traditional populations of the Brazilian Amazon. *American Journal of Physical Anthropology, 14*, 132.

Silva, H.P. (2002b). Aspectos demográficos e médico-epidemiológicos dos residentes na Floresta Nacional de Caxiuanã, Melgaço, Pará. In P. L. B. Lisboa (ed.), *Caxiuanã*. Belém: Museu Paraense Emílio Goeldi/MCT.

Silva, H.P. (2004a). Impactos da degradação ambiental na saúde humana: Desafios para o século XXI. *Sociedade Médica em Revista, 11*, 8–11.

Silva, H.P. (2004b). *Saúde e doença em relação ao meio ambiente: Desafios para as populações em situação de vulnerabilidade* (pp. 108–10). Belo Horizonte: PUC Minas/IDHS/PNUD.

Silva, H. P., & Eckhardt, R. B. (1994). Westernization and blood pressure variation in four Amazonian populations. *Collegium Anthropologicum, 18*, 81–7.

Silva, H. P., & Crews, D. E. (1995). Growth and development of Cofán children from Zabalo, Ecuador. *American Journal of Human Biology, 7*, 134–5.

Silva, H. P., & Padez, C. (2006). Secular trends in age at menarche among Caboclo populations from Pará, Amazonia, Brazil: 1930–1980. *American Journal of Human Biology, 18*, 83–92.

Silva, H. P., Almeida, S., & Lisboa, P. L. B. (1995a). 'Ferreira Penna' Research Station: A new opportunity for research in the Lower Amazon Basin, Caxiuanã, Pará, Brazil. *Antropologia Biológica, 3*, 59–68.

Silva, H. P., Crews, D. E., & Neves, W. A. (1995b). Blood pressure variation in two rural Amazonian populations from Brazil. *American Journal of Human Biology, 7*, 535–42.

Silva, H. P., James, G. D., & Crews, D. E. (2006). Blood pressure, seasonal body fat, heart rate, and ecological differences in caboclo populations of the Brazilian Amazon. *American Journal of Human Biology, 18*, 10–22.

Silveira, I. M., Quaresma, H. D. B., Guapindaia, V. L., & Machado, A. L. (1997). As populações pré-históricas e atuais. In P. L. B. Lisboa, (ed.), *Caxiuanã*. Belém: Museu Paraense Emílio Goeldi/MCT.

Siqueira, A. D. (1997). *Ecology of food and nutrition: Patterns of land use and nutritional status among caboclo populations on Marajó Island, Pará, Brazil*. Unpublished Ph.D. Dissertation, Department of Anthropology, Indiana University, Bloomington.

Smith, S. G. (1970). *Intestinal helminthiasis of an indigenous Cofán population in the Amazon Basin*. Unpublished M.A. Dissertation, Auburn University, Auburn.

Smith, T., Earland, J., Bhatia, K., Heywood, P., & Singleton, N. (1993). Linear growth of children in Papua New Guinea in relation to dietary, environmental and genetic factors. *Ecology of Food and Nutrition, 31*, 1–25.

Sponsel, L. E. (1986). Amazon ecology and adaptation. *Annual Review Anthropology, 15*, 67–97.

Sponsel, L. E. (Ed.). (1995). *Indigenous peoples and the future of Amazonia: An ecological anthropology of an endangered world.* Tucson: The Univ. of Arizona Press.

Tanner, J. M. (1990). *Fetus into man: Physical growth from conception to maturity*. Cambridge: Harvard University Press.

Ulijaszek, S. J., & Strickland, S. S. (1993). *Seasonality and human ecology*. Cambridge: Cambridge Univ. Press.

Vasconcelos, J. C. (1981). Bactérias enteropatogênicas de diarréia infantil aguda em Tucuruí, Pará. *Acta Amazonica, 11*, 527–38.

Wagley, C. (1963). *An introduction to Brazil.* New York: Columbia University Press.

Wagley, C. (Ed.). (1974). Man in the Amazon. Gainesville: University Press of Florida.

Waterlow, J. C. (1973). Note on the assessment and classification of protein-energy malnutrition in children. *Lancet, 1*, 87–9.

Weiner, J. S., & Lourie, J. A. (1969). *Human biology: A guide to field methods.* Oxford: Blackwell Scientific Publications.

Weiner, J. S., & Lourie, J. A. (Ed.). (1981). *Practical human biology*. New York: Academic Press.

WHO. (1995). *Physical status: The use and interpretation of anthropometry*. Genebra: World Health Organization.

World Bank. (1993). *Investing in health*. Washington: The World Bank.

Conclusion

Rui Murrieta, Walter Neves, Henyo Barretto Filho and Cristina Adams

Abstract In the conclusion, Rui Murrieta, Walter Neves, Henyo Barreto Filho and Cristina Adams build their final considerations taking as a starting point the contributions of the many authors, as they seek to set up a research agenda for the Amazonian *Caboclo* Societies.

Keywords History · Anthropology · Identity · Landscape · Health and diet

Thorough analysis of the structuring of modernity in the nineteenth and twentieth centuries among different Third-world populations has unveiled highly complex and dynamic facets of what is a far from homogeneous or linear process (Appadurai, 1996; Escobar, 1995). Although the countless attempts at explaining this phenomena that have appeared over the last one hundred years are undoubtedly important (especially when we consider the foundational character of many of them), they still do not come close to encompassing and understanding the multiplicity of local manifestations and responses to what we have come to call modernization. The articles in this book have sought to contribute to creating a more nuanced and multidimensional panorama for this advancement and its ramifications for the caboclo societies of the Amazon. In what follows we will present, by way of a provisional synthesis, a run through the thematic interfaces among the various contributions whilst assuming a critical stance before them.

History and Identity

One of the common points to all of the articles presented in this book is the need to cut to the historical root of what we have called the ideological invisibility of the caboclos. Even acknowledging that the academic community went some way toward rupturing that invisibility with the numerous research projects carried out during the 1990s in diverse fields of knowledge about these groups (Schmink, 2004; see the introduction to this volume), as well as through the

C. Adams et al. (eds.), *Amazon Peasant Societies in a Changing Environment*,
DOI 10.1007/978-1-4020-9283-1_15, © Springer Science+Business Media B.V. 2009

exponential growth in interventionist programs directed toward them, the present edited volume shows that our knowledge of the history and socio-ecological diversity of this process is still riddled with gaps. For Nugent (1993, 1994, and in this volume), the engineering of a homogeneity of landscapes represented by an all-dominating and obliterating supernature is one of the central points in the construction of the political invisibility of the caboclo. If the generalized figure of the Indian blends easily enough into this natural world, even if through naïve representations based upon a supposed pre-disposition to conservation, the same cannot be said of those who make up the historical Amazonian peasantry: the caboclos. Nugent pinpoints as one of the factors behind our difficulty in satisfactorily categorizing these groups, whether within a wholly natural landscape or in capitalized society, as residing in the fact that the caboclo, the mestiço or river-dweller are social categories too uncomfortably limitrophic for such an intentionally homogenizing process as modernization. Nugent sagaciously refers to the hybrid, volatile and 'opportunistic' nature of caboclo social relations and economic practices (also highlighted by other authors, such as Harris, 2001, in this volume; Pace, 1998; Lima, 1992; Lima & Alencar, 2001, in this volume) that mix 'purely' capitalist roles, especially in their integration with the global market, with recognizably pre-modern (or, perhaps non-modern would be more accurate) quotidian modes of social organization and subsistence.

Following this line of reasoning, the very term caboclo came in for criticism and questioning. During the last few years various researchers have endorsed the view that the word caboclo is inappropriate to refer to such a heterogeneous population (Lima, 1992; Pace, 1998; Harris, 2001). Add to that the pejorative use of the term, its ambiguous and relational application, and its potentially racist and segregationist overtones, and the word is met with even sterner rejection in the academic world (see Lima, 1992; Pace, 1998; Harris, 2001). However, can we really say that these are the only dimensions to the use of the term? One of the aspects most often neglected, particularly by foreign researchers, is the important role the term 'caboclo' has played as a marker and distinguisher for the identity of the Amazonian mestiço in structuring a counter-discourse on the region's social, political and artistic spheres. The political side can be seen in the way regional politicians identify themselves as 'caboclo' in a bid to stave off, disqualify or neutralize criticism, opinions or attempts at intervention coming from 'others' (read: non-locals, non-Amazonians, etc.), or in the forging of regional literary types (as in the works of Dalcídio Jurandir or Haroldo Maranhão). Our intention here is not to substitute one vision (pejorative, discriminatory) for another (alternative and self-congratulatory), but simply to show that both exist in the common acceptations of the term, displaying all the ambiguity and diversity of meaning the word 'caboclo' contains. It is not by ideologically 'cleansing' the terms historically used to refer to these populations that we will make the worldviews they express disappear. And so, despite the lack of consensus among the authors in this volume (see Brondízio, Harris and Lima), we have maintained the term 'caboclo' as a concept in reconstruction and re-signification, which is capable of encapsulating more than a history of exclusion, but of incorporating a positive identification of a way of life intimately linked to the socio-ecological landscape of the Amazon.

The tension between the extremes of meaning mentioned above was already present in the first attempts to homogenize Amazonian society by the modern Portuguese state. Décio Guzman's article in this volume identifies such a project in the form of the decree of 1755 prohibiting the use of the word 'caboclo' with reference to the descendents (and new subjects of the Portuguese Crown) from marriages between Indians and whites. This conflict of the ideological agenda of the state, which needed to obliterate potential regional identities and idiosyncrasies, for better or for worse, in its bid to crystallize a single national identity, would recur on various occasions throughout the regional history. This dialectic is a constant in Amazonian history and set the tone for many of the social movements of the nineteenth century, such as the *Cabanagem* movement in Pará (Cleary, 1993), and for all the contradictions of the extractivist system of the rubber era (Santos, 1980; Weinstein, 1993). It is evident that the characteristics the modernist mindset of society identifies as primitive and retrograde surged and structured in response to a socio-ecological context engendered by the colonial system itself. This leads us to conclude that this way of life, like that of various peasant groups throughout Latin America, is more than just the degenerated remnant of riverine indigenous culture, but the product of the implantation of a modern, globalized mercantile system, that began in the fifteenth Century (Braudel, 1981; Gruzinski, 2003; Wallerstein, 1974; Wolf, 1982). In other words, the caboclos are, by definition, modern, though they belong to a peripheral landscape seen as undesirable to the modern capitalist configuration implied by the dominant political economy.

It is on the flashpoints between the structures of a globalized market economy and the national state, as well as the flexibility of forms of negotiation in the caboclo quotidian, that Harris makes his biggest contribution to this volume. He underscores the fact that the same contradiction as expressed in the attempt to integrate a wider capitalist system with pre-modern modes of socio-economic organization, such as parentage and patronage, is also expressed in the manifestations of religiosity and relationship with the dominant Catholicism. The survival of a cult of saints alongside a belief in a world of spells and enchanted beings expresses another dimension of this same capacity for the negotiation and accommodation of apparently conflicting categories and worldviews in the daily life of these people. The crux of that ability resides in their capacity to live in an ever-changing present in which history or the collective past plays at most a negligible role. Like the forms of social structuring and socio-economic organization, the past is remembered and translated through the existence and history of the family unit, whether nuclear or extended (Lima & Alencar, 2001). The point of ontological reference for the past as much as the present is always the family and the personal relationships assimilated as the logic of kinship. Some ethnographic examples in this volume perfectly illustrate this aspect.

Landscape and the Use of Natural Resources

Of enormous importance to our final discussion are the effects of this socio-political invisibility upon the material dimensions of the caboclo economy and ecology. We have already seen how the complete naturalization of the Amazonian landscape as a vast and pristine forest is one of the core elements that led to that invisibility (Nugent, 1993, 1990, this volume). This same interpretation also impedes the identification and recognition of forms of use and management of forest resources that do not fit into the classical passive extractivist model perpetuated by the anthropological literature on the region, or into a monocultural and extensive model of modern cultivation (Bunker, 1985; Weinstein, 1993). As a result, most of the subsistence activities of the caboclo, located in a continuum between these extremes, are simply ignored, underestimated or disqualified in their productive capacity. Once again, we find ourselves debating with historically constructed value-judgments and images of what 'caboclo economy' and Amazonian landscape are and how they are described and defined. In this volume Brondízio describes the intense process of managing the açaí stands of Ponta de Pedras and argues that it fits much more comfortably with agroforest cultivation than with extraction or gathering. To think of the riverine economy as based upon a system of intensive cultivation that is efficiently integrated with the regional and national markets, albeit whilst still rooted in family-run smallholding, appears absurd to those who associate riverine communities with 'primitive' extractivism and 'snatch-and-grab' forms of exploiting forest produce. Without in any way wishing to create an essentialized notion of the 'noble savage' or 'naturally conservationist man', we believe that the forms of cultivation used among the historical peasantry have proved efficient within their demographic proportions in dealing with the chronic instability of the mode of hybrid production established in the region (see Bunker, 1985; and Weinstein, 1993). Brondízio's comparison between caboclos and settlers clearly illustrates the points of intersection between these two subsistence models, especially when it comes to the mixed employment of intensive and extensive strategies of forest resource use and cultivation, their flexible integration within an always unstable market and their dependence on family-based social capital. However, as already mentioned, these are traits that are incompatible with the agribusiness model of economic productivity and efficiency. What is most interesting about this is that this apparent antithesis is manifest in the very configuration of the landscape of these forms of cultivation, which seem to blend into the 'natural' forest, opposing themselves disconcertingly to the monolithic, geometrical and rigorous aesthetic of modern monoculture.

It would seem quite clear that such stereotypes are not only directly linked to ignorance of these forms of landscape use and stewardship, but also of the forms of organizing the work, distributing the resources and the ways the domestic units occupy the geography. We already know that micro-ecological variations in the riverine environments and the forms of exploration associated with them are far more complex and numerous than was supposed a decade ago (for some interesting examples, see Hiraoka, 1985; Moran, 1993; Murrieta & WinklerPrins, in this volume; Murrieta, Dufour, & Siqueira, 1999; Raffles, 1999; Raffles & WinklerPrins, 2003; WinklerPrins, 2001, 2002). In the wake of this trend, Célia Futemma presents highly illustrative data on the organizational aspects of the Patos community, which is located in an area of considerable ecological and landscape variability. Divided between the floodplains and the uplands, the community developed its own codes for access not only to specific zones, but to private resources. The key here is not simply to create customary forms of access to resources, but, above all, to use the social capital already there in the local kinship system to define them. In this manner a certain equity in resource distribution is achieved, as is fair access to the complementary micro-environments essential to the subsistence of the population (also see WinklerPrins, 2001, 2002). The importance of these unofficial, 'alternative arrangements' compensates for the negligence of public initiatives (especially in terms of agrarian reform) that ignore the critical role of this environmental complementarity. The functionality of these overlappings between social capital and economic/ecological capital is perhaps the least known aspect of caboclo social organization, but potentially one of the most crucial. This importance does not exclusively refer to the normative aspect of these communal codes and practices, but also their impact on current interventionist policy, whether of the state or non-governmental organizations.

Subsistence, Nutrition and Health

As with the other areas of caboclo social organization and structure, the analysis of the riverine diet in the Amazon has also suffered from a simplifying, bias pejorative bias. The anthropological literature has always considered the cassava/fish binomial to be a poor and unsatisfactory dietary repertoire from a nutritional point of view (e.g., Castro,1985), as it resolutely revolves around cassava and its direct derivatives. Though the protein content in cassava is in fact negligible, which could, in principle, have become a nutritional impediment to the development and functioning of human physiology, two factors make this cultivar a strategic item within the traditional Amazonian context: its extremely high caloric content, especially when we consider the region's well-known environmental limitations, and its potential for stockpiling, unrivalled among the traditional Amazonian cultivars. Once again, another erroneous judgment on a dimension of riverine life took as its reference the agricultural standards of the Old World, whose intensive monocultural base since the Middle Ages has been unfavorable to the notion of complementarity between items often inconspicuous or hardly visible in the landscape. Obviously, this is justified by the characteristics of the agricultural produce of the Old World, which combine unequalled nutritional qualities, especially in terms of protein content, derived from an artificial process of natural selection worked over millennia (Braudel, 1981; Harris, 1977;

Diamond, 2003). The adoption of an European monocultural model by caboclo societies with huge technological limitations would have run against the ecological grain of the tropical landscape and the subsistence strategies developed by the Amerindian cultures.

In the humid tropics of the Amazon, the nutritional shortcomings of cassava were efficiently compensated for by the composition of a diverse repertoire of cultivars often mixed on the same plot and by the exploration of the region's most frequent source of protein: fish. In this volume, Adams and collaborators use finely-tuned quantitative studies to illustrate the considerable dependence of the riverine communities of the islands of Marajó and Ituqui on cassava and fish. The authors also identified a certain instability in the caloric consumption of these populations during the annual cycle, which would seem to be directly connected with the problems and limitations of cassava cultivation vis-a-vis the traditional technologies and the manpower limitations that go with the caboclo reliance on family-based labor. In other correlate work to that presented here, Adams, Murrieta and collaborators (Adams, 2002; Adams, Murrieta, & Sanches, 2005; Murrieta & Dufour, 2004) detected a moderate deficit in individual calorie intake among the domestic units of Ituqui and other riverine regions. Surprisingly, no deficit was observed for protein ingestion.

The implications of this data surpass the domain of applied research, as they run up against theoretical questions that have held a central place in the anthropological debate on the Amazon over the last six decades. Despite the tumultuous historical and political environment that has characterized the last 500 years of Amazonian history, none of it has been able to dislodge the manioc/fish combination from the core of the region's material sustenance. The ubiquity and resilience of this combination leads us to suggest that the ecological characteristics of the hylea, as argued by Betty Meggers for the last five decades, do limit the agricultural possibilities within this technologically restricted context, whether for edaphic reasons on the interfluvial zones, or because of the unpredictability of the flood cycles in the varzea. Taking into account the use of multiple micro-environments, energy intake derived from carbohydrates continues to be the most crucial aspect of caboclo survival in the Amazon, as extensively defended by Meggers. Even in situations that allow these populations to align more effectively with the regional markets and therefore enable them to generate more income, the money obtained is largely used to buy flour, or other calorie-rich foods (Murrieta & Dufour, 2004).

Recent studies have strongly suggested that the focus of research and interventions in relation to nutritional quality for the Amazonian riverine populations should be shifted from production and food intake to the more basic aspects of sanitation and medical care. As Silva clearly demonstrates in this volume, Amazonian caboclo populations suffer from a staggering frequency of various intestinal parasitological infections. Even if they were ingesting satisfactory quantities of nutrients, as would seem to be the case on Marajó and Ituqui, such high levels of polyparasitosis could easily compromise the adequate development of the human organism, especially in the first years of life (Silva, 2001; Wilson et al,

1999). The author's findings on height-per-age measurements among various caboclo groups leave no doubt as to the existence of relatively high levels of chronic malnutrition among some of these populations. We can therefore conclude that the most important stress factors on these populations are epidemiological in nature as opposed to food shortages.

This information – especially if confirmed in the future among other caboclo populations in less benign ecosystems in terms of natural productivity – denounces the notion that these populations are always suffering from nutritional deficiencies due to the scarcity of food. This misconception becomes especially perverse in the implementation of state or non-governmental intervention programs for these populations. Normally devoid of any acceptable empirical base, these programs end up foisting cultivars and strategies (vegetable gardens) upon caboclo communities that are utterly alien to their local alimentary repertoires, and therefore doomed to failure (Adams et al., 2005). Such equivocal intervention – whether for scientific ignorance or political opportunism – stems from the lack of even the most rudimentarily acceptable diagnostics of the critical factors to which these caboclo populations are subject.

Environmental Policy, Sustainability and Traditional Populations

Another controversy concerning the Amazonian caboclo societies that has been exhaustively discussed is that of the compatibility of the permanence of human groups inside protected areas in relatively well-conserved Brazilian ecosystems. In this volume, Lima and Barreto Filho dealt with some of the most urgent issues on this topic. Firstly, Lima staunchly defends a position that is far from orthodox among Brazilian anthropologists, stressing the exclusionary aspect of the sustainable development model when applied as social engineering in real peripheral contexts. In so doing, the author aligns herself with a recent international trend that questions the prescription and ramifications of the concept of sustainable development that was simply hurled upon the developing countries without its theoretical logic having first been subjected to acceptable scrutiny in the light of human history. The pressing need of these countries to carve out meaningful political space for themselves on the international conservation and development scene meant they embraced this political agenda without properly analyzing the contradictions inherent to the concept of sustainable development. The urgency to incorporate the symbolic and ideological capital of this concept created a kind of mantra repeated *ad nauseam* in the hope of convincing the planet of its feasibility. Following this line of reasoning Lima uses an economic analysis of the domestic economies in Mamirauá to show how the implantation of 'self-sustainable' development models is socially unjust, as it freezes the aspirations of these groups in a state of material poverty. Such expectations would be unacceptable if applied to urban middle-class groups in emerging nations, or indeed in developed countries, who would never accept any limits imposed upon their levels of consumption. In short, what emerges is a social contract that imprisons these subjects to a galvanized low profile in time. A social contract of extraordinarily high long-term risk, because the 'contracting' public and private institutions capture the 'contractee' in an a-historical freeze-frame, assuming their levels of consumption and local growth will remain constant over time. The frustration of this expectation could lead to the rescission of the contract, resulting in the loss of benefits already gained.

In his contribution, Barretto Filho discusses the concept of traditional populations and underlines its incompatibility with sustainable development, especially when applied inside conservation areas. The author argues that, if only the traditional populations living inside protected areas are taken into consideration by the nation's park planners and environmental managers when demarcating these areas, it will create a clear asymmetry between those who, by some as yet arbitrary and tenuous definition, are considered 'traditional' and other populations seen as newcomers. The author's position is that all groups that happen to occupy an area designated for environmental protection should be cause for the same social concerns normally reserved for those considered 'traditional'. The non-incorporation of this strategy has on various occasions sparked scenes of social conflict in Brazil. However, although Barretto Filho's concern with adopting a more encompassing concept of social justice in terms of maintaining human societies within protected areas is laudable, the doubt is whether an attempt to accommodate a broader range of subjects would not confront environmental planners with an unviable development scheme from the political point of view, vis-a-vis restrictions imposed by the sources of funding normally available to the environmental protected areas. Suffice it to say that the mere guarantee of permanence for traditional societies inside protected areas is a recent conquest, given the heavy constraints of national and international environmental organs on human permanence inside such areas.

Fabio de Castro's chapter also contributes to the discussion on the difficulties in applying sustainable development models (such as the management of communitarian lakes) by showing that the use of regional socio-economic categories like that of 'traditional population' entails a homogenization that might be very useful from the discursive perspective, but which does not correspond to the social and ecological reality of the Amazonian communities. As such, so long as socio-environmental diversity is not truly recognized and incorporated into public political strategies for Amazonian development, there is little chance of achieving an alternative model to deforestation or of easing the pressures of agribusiness.

Gender Relations and Daily Life

The reconstruction and explanation of the subsistence systems of caboclo societies necessarily involves understanding more precisely the local social capital, also seen through the construction of gender and kinship relations in daily life. As already mentioned, the basic language that permeates the relationships between these human groups and the material bases of their sustenance resoundingly rests on the precedence of immediate local relationships over the impersonal ties of the modern capitalist model. In their contributions to this volume, Murrieta, WinklerPrins and Siqueira demonstrate, in convincing fashion, the different spheres of these relations. The point of intersection between the two authors is the rich and dynamic social universe of the caboclo domestic unit, where the most compartmentalized productive activities, which occur outside the household, are shot through with affectively constructed desires and motivations. Gender relations occupy an important place in the forge of this complex mesh of relations. This phenomenon is made clear by Murrieta and WinklerPrins' ethnographic analysis, which takes the spatial reference of the vegetable gardens and household yards of Ituqui as a platform for understanding this process. These gardens and yards, privileged vehicles of re-signification, function as the arena in which conflicts are resolved, affections are reaffirmed and the spaces of masculine and feminine power are demarcated. They also help fix successful productive experiments. Once again, caboclo societies show us how observed human behaviors (and probably their motivations) derive from a negotiated amalgamation of material needs, emotional bonds and meanings, which, in relation to the social sciences, leads us to question those analytical models that place all their stress on just one of these components. It is obvious that even issues related to biological adaptability and resilience can only be sufficiently understood through the juxtaposition of these elements of social life.

Siqueira, in her contribution, shows that the opposite should also be observed if we are to understand the bare minimum of acceptability in our explanations of these social phenomena. She shows that women's power to influence the decisions taken by the riverine domestic unit is correlated to their economic participation in the family economy. Besides economic importance, another factor that increases the participation of riverine women in decision-making is their level of involvement in extra-local institutional networks. Hence the motivation toward institutional engagement has a dual base: a symbolic political aspect, and a pragmatic/economic bent.

Final Considerations

Having come to the end of this book, the reader will certainly not have passed unfazed through the new critical approaches to the study of the historical Amazonian peasantry it presents. We hope that the adoption of a diachronic and prismatic view, which includes these societies within the context of western expansionism and all of the resistance, tensions and negotiations that go with it, will have broadened the interpretation of the Amazonian social reality.

We believe that the new theoretical approaches presented here will not only contribute to the discussion on the type of development model intended for the twenty-first century Amazon, but also encourage new researchers to devote themselves to studying this socio-environmental reality.

References

Appadurai, A. (1996). *Modernity at large*. Minnesota: University of Minnesota Press.

Adams, C. (2002). *Estratégias adaptativas de duas populações caboclas (Pará) aos ecossistemas de várzea estuarina e estacional: Uma análise comparativa*. Unpublished PhD Dissertation, Departamento de Ecologia, Universidade de São Paulo, São Paulo.

Adams, C., Murrieta, R. S. S., & Sanches, R. (2005). Agricultura e alimentação em populações ribeirinhas das várzeas do Amazonas: Novas perspectivas. *Ambiente & Sociedade*, *8*, 65–86, 2005.

Braudel, F. (1981). The Structures of Everyday life: The capitalism and civilisation, 15th to 18th century. London: Collins.

Bunker, S. (1985). *Underdeveloping the Amazon*. Urbana: University of Illinois Press.

Castro, J. (1985). *Geografia da fome. O dilema brasileiro: Pão ou aço*. São Paulo: Brasiliense.

Cleary, D. (1993). After the frontier: Problems with political economy in the Modern Brazilian Amazon. *Journal of Latin American Studies*, *25*, 331–49.

Diamond, J. (2003). *Armas, germes e aço: Os destinos das sociedades humanas*. São Paulo: Editora Record.

Escobar, A. (1995). *Encountering development: The making and unmaking of development*. Princeton: Princeton University Press.

Gruzinski, S. (2003). *O pensamento mestiço*. São Paulo: Companhia das Letras.

Harris, M. (1977). *Cannibals and kings*. New York: Random House.

Harris, M. (2001). *Life on the Amazon: The anthropology of a Brazilian peasant village*. London: British Academy.

Hiraoka, M. (1985). Cash cropping, wage labour, and urbanward migrations: Changing floodplain subsistence in the Peruvian Amazon. In E. P. Parker (Ed.), *The Amazon caboclo: Historical and contemporary perspectives* (pp. 199–242). Williamsburg: College of William and Mary.

Lima, D. (1992). *The social category caboclo: History, social organisation and outsider's social classification of the rural population of the Amazonian Region*. Unpublished Ph.D. Dissertation, University of Cambridge, Cambridge.

Lima, D., & Alencar, E. (2001). A Lembrança da história: Memória social, ambiente e identidade nas várzeas do Solimões. *Lusotopie*, *2001*, 27–48.

Moran, E. F. (1993). *Through Amazonian eyes*. Iowa City: University of Iowa Press.

Murrieta, R. S. S., & Dufour, D. (2004). Fish and farinha: Protein and energy consumption in Amazonian rural communities on Ituqui Island, Brazil. *Ecology of Food and Nutrition*, *43*, 231–55.

Murrieta, R. S. S., Dufour, D. L., & Siqueira, A. D. (1999). Food consumption and subsistence in three caboclo populations on Marajó Island, Amazonia, Brazil. *Human Ecology*, *27*, 455–75.

Nugent, S. (1993). *Amazonian caboclo society: An essay on invisibility and peasant economy*. Oxford: Berg.

Nugent, S. (1994). *Big mouth: The Amazon speaks*. San Francisco: BrownTrout Publishers.

Pace, R. (1998). *Amazon town: Gurupá revisited*. London/Boulder: Lynne Rienner.

Raffles, H. (1999). 'Local Theory': Nature and the making of an Amazonian place. *Cultural Anthropology*, *14*, 323–60.

Raflles, H., & Winklerprins, A. M. G. A. (2003). Further reflections on Amazonian environmental history: Transformations of rivers and streams. *Latin American Research Review*, *38*(3), 165–87.

Santos, R. (1980). *História econômica da Amazônia*. Belém: Sejup.

Schmink, M. (2003). No longer invisible, but still enigmatic: Amazonian peasant identities and cosmographies. *Reviews in Anthropology*, *32*, 223–37.

Silva, H. P. (2001). *Growth, development, nutrition and health in caboclo populations from the Brazilian Amazon*. Unpublished Ph.D. Dissertation, Department of Anthropology, Ohio State University, Columbus.

Wilson, W., Dufour, D., Staten, L. K., Barac-Nieto, M., Reina, J. C., & Spurr, G. B. (1999). Gastrointestinal parasitic infection, anthropometric status, and physical work capacity in Colombian boys. *American Journal of Human Biology*, *11*, 763–71.

WinklerPrins, A. M. G. A. (2001). Why context matters: Local soil knowledge and management among an indigenous peasantry on the Lower Amazon floodplain, Brazil. *Etnoecológica*, *5*, 6–20.

WinklerPrins, A. M. G. A. (2002). Recent seasonal flooplain-upland migration along the Lower Amazon River. *The Geographical Review*, *92*, 415–31.

Weinstein, B. A. (1993). *Borracha na Amazônia: Expansão e decadência 1850–1920*. São Paulo: Hucitec/Edusp.

Wolf, E. (1982). *Europe and people without history*. Berkeley: University of California Press.

Wallerstein, I. (1974). *The modern world-system: Capitalist agriculture and the origins of the European world-economy in the sixteenth century*. New York: Academic Press.

Index